Ordinary Differential Equations in Theory and Practice

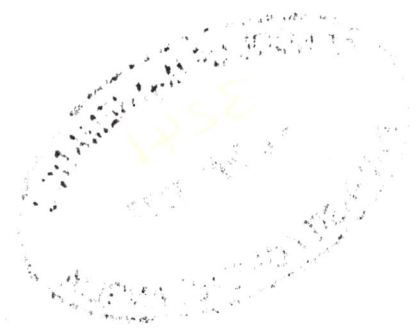

Ordinary Differential Equations in Theory and Practice

R. M. M. Mattheij and J. Molenaar
TU Eindhoven, The Netherlands

JOHN WILEY & SONS
Chichester · New York · Brisbane · Toronto · Singapore

Other Wiley Editorial Offices

John Wiley & Sons, Inc., 605 Third Avenue,
New York, NY 10158-0012, USA

Jacaranda Wiley Ltd, 33 Park Road, Milton,
Queensland 4064, Australia

John Wiley & Sons (Canada) Ltd, 22 Worcester Road,
Rexdale, Ontario M9W 1L1, Canada

John Wiley & Sons (Asia) Pte Ltd, 2 Clementi Loop #02-01,
Jin Xing Distripark, Singapore 0512

British Library Cataloguing in Publication Data

A catalogue record for this book is available from the British Library

ISBN   0 471 95674 0; 0 471 96530 8 (pbk.)

Produced from camera-ready copy supplied by the authors using TeX
Printed and bound in Great Britain by Biddles Ltd, Guildford and King's Lynn
This book is printed on acid-free paper responsibly manufactured from sustainable forestation, for which at least two trees are planted for each one used for paper production.

Contents

To our wives, Marie-Anne and Gerda

Preface

Ordinary differential equations (ODE) play an important rôle in the solution of many problems. This makes them an essential topic in any mathematics and/or science curriculum. Often, the treatment of the subject is scattered over various courses. Indeed, apart from their inclusion in analytical courses on differential equations (where the numerical analysts might be poorly explained or justified, and its use haphazard), they appear again in most numerical courses, which quite often contain a small crash course with an analytical introduction to the subject. Also some parts of the theory appear to be introduced just as a vehicle for some kind of application or justification of material, not being related to the subject otherwise. This scatter is quite well reflected in available textbooks.

Because of the aforementioned overlap of courses, but even more in order to emphasise their relationships and cohesion, we think it helpful to have a comprehensive and integrated treatment of analytical and numerical aspects in combination with the modelling of relevant problem classes. Overall it may also improve efficiency in teaching ODE. This text is therefore based on the following thoughts: to provide enough insight into qualitative aspects of ODE, such as the existence and uniqueness of solutions, to offer a thorough account of quantitative methods for approximating solutions numerically and, finally, to acquaint the reader with mathematical modelling where such ODE often play a significant rôle. Through such a combination the continuous (analytical) and the discrete (numerical) cases can be dealt with more efficiently, since the former may be considered as a limiting case of the latter, and the latter as an approximation of the former.

The theory of ODE is quite extensive. Despite the fact that it is an old and well-established area of analysis, it is still enjoying a lot of research interest, for example in dynamical systems, bifurcation, chaos, singular perturbations, differential-algebraic equations and waveform relaxation (just to mention a few topics). So, we had to make a selection; this was inspired mainly by didactical and practice-oriented objectives. A fairly obvious choice was to put a lot more emphasis on initial value problems than on boundary value problems (and thus on the variegated notion of stability, a typical evolutionary concept).

There are clearly related theories for the latter class of problems which can be treated well within such a framework, nevertheless. Also some of the other, more specialised, topics are not covered in much detail, but hopefully well enough to give some readers more appetite for further study.

Roughly speaking, the book consists of four parts. First we have a general introduction. This is intended for the reader to get acquainted with some basic concepts, but even more to acquire the proper feel for interpreting the equations, such that they will not be solely mathematical objects without relation to other areas of science. We have opted for an interpretation in terms of (classical) mechanics, because this field can claim the oldest rights. The second part of the book is related mainly (though not exclusively!) to analytical theory, viz. Chapters II (existence and uniqueness), IV (linear equations), V (stability), VI (chaos theory), the first half of Chapter VIII (singular perturbations), and the first parts of Chapters IX (differential-algebraic equations) and X (boundary value problems). The third part includes numerical aspects, as in Chapters III (one-step methods) and VII (multistep methods), the second half of Chapter VIII (stiff equations) and the second half of Chapters IX and X. The fourth part, consisting of Chapters XI and XII, makes up an extensive and essential part of this book and is added both for illustration and motivation. First, in order to emphasise the importance of ODE in applications, their rôle in classical mechanics is described in Chapter XI. In Chapter XII it is then shown how ODE play a rôle in mathematical modelling for a range of phenomena. As modelling requires more than just knowledge of analytical and numerical methods, the first sections of Chapter XI are devoted to the 'art' of modelling. Our original plan was to involve every concept, as introduced in Chapters I–XI, in one of the models in Chapter XII; we soon found out that this was too ambitious. Therefore we have highlighted only the most important aspects, such as dimensional analysis, stability analysis of equilibria, linearisation and global phase plane analysis, and show which conclusions are to be drawn from them. Apart from these more elaborate cases, most concepts are illustrated throughout the text by examples. We hope that both the examples and the models will inspire the reader to look for similar cases and to work them out along similar lines.

The numerical results in this book have been obtained by using the PASCAL procedures Fehlberg, Adams and Gear, which were written on the basis well-known public domain software. The interested reader may obtain these, as well as the (FORTRAN) boundary value problem code MUSN, through anonymous ftp through win.tue.nl, in the directory pub/local/numerical. An alternative way to transfer these programs is through WWW at http://www.win.tue.nl/win/math/an/ftp.

This book is intended to be used as a textbook for both undergraduate and graduate courses, but we hope that it will also be useful as a source of reference and inspiration for students and researchers alike. Although we have tried to integrate analytical and numerical aspects as far as we deemed it useful, the division into chapters is such that the book can also be used merely as an (analytical) introduction to ODE. This possibility – though even less intentional – also holds for a more typical numerical treatment of the subject.

In writing this book we have benefited from the ideas and support (in various ways) of a number of persons. In particular we would like to mention Prof. van Groesen who provided us with advice on a very early Dutch version of this text. We want to give a special acknowledgement to H. Willemsen, to whose credit it is that we have such fine displays of tables and figures and who reworked the text over and again. We wish to express our gratitude also to Mrs. A. Klooster, who prepared the Latex manuscript in an exeedingly accurate manner. Finally we would like to thank students for their positive criticism of the earlier drafts of the book.

R.M.M. Mattheij   Eindhoven
J. Molenaar   October 1995

I

Introduction

In §1 we introduce initial value problems through an example from mechanics; here we employ elements from the theory of ordinary differential equations (ODE) that will be worked out in more detail in later sections and chapters. Then, in §2, we introduce vector fields and systems of first order differential equations. Their classification and some quite general properties are treated in §3. In §4 we show how higher order differential equations can be reduced to systems of first order; this then justifies the exclusive treatment of the first order equations at a later stage. The discrete analogue of a differential equation, viz. the difference equation, is defined and described in §5. In §6 we indicate how the solution of an ODE can be approximated through discretisation, thus giving a difference equation. We also introduce the notion of consistency for measuring the discrepancy between the solution of an ODE and that of its discretised counterpart.

1. Introduction

The study of *ordinary differential equations* (ODE) goes back to times when classical mechanics was being developed. This has permeated both notation and terminology. This appears most clearly in the independent variable t, which often corresponds to the physical concept of *time*, although also space or other notions play the rôle of independent variable occasionally. Mechanical systems can often be quite helpful in interpreting results or even inspiring methods. Therefore we should like to start this chapter by introducing important ODE concepts with this interpretation in mind. In Chapter XI we shall deal with mechanical systems in detail.

Consider a particle, with unit mass, and denote:

t : *time*

$\mathbf{y}(t)$: vector of spatial coordinates, i.e. *position*

$\dot{\mathbf{y}}(t)$: time derivative of $\mathbf{y}$, i.e. *velocity*.

If we assume that we can force the particle to have a prescribed velocity $\mathbf{v}(t, \mathbf{y})$ at any time t and position $\mathbf{y}$, then the relation

(1.1a) $\dot{\mathbf{y}} = \mathbf{v}(t, \mathbf{y})$

will describe trajectories of such particles. Since it contains a first derivative and not a higher one, it is called a *first order ODE*. A relevant question then is which specific trajectory is defined by this differential equation, if we prescribe the initial position at $t = t_0$, say

(1.1b) $\mathbf{y}(t_0) = \mathbf{y}_0$.

The relations (1.1a,b) constitute a so-called *initial value problem* (IVP). In most mechanical situations the actual velocity field is not known a priori. However, one can often find an expression for the *force* $\mathbf{F}(t, \mathbf{y})$, exerted on the particle at time t and position $\mathbf{y}$; it causes an acceleration of the particle, which is the second derivative of the position. Hence we find

(1.2a) $\ddot{\mathbf{y}} = \mathbf{F}(t, \mathbf{y})$,

which is essentially *Newton's second law*. We should now provide more information for the solution, namely the position **and** velocity at t_0, denoted by $\mathbf{y}_0$ and $\dot{\mathbf{y}}_0$ respectively:

(1.2b) $\begin{cases} \mathbf{y}(t_0) := \mathbf{y}_0 \ , \\ \dot{\mathbf{y}}(t_0) := \dot{\mathbf{y}}_0 \ . \end{cases}$

Clearly (1.2a,b) also form an IVP. Since we have a second derivative in (1.2a), we call it a *second order ODE*.

By introducing new quantities $\mathbf{x}$, $\mathbf{f}$ and $\mathbf{x}_0$, defined as

(1.3) $\mathbf{x} := \begin{bmatrix} \mathbf{y} \\ \dot{\mathbf{y}} \end{bmatrix}$, $\mathbf{f} := \begin{bmatrix} \dot{\mathbf{y}} \\ \mathbf{F} \end{bmatrix}$, $\mathbf{x}_0 := \begin{bmatrix} \mathbf{y}_0 \\ \dot{\mathbf{y}}_0 \end{bmatrix}$,

we obtain the IVP

(1.4) $\begin{cases} \dot{\mathbf{x}}(t) = \mathbf{f}(t, \mathbf{x}) \ , \\ \mathbf{x}(t_0) = \mathbf{x}_0 \ . \end{cases}$

We note that (1.4) has the same (first order) form as (1.1), and is equivalent to (1.2). We say that the IVP (1.4) is in *standard* form. Throughout this book we shall use a notation like (1.4). Since the ODE is characterised by $\mathbf{f}$ and the initialisation by t_0 and $\mathbf{x}_0$, it is sometimes convenient to indicate problem (1.4) by IVP $(\mathbf{f}, t_0, \mathbf{x}_0)$. A vector function $\mathbf{x}$ of t satisfying IVP $(\mathbf{f}, t_0, \mathbf{x}_0)$ is called its *solution*. Important questions for such an IVP are:

– which requirements should $\mathbf{f}$ satisfy in order to guarantee the existence and uniqueness of $\mathbf{x}$ on a certain time interval?

– which properties does the solution **x** have?

– how can we actually calculate this solution?

Example I.1.

(i) Consider a particle which can only move along a line, the x-axis say. We might think of a bead on a straight spindle. The particle is not subject to forces. This means that in (1.2) the (scalar) force F vanishes, so that (1.2a) reads

$$\dot{\mathbf{x}} = \begin{bmatrix} 0 & 1 \\ 0 & 0 \end{bmatrix} \mathbf{x} , \quad \text{where} \quad \mathbf{x} := \begin{bmatrix} x \\ \dot{x} \end{bmatrix} .$$

Straightforward integration yields the solution

$$x(t) = x_0 + \dot{x}_0\, t .$$

This solution can be read as an expression of *Newton's first law*: a mass not subject to any force moves with constant - possibly vanishing - velocity.

(ii) Consider a one-dimensional *harmonic oscillator*: a particle which moves along the x-axis and is attracted to a fixed position, the origin say, with a force proportional to the distance from that position. Newton's second law gives (cf. (1.2a))

$$m\ddot{x} = -kx ,$$

with m the mass and $k > 0$ the spring constant. Rewritten as a first order ODE this ODE reads

$$\dot{\mathbf{x}} = \begin{bmatrix} 0 & 1 \\ -k/m & 0 \end{bmatrix} \mathbf{x} .$$

One may verify that the solution is given by

$$x(t) = x_0 \cos \omega(t - t_0) + \frac{\dot{x}_0}{\omega} \sin \omega(t - t_0) ,$$

where $\omega := \sqrt{k/m}$.

2. First order ODE

Let $\Omega \subset I\!\!R^n$ be an open, non-empty set. A mapping $\mathbf{f} : \Omega \to I\!\!R^n$ is called a *vector field* on Ω, usually written as

$$(2.1) \qquad \mathbf{f}(\mathbf{x}) := \begin{bmatrix} f_1(x_1, ..., x_n) \\ \vdots \\ f_n(x_1, ..., x_n) \end{bmatrix} .$$

We shall assume throughout that $\mathbf{f}$ is continuous. If, in addition, $\mathbf{f}$ is differentiable, the derivative of $\mathbf{f}$ is given by

$$(2.2) \qquad \mathbf{J}(\mathbf{x}) := \frac{\partial \mathbf{f}}{\partial \mathbf{x}} := \begin{bmatrix} \dfrac{\partial f_1}{\partial x_1} & \cdots & \dfrac{\partial f_1}{\partial x_n} \\ \vdots & & \\ \dfrac{\partial f_n}{\partial x_1} & \cdots & \dfrac{\partial f_n}{\partial x_n} \end{bmatrix} .$$

This derivative is called the *Jacobian matrix* or *functional matrix*. Its determinant $\det(\mathbf{J}(\mathbf{x}))$ is called the *Jacobian*.

If $n = 1$, f is a *scalar field*. We shall include this case even though we shall (generally) talk about a vector field. A vector field may also depend on an additional parameter t, i.e. $\mathbf{f} := \mathbf{f}(t, \mathbf{x})$. This parameter t will play the rôle of an independent parameter, while $\mathbf{x}$ will be assumed to depend on t. The set Ω from which $\mathbf{x}(t)$ has its values is called the *state space* or *phase space*. The notion of phase space is often reserved for spaces consisting of positions and velocities as in (1.3). If t takes its values from an interval I, we call $I \times \Omega$ the *time-state space*.

For a given $\mathbf{f}(t, \mathbf{x})$ defined on $I \times \Omega$ the associated IVP reads:

$$(2.3a) \qquad \dot{\mathbf{x}} = \mathbf{f}(t, \mathbf{x}) ,$$

$$(2.3b) \qquad \mathbf{x}(t_0) = \mathbf{x}_0 .$$

In Chapter II we shall show that this IVP has a unique solution on an interval $J \subset I$ around t_0, if $\mathbf{f}$ satisfies rather mild conditions. Initial value (or *Cauchy*) problems are also called *evolution problems*, because they deal with problems that evolve in time. To express the dependency on its initial value we shall sometimes denote the solution as (cf. Fig. I.1)

$$(2.4) \qquad \mathbf{x}(t) = \Psi(t \, ; \, t_0, \mathbf{x}_0) .$$

We now introduce the following useful concepts:

– The *orbit* or *trajectory* of IVP $(\mathbf{f}, t_0, \mathbf{x}_0)$ is the curve

$$\{ \Psi(t \, ; \, t_0, \mathbf{x}_0) \, | \, t \in J \}$$

in the state space. The corresponding *positive orbit* is obtained by taking the part with $t \geq t_0$.

– The *solution curve* or *integral motion* of IVP $(\mathbf{f}, t_0, \mathbf{x}_0)$ is the curve

$$\left\{ \begin{bmatrix} t \\ \Psi(t \, ; \, t_0, \mathbf{x}_0) \end{bmatrix} \, \middle| \, t \in J \right\}$$

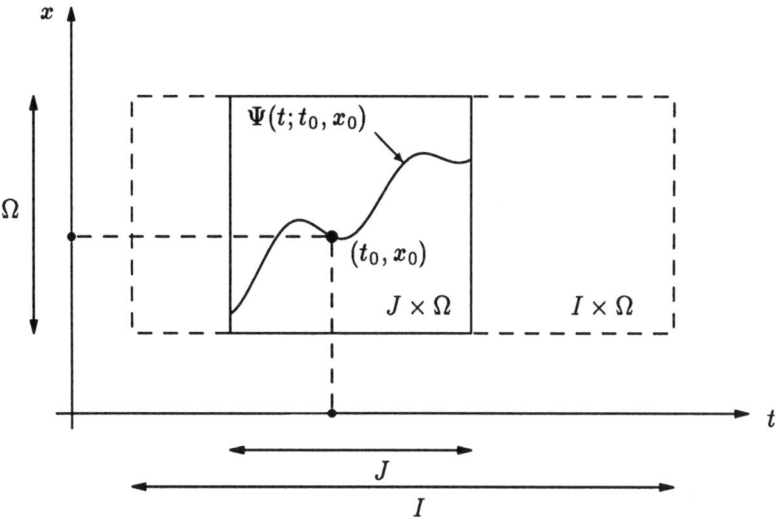

Figure I.1

in the time-state space ($\subset \mathbb{R}^{n+1}$). For $n = 1$ it is simply the graph of x as a function of t.

– The *direction field* of an ODE is the vector field

$$\begin{bmatrix} 1 \\ \mathbf{f}(t, \mathbf{x}) \end{bmatrix}.$$

Note that the orbit is tangent to the vector field at any point, while the solution curve is tangent to the direction field at any point.

Example I.2.
The vector field of Example I.1(i) is given by

$$\mathbf{f} = \begin{bmatrix} \dot{x} \\ 0 \end{bmatrix}.$$

In Fig. I.2, we have indicated $\mathbf{f}$ by arrows, which indicate both direction and magnitude of the force in the phase space. Clearly the trajectories are horizontal lines. The solution curves are straight lines in $\mathbb{R}^3$. For fixed $\dot{x}_0$ we can draw x as a function of t; in Fig. I.3 this is done for various x_0, a fixed value of $\dot{x}_0 > 0$, and $t \geq t_0$.

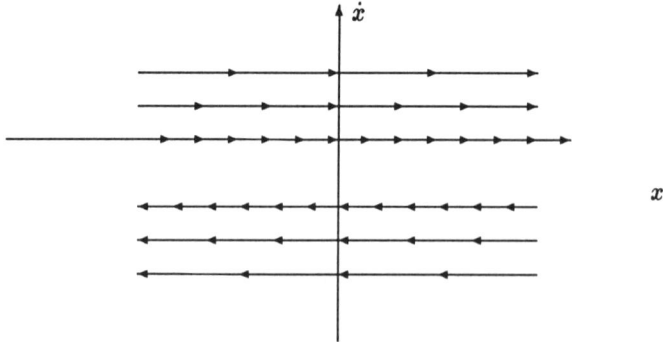

Figure I.2   Phase space and vector field for Example I.2.

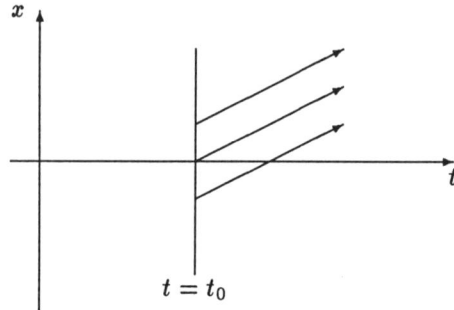

Figure I.3   Solution curves for Example I.2 for three x_0 values, a fixed value of $\dot{x}_0 > 0$, and $t \geq t_0$.

Example I.3.
The vector field corresponding to the one-dimensional harmonic oscillator in Example I.1 (ii) is given by

$$\mathbf{f} := \left[\begin{array}{c} \dot{x} \\ -\dfrac{k}{m}\,x \end{array} \right] ,$$

as drawn in Fig. I.4 for $k = m$. For these parameter values the orbits are circles. In general the orbits are ellipses.
A solution curve is sketched in Fig. I.5.
Of special importance are the points $\hat{\mathbf{x}} \in \Omega$, where

$$(2.5) \qquad \mathbf{f}(t, \hat{\mathbf{x}}) = \mathbf{0} , \qquad t \in I .$$

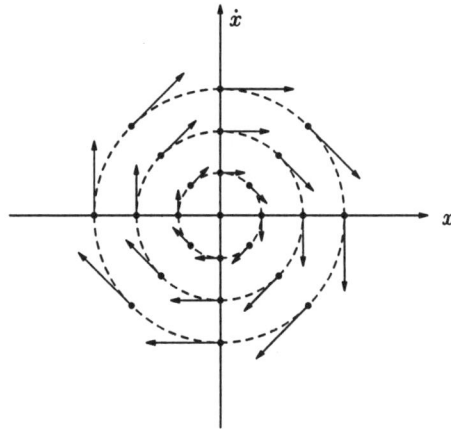

Figure I.4 Phase space and vector field for Example I.3.

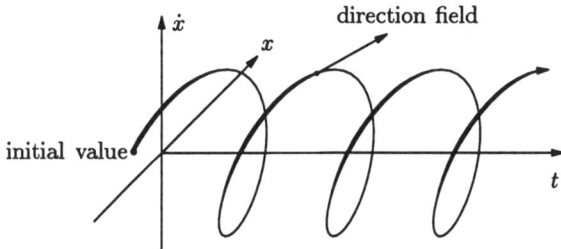

Figure I.5 Solution curve for Example I.3.

In the literature, they are referred to as *stationary, equilibrium, singular, fixed, rest,* or *critical* points. Here, we prefer to use the first term. A stationary point is a solution of the IVP $(\mathbf{f}, t_0, \hat{\mathbf{x}})$ for any $t_0 \in I$. The orbit is then contracted on one point.

A subset $\hat{\Omega} \subset \Omega$ is called an *invariant set* if for any $\mathbf{x}_0 \in \hat{\Omega}$ the solution of the IVP $(\mathbf{f}, t_0, \mathbf{x}_0)$ remains within $\hat{\Omega}$ for all $t \in J$.

Example I.4.

(i) All solutions of the free particle in Example I.1(i) are stationary if $\dot{x}_0 = 0$. The x-axis is an invariant set in the state space.

(ii) The harmonic oscillator in Example I.1(ii) has a stationary solution if $x_0 = \dot{x}_0 = 0$. The invariant sets in the state space are ellipses.

In the following we shall often use the equivalence between the IVP $(\mathbf{f}, t_0, \mathbf{x}_0)$ and a (Volterra) integral equation:

Property 2.6. *A function* $\mathbf{x}$ *of* t *is a solution of the IVP* $(\mathbf{f}, t_0, \mathbf{x}_0)$ *if and only if* $\mathbf{x}$ *is a solution of the integral equation*

$$(2.6a) \qquad \mathbf{x}(t) = \mathbf{x}_0 + \int_{t_0}^{t} \mathbf{f}\Big(s, \mathbf{x}(s)\Big) \, ds \ .$$

Proof: If we integrate (2.3a) on (t_0, t) and use (2.3b), we obtain

$$\int_{t_0}^{t} \dot{\mathbf{x}}(s) \, ds = \mathbf{x}(t) - \mathbf{x}_0 = \int_{t_0}^{t} \mathbf{f}\Big(s, \mathbf{x}(s)\Big) \, ds \ .$$

Differentiation of (2.6a) gives (2.3a), while substitution of $t = t_0$ in (2.6a) gives (2.3b).   ☐

Instead of prescribing conditions for $\mathbf{x}(t)$ at one particular point t_0, we may as well do this at more than one point. A particular case of such *multipoint conditions* are *two-point boundary conditions*, where the solution is partially prescribed at $t = t_0$ and partially at $t = t_1 \neq t_0$, say. The resulting *boundary value problem* (BVP) then is to find the solution of (2.3) on (t_0, t_1). Quite often the boundary conditions (BC) are *separated*, i.e. the ones at $t = t_0$ are decoupled from the ones at $t = t_1$. An important case of non-separated BC are *periodic* BC, i.e.

$$(2.7) \qquad \mathbf{x}(t_0) = \mathbf{x}(t_1) \ .$$

Although this book is mainly concerned with IVP, we have devoted a separate chapter (Chapter X) to BVP. The existence of solutions of IVP only depends on fairly weak smoothness assumptions on $\mathbf{f}$. For BVP the matter is more complicated, as the following example shows.

Example I.5.
Consider again the harmonic oscillator of Example I.1(ii) with $k = m$:

$$\ddot{x} = -x \ .$$

As remarked before, the general solution is given by

$$x(t) = a \cos t + b \sin t \ .$$

Requiring $x(0) = 0$ results in $a = 0$. In particular this implies that $x(\pi) = 0$ as well. Hence we cannot prescribe $x(\pi)$ arbitrarily. Therefore the BVP

$$\begin{cases} \ddot{x} = -x \\ x(0) = 0 \\ x(\pi) = c, \quad c \in \mathbb{R} \end{cases}$$

will not have a solution in general. If $c = 0$, there is one; but it is not unique, since any multiple of $\sin t$ then satisfies, too.

3. Classification and properties

In this section we shall further classify our problem setting. The ODE

(3.1) $\dot{\mathbf{x}} = \mathbf{f}(t, \mathbf{x})$

is called *linear* if $\mathbf{f}$ has the form

(3.2) $\mathbf{f}(t, \mathbf{x}) := \mathbf{A}\,\mathbf{x} + \mathbf{b}\,,$

where $\mathbf{A}(t) \in \mathbb{R}^{n^2}$ and $\mathbf{b}(t) \in \mathbb{R}^n$ for all t. A linear ODE is called *homogeneous* if $\mathbf{b}(t) \equiv 0$. If $\mathbf{A}(t) \equiv \mathbf{A}$ for some constant matrix $\mathbf{A}$, we call the ODE *linear with constant coefficients*. We remark that a vector field of the form (3.2) should rather be called '*affine*'; strictly speaking only homogeneous vector fields are linear in the sense that they satisfy the condition $\mathbf{f}(c_1\mathbf{x}_1 + c_2\mathbf{x}_2) = c_1\,\mathbf{f}(\mathbf{x}_1) + c_2\,\mathbf{f}(\mathbf{x}_2)$ for scalar c_1, c_2. However, we prefer to follow the usage common in the literature.

The ODE (3.1) is called *autonomous* if $\mathbf{f}$ does not depend on t explicitly, i.e.

(3.3) $\dot{\mathbf{x}} = \mathbf{f}(\mathbf{x})\,.$

For autonomous ODE we have the following *translation property*.

Property 3.4. *If $\mathbf{x}(t)$ is a solution of (3.3) on an interval (a, b), then for any $s \in \mathbb{R}$, $\mathbf{x}(t + s)$ is a solution of (3.3) on $t \in (a - s, b - s)$.*

Note that the solution curves in the time-state space corresponding to $\mathbf{x}(t + s)$ can be found from those corresponding to $\mathbf{x}(t)$ by shifting the latter over a distance s in the negative time direction. The orbits in state space remain the same, however. Using the notation in (2.4), we can write the translation property as

(3.5) $\Psi(t\,;\,t_0, \mathbf{x}_0) = \Psi(t - t_0; 0, \mathbf{x}_0)$ for all $t_0 \in I\,.$

This is illustrated in Fig. I.6.

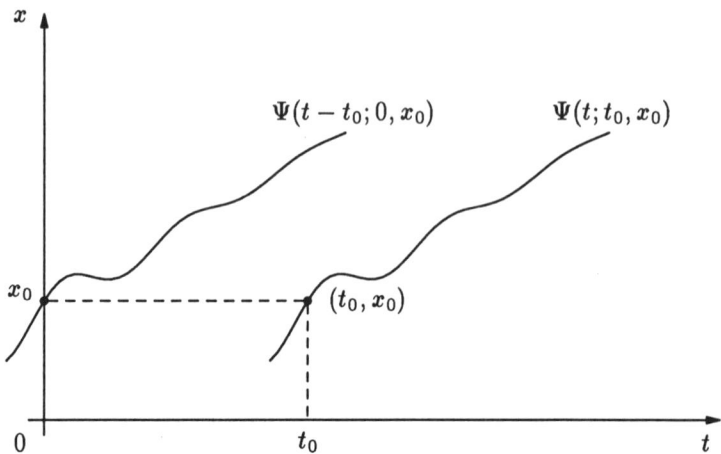

Figure I.6   Shifted solution of the autonomous ODE (3.3).

The trajectory is now completely determined by $\mathbf{x}_0$, of course. We conclude that it is not restrictive to take always $t_0 = 0$ for an autonomous system. The solution may then be written more simply as

$$\Psi(t \,;\, \mathbf{x}_0) \equiv \Psi(t \,;\, 0, \mathbf{x}_0) \;.$$

Example I.6.
Gravity may be assumed time independent for smaller scale situations. When a particle is dropped, it will follow a trajectory which is independent of the time it was being dropped. Now consider a unit mass under influence of a gravitational force $-g$, where g is the gravitational acceleration. If the distance z is measured in a positive direction from the earth's surface, we find, according to Newton's second law,

$$\ddot{z} = -g \;.$$

In the first order vector form the corresponding vector field reads

$$\mathbf{f}(\mathbf{x}) = \begin{bmatrix} 0 & 1 \\ 0 & 0 \end{bmatrix} \mathbf{x} - \begin{bmatrix} 0 \\ g \end{bmatrix} \;,$$

where $\mathbf{x} := (z, \dot{z})^T$.
Clearly, the system is linear and autonomous. If we prescribe z_0 and $\dot{z}_0$ at $t_0 = 0$, we obtain the solution

$$z(t) = z_0 + \dot{z}_0\, t - \tfrac{1}{2}\, g\, t^2 \ .$$

If is often convenient to have $t_0 = 0$ as a default. We can always achieve this by redefining the vector field by

$$\bar{\mathbf{f}}(t, \mathbf{x}) := \mathbf{f}(t + t_0, \mathbf{x}) \ .$$

Then, we have IVP $(\mathbf{f}, t_0, \mathbf{x}_0) = $ IVP $(\bar{\mathbf{f}}, 0, \mathbf{x}_0)$. Yet, since the solution of a non-autonomous solution depends on the choice of t_0, we sometimes need to show the explicit dependence on t_0.

A particular case of a non-autonomous ODE is a *periodic* ODE. Then we have for some $T > 0$ and all $t \in I$

$$(3.6) \qquad \mathbf{f}(t + T, \mathbf{x}) = \mathbf{f}(t, \mathbf{x}) \ .$$

The smallest value of T for which this holds is called the *period*. From (3.6) we see that the form of the solutions is not affected if we shift t_0 to $t_0 \pm nT$, $n \in I\!N$. This property can be expressed as follows:

$$(3.7) \qquad \Psi(t \pm nT\,;\, t_0 \pm nT, \mathbf{x}_0) = \Psi(t\,;\, t_0, \mathbf{x}_0) \ .$$

A solution $\mathbf{x}(t)$ is called *periodic* if there exists a period $T > 0$ such that $\mathbf{x}(t+T) = \mathbf{x}(t)$ for all t in the existence interval. We remark that a solution of a periodic ODE need *not* be periodic itself. However, if the solution happens to be periodic we have the following property:

Property 3.8. *If a solution of a periodic ODE is periodic, then it has the same period as the vector field.*

Proof: Assume the vector field $\mathbf{f}(t, \mathbf{x})$ to have period T and the solution $\mathbf{x}$ period $\tilde{T}$ with $\tilde{T} \neq T$. Because $\mathbf{x}$ is periodic, this also holds for its derivative $\dot{\mathbf{x}}$. Since $\dot{\mathbf{x}}(t) = \mathbf{f}(t, \mathbf{x})$ we conclude that $\mathbf{f}$ must be periodic, not only with period T, but also with period $\tilde{T}$. Thus we have $\mathbf{f}(t + \tilde{T}, \mathbf{x}) = \mathbf{f}(t, \mathbf{x})$. Because of the T-periodicity of $\mathbf{f}$ we may write for any $n \in I\!N$: $\mathbf{f}(t + \tilde{T} - nT, \mathbf{x}) = \mathbf{f}(t, \mathbf{x})$. By choosing n such that $0 < \tilde{T} - nT < T$, we may conclude that $\mathbf{f}$ is periodic with a period smaller than T. Because this is inconsistent with the assumptions, we conclude that $T = \tilde{T}$. $\square$

Example I.7.
For the harmonic oscillator in Example I.1(ii) we take a driving force equal to 2 cos t. For $k = m$ we obtain the periodic ODE

$$\dot{\mathbf{x}} = \begin{bmatrix} 0 & 1 \\ -1 & 0 \end{bmatrix} \mathbf{x} + \begin{bmatrix} 0 \\ 2\cos t \end{bmatrix} \ .$$

If $\mathbf{x}(0) = 0$, we find the apparently non-periodic solution

$$\mathbf{y}(t) = \begin{bmatrix} t \sin t \\ \sin t + t \cos t \end{bmatrix}.$$

The class of periodic solutions is a subset of the class of quasi-periodic solutions. A solution is called *quasi-periodic* if it contains at least two frequencies which have an irrational ratio. An example is the function

$$x(t) := \sin(\omega_1 t) + \sin(\omega_2 t).$$

If ω_1/ω_2 is irrational, this function is not periodic, although it has a regular structure.

We conclude this section by dwelling upon the shift Property 3.4 for autonomous ODE. For this we consider an autonomous IVP $(\mathbf{f}, 0, \mathbf{x}_0)$ that has a solution for any $\mathbf{x}_0 \in I\!\!R^n$. As before we denote this solution by $\Psi(t\,; \mathbf{x}_0)$. It is assumed that the solutions exist for all t. We now take t as parameter and consider the map $\Psi(t\,; \cdot) : I\!\!R^n \to I\!\!R^n$. In order to emphasise that it is in fact a transformation of $I\!\!R^n$, we shall write

$$(3.9) \qquad \Psi_t(\cdot) := \Psi(t\,; \cdot).$$

Varying t we obtain a family of such transformations, which calls for the intuitive idea of 'flow'. Hence we call (3.9) the *phase flow*. We note that the following composition rule holds:

$$\Psi_t \circ \Psi_s(\mathbf{x}_0) := \Psi_t\Big(\Psi_s(x_0)\Big) = \Psi\Big(t\,; \Psi(s\,; \mathbf{x}_0)\Big) = \Psi_{t+s}(\mathbf{x}_0),$$

so

$$(3.10) \qquad \Psi_t \circ \Psi_s = \Psi_{t+s} = \Psi_s \circ \Psi_t.$$

Hence the phase flow satisfies the *group properties*

$$(3.11a) \qquad \Psi_0 = \mathbf{I} \qquad \text{(identical mapping in } I\!\!R^n)$$

$$(3.11b) \qquad \Psi_t^{-1} = \Psi_{-t}.$$

Moreover, Ψ_t is differentiable as a function of t. Since

$$(3.11c) \qquad \frac{d}{dt}\Psi_t(\mathbf{x})\Big|_{t=0} = \mathbf{f}(\mathbf{x}) \qquad \text{for all } \mathbf{x} \in I\!\!R^n,$$

we call $\mathbf{f}$ the *infinitesimal generator* of the group. This generator is typical for Ψ_t. Indeed, also the converse holds, as can easily be checked.

Property 3.12. *If Ψ_t, $t \in I\!\!R$, is a one-parameter group of transformations from $I\!\!R^n \to I\!\!R^n$, being differentiable with respect to t, then $\Psi_t(\mathbf{x})$ is the solution of the IVP $(\mathbf{f}, 0, \mathbf{x})$, where $\mathbf{f}$ is given by*

$$\mathbf{f} = \frac{d}{dt}\Psi_t\Big|_{t=0}.$$

Example I.8.

(i)   A simple autonomous scalar IVP is given by

$$\begin{cases} \dot{x} = x \\ x(0) = x_0 \ , \end{cases}$$

with solution

$$x(t) = \Psi(t\,;\,x_0) := x_0\,e^t \ .$$

The group properties are trivial here. One immediately sees e.g.

$$\Psi_s \circ \Psi_t(x_0) = x_0\,e^{s+t} = \Psi_{s+t}(x_0) \ .$$

(ii) The general solution of the harmonic oscillator in Example I.1(ii) with $k = m$ is

$$\mathbf{x}(t) = \begin{bmatrix} \cos t & \sin t \\ -\sin t & \cos t \end{bmatrix} \mathbf{x}_0 =: \mathbf{Q}(t)\,\mathbf{x}_0 \ .$$

The (orthogonal) transformation $\mathbf{Q}$ satisfies

$$\mathbf{Q}(s)\,\mathbf{Q}(t) = \mathbf{Q}(s+t) \ ,$$

from which all group properties follow.

4. Higher order ODE

As indicated in §1, the notion of *order* relates to the highest order derivative appearing in the equations. In this section we restrict ourselves to scalar problems. Generalisations to systems of vector-valued higher order ODE are obvious.

Let $y(t)$ be a scalar function, which is n times continuously differentiable on an interval $I \subset \mathbb{R}$. Let us denote derivatives as

$$\frac{d^k}{dt^k}\,y(t) = y^{(k)}(t) \ , \qquad k = 0, 1, ..., n \ ,$$

or

$$y = y^{(0)}, \quad \dot{y} = y^{(1)}, \quad \ddot{y} = y^{(2)}, \quad ... \ .$$

In general we call a relation between $y^{(0)}, ..., y^{(n)}$, on an interval $I \subset \mathbb{R}$, say

$$(4.1) \qquad F\Big(t, y(t), y^{(1)}(t), ..., y^{(n)}(t)\Big) = 0 \ ,$$

an *ODE of order n*. In the form (4.1) this higher order ODE (HODE) is *implicit*. Quite often we have an *explicit* form

$$(4.2) \qquad y^{(n)} = G\Big(t, y(t), y^{(1)}(t), ..., y^{(n-1)}(t)\Big) \ .$$

The transition from the implicit form (4.1) to the explicit form (4.2) is possible at least in the vicinity of points in the time-state space where $\partial F/\partial y^{(n)} \neq 0$. There exist interesting cases where the latter condition is not satisfied, see Chapter IX.

Explicit HODE can be written as a first order system. This transformation is not unique. As in the transition from (1.2) to (1.3) we define a vector $\mathbf{x}$, with components x_i, $i = 1, ..., n$, by

$$(4.3a) \qquad x_i(t) := y^{(i-1)}(t) ,$$

and a vector field $\mathbf{f}(t, \mathbf{x})$ by

$$(4.3b) \qquad \mathbf{f}(t, \mathbf{x}) := \Big(x_2, x_3, ..., x_n, G(t, x_1, ..., x_{n-1})\Big)^T .$$

Then the HODE (4.2) is equivalent to the first order system

$$\dot{\mathbf{x}} = \mathbf{f}(t, \mathbf{x}) .$$

Sometimes a different transformation may be more appropriate, as the following example shows:

Example I.9.
Consider the so-called Liénard equation,

$$\ddot{y} + f(y)\,\dot{y} + g(y) = 0 ,$$

for some given functions f and g. Applying (4.3) we obtain

$$\dot{x}_1 = x_2$$
$$\dot{x}_2 = -f(x_1)\,x_2 - g(x_1) .$$

An alternative can be given if a primitive of f is to hand. So let

$$F(y) := \int_0^y f(s)\,ds.$$

Since

$$\ddot{y} + f(y)\,\dot{y} = \frac{d}{dt}\Big(\dot{y} + F(y)\Big) ,$$

we obtain the commonly used first order form

$$\dot{x}_1 = x_2 - F(x_1)$$
$$\dot{x}_2 = -g(x_1) .$$

A linear explicit HODE has the form

(4.4) $y^{(n)} + a_{n-1}(t)\, y^{(n-1)} + ... + a_0(t)\, y = b(t)$,

where $a_{n-1}, ..., a_0$ and b are given functions. Implicit HODE that are linear can always be made explicit. The transformation (4.3) yields the first order form $\dot{\mathbf{x}} = \mathbf{A}\,\mathbf{x} + \mathbf{b}$ with

(4.5a) $\mathbf{b}(t) := [0, 0, ..., 0, b(t)]^T$,

and

(4.5b) $\mathbf{A}(t) := \begin{bmatrix} 0 & 1 & 0 & \cdots & & 0 \\ \vdots & \vdots & \ddots & & & \vdots \\ \vdots & \vdots & & \ddots & & 0 \\ 0 & 0 & & & & 1 \\ -a_0 & -a_1 & \cdots & & \cdots & -a_{n-1} \end{bmatrix}$.

This matrix $\mathbf{A}(t)$ is called the *companion matrix*.

Naturally, we see from the transformation of an n-th order HODE to a first order system that n initial values are needed for having a meaningful initial value problem. In the next chapter we shall dwell on questions like existence and uniqueness.

5. Difference equations

In the foregoing sections we focused on ODE, which contained the derivative $\dot{\mathbf{x}}$ of the state vector. This implies that we tacitly assumed both the state vector and the independent variable t to be continuous. For some processes such a continuous description is not possible or not appropriate. One may think of population models or models in economies, where either the dependent or independent variables are taken from a discrete set of values. Hence, rather than differential quotients one is bound to use *difference quotients*. This leads to so-called *difference equations* (Δ-equations); the most important class of Δ-equations in this book will be that of numerical schemes arising from ODE. Besides these origins, difference equations have a place of their own in analysis with many analogues to differential equations [2]. It is not our intention here to consider Δ-equations exhaustively. Rather we should like to indicate a few useful properties which will play a rôle later on.

Instead of the continuous variable t in ODE we shall use the index i in Δ-equations, which is supposed to run from 0 to $N \in I\!N$; the latter index set is indicated by J. Let a sequence of vector fields $\mathbf{f}_i(\mathbf{x}) : \Omega \to I\!R^n$ be given, where $i \in J$ and $\Omega \subset I\!R^n$. Then a *first order* Δ-equation has the form

(5.1a) $\mathbf{x}_{i+1} = \mathbf{f}_i(\mathbf{x}_i)$, $\mathbf{x}_i \in \Omega$, $i \in J$.

By defining

(5.1b) $\mathbf{x}_0 = \mathbf{v}_0$, $\mathbf{v}_0 \in I\!\!R^n$,

we obtain a discrete analogue of an IVP. As for ODE, we can define *orbit* or *trajectory*, and *solution curve* or *integral motion* in an obvious way; they are the discrete point sets within $\Omega \subset I\!\!R^n$ and $J \times \Omega \subset I\!\!R^{n+1}$ respectively. A solution will be denoted by $\{\mathbf{x}_i\}_{i \in J}$. In Fig. I.7 we have drawn a solution for the scalar case $n = 1$.

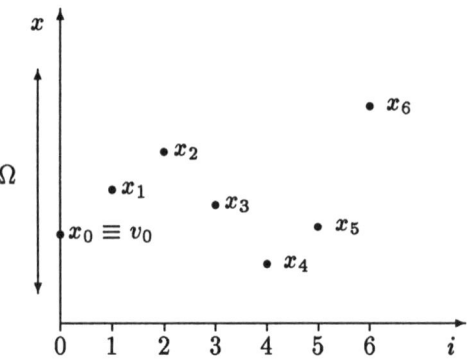

Figure I.7   Discrete solution curve.

Because of the recursive character (5.1a) is also called a (*one-step*) *recursion*. A *stationary solution* of (5.1a) is a constant solution $\hat{\mathbf{x}}$ which satisfies

(5.2) $\hat{\mathbf{x}} = \mathbf{f}_i(\hat{\mathbf{x}})$, $i \in J$.

If the $\{\mathbf{f}_i\}$ are independent of i, the Δ-equation (5.1a) is called *autonomous*. The Δ-equation is called *linear* if it has the form

(5.3) $\mathbf{x}_{i+1} = \mathbf{A}_i \mathbf{x}_i + \mathbf{b}_i$, $i \in J$,

where $\mathbf{A}_i \in I\!\!R^{n \times n}$ and $\mathbf{b}_i \in I\!\!R^n$. This recursion is *homogeneous* if $\mathbf{b}_i \equiv 0$. In particular, for $n = 1$ we have a *scalar* recursion of the form

(5.4) $x_{i+1} = a_i x_i + b_i$, $i \in J$.

Its solution is given by

(5.5) $x_i = \left(\prod_{j=0}^{i-1} a_j \right) x_0 + \sum_{j=0}^{i-1} \left(\prod_{l=j+1}^{i-1} a_l \right) b_j$, with the conventions that empty sums are 0 and empty products are 1. For convenience we restrict ourselves to scalar equations in the rest of this section.

A scalar implicit k-th order Δ-equation is given by

$$(5.6) \qquad F_i(x_{i+1}, ..., x_{i+1-k}) = 0 , \qquad i = k-1, k, ... ,$$

for which $x_0, ..., x_{k-1}$ need be given in order to have a well-defined IVP. A Δ-equation like (5.6) is also called a k-step recursion. If it can be made explicit, it reads as

$$(5.7) \qquad x_{i+1} = G_i(x_i, ..., x_{i+1-k}) , \qquad i = k-1, k, ...$$

In analogy to the differential operator one may introduce the *shift operator* E, defined by

$$(5.8a) \qquad E\, x_i := x_{i+1} .$$

The so-called *forward difference* is then defined by

$$(5.8b) \qquad \Delta\, x_i := (E - I)\, x_i ,$$

and the *backward difference* by

$$(5.8c) \qquad \nabla\, x_i := (E - I)\, x_{i-1} = (I - E^{-1})\, x_i .$$

The k-step Δ-equation (5.6) can then be written as

$$(5.9) \qquad F_i(E^k\, x_{i-k+1}, E^{k-1}\, x_{i-k+1}, ..., E\, x_{i-k+1}, x_{i-k+1}) , \quad i = k-1, k, ...$$

In the linear, explicit case the k-th order recursion reads typically as

$$(5.10) \qquad x_{i+1} = \sum_{j=1}^{k} a_{i,j}\, x_{i-j+1} + b_i .$$

We can write (5.10) as a first order system $\mathbf{x}_{i+1} = \mathbf{A}\mathbf{x}_i + \mathbf{b}_i$, $i \geq 0$, by defining

$$(5.11) \qquad \mathbf{x}_i := \begin{bmatrix} x_i \\ \vdots \\ x_{i+k-1} \end{bmatrix} , \qquad \mathbf{b}_i := \begin{bmatrix} 0 \\ \vdots \\ 0 \\ b_{i+k} \end{bmatrix} ,$$

$$\mathbf{A}_i := \begin{bmatrix} 0 & 1 & 0 & \cdots & & 0 \\ \vdots & \vdots & \ddots & & & \vdots \\ \vdots & \vdots & & \ddots & & 0 \\ 0 & 0 & & & & 1 \\ a_{i+k-1,k} & \cdots & \cdots & \cdots & & a_{i+k-1,1} \end{bmatrix} .$$

If we have $a_{i,j} \equiv a_j$ and $b_i \equiv 0$, i.e. the equation is autonomous and homogeneous, we may write (5.10) in the form

$$(5.12) \qquad \left(E^k - \sum_{j=1}^{k} a_j E^{k-j} \right) x_{i-k+1} = 0 \ .$$

Trying solutions of the form $x_i = \lambda^i$ for some λ, we find the *characteristic polynomial*

$$(5.13) \qquad \rho(\lambda) := \lambda^k - \sum_{j=1}^{k} a_j \lambda^{k-j} \ .$$

If λ is a simple root of $\rho(\lambda) = 0$, (5.12) has indeed a solution $\{x_i\}$ with $x_i = \lambda^i$. If λ is l-fold, there are such solutions with $x_i = \lambda^i, x_i = i\,\lambda^i, ..., x_i = i^{l-1}\,\lambda^i$; see also Appendix D. The general solution is then a linear combination of such solutions. The polynomial $\rho(\lambda)$ is precisely the characteristic polynomial of the matrix $\mathbf{A}$ in (5.11) with $a_{i,j} \equiv a_j$.

Example I.10.
Consider the Δ-equation named after Fibonacci:

$$x_{i+1} = x_i + x_{i-1} \ .$$

The characteristic polynomial is given by

$$\rho(\lambda) = \lambda^2 - \lambda - 1 \ ,$$

with roots $\lambda_1 = \frac{1}{2} + \frac{1}{2}\sqrt{5}$ and $\lambda_2 = \frac{1}{2} - \frac{1}{2}\sqrt{5}$. The general solution is given by

$$x_i = c_1 \lambda_1^i + c_2 \lambda_2^i \ ,$$

with the coefficients c_1, c_2 to be determined from the initial values x_0 and x_1.

6. Discretisations

In this section we shall give an introduction to numerical concepts met when approximating solutions of IVP numerically. The starting point is that one needs values of the solution at certain time points only, e.g. a set sufficient to produce a graph. As will turn out this is also a prerequisite for the numerical methods in this book, which will be based on Δ-equations. We restrict ourselves here to scalar equations.

Let $t_0, ..., t_N$ be a set of time points and assume we like to approximate the solution values $x(t_i)$, $i = 0, ..., N$. The set $\{x(t_i)\}_{i=0}^{N}$ is called a *discretisation* of the interval $\{x(t) \mid t_0 \leq t \leq t_N\}$. The set $\{t_i\}_{i=0}^{N}$ is called the *grid*. Because all quantities are calculated at the grid points only, we have to replace the differential operator by a differential quotient. This replacement is called a *discretisation* of the operator. Given the grid this discretisation is not unique. For example, a simple discretisation of $\dot{x}(t_i)$ is

$$\dot{x}(t_i) \doteq \frac{x(t_{i+1}) - x(t_i)}{t_{i+1} - t_i} .$$

Here, the notation $\doteq$ denotes that equality only holds if a limiting procedure is applied, namely $t_{i+1} \to t_i$. The ODE $\dot{x} = f(t, x)$ could be discretised as

$$(6.1) \qquad \frac{x(t_i + h_i) - x(t_i)}{h_i} \doteq f\Big(t_i, x(t_i)\Big) , \quad h_i := t_{i+1} - t_i .$$

Since we know $x(t_0)$, we may use (6.1) to compute an approximation of the solution. Another way to obtain approximations is to rewrite the ODE as an integral equation first (cf. (2.6a)); so

$$(6.2) \qquad x(t) = x(t_0) + \int_{t_0}^{t} f\Big(s, x(s)\Big) ds .$$

We can now *discretise* the integral. This leads to so-called *quadrature formulae*. First, we write

$$(6.3) \qquad x(t_{i+1}) = x(t_i) + \int_{t_i}^{t_{i+1}} f\Big(s, x(s)\Big) ds .$$

The simplest way to discretise the integral is to approximate f on the interval $[t_i, t_{i+1}]$ by its value at t_i. This yields the *Euler forward formula*

$$(6.4) \qquad x(t_{i+1}) \doteq x(t_i) + h_i f\Big(t_i, x(t_i)\Big) .$$

It is a so-called *explicit, one-step recursion*. In general, one-step methods are given by

$$(6.5) \qquad x(t_{i+1}) \doteq x(t_i) + h_i \Phi(t_i, x(t_i), x(t_{i+1}), h_i) .$$

If Φ is *not* dependent on $x(t_{i+1})$ we have an *explicit* method, as in (6.4). Otherwise, (6.5) is called an *implicit* one-step method. We shall dwell on one-step methods in Chapter III. One can also find k-step methods to approximate a solution of an ODE numerically. This will be investigated in Chapter VII.

Here, we should like to work out how well an ODE is approximated by some discretisation with step h. Consider the standard ODE

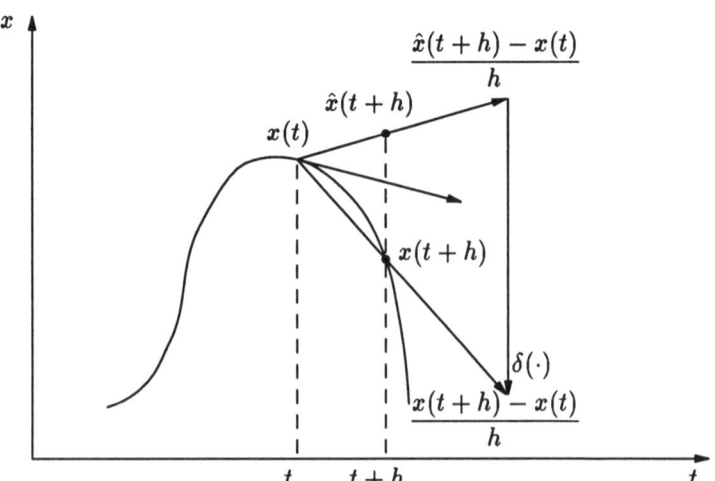

Figure I.8 Local error.

$$(6.6) \qquad \dot{x} = f(t, x) \ .$$

Let the value $x(t)$ at time t of some solution be known. If we do one step in parameter space, say from t to $t + h$, we should follow the solution arriving at the point $x(t + h)$ (see Fig. I.8). Doing one step with (6.5) would give a point $\hat{x}(t+h)$, say, where $\hat{x}(t+h)$ is different from $x(t+h)$ in general. Now note that it is *not* sufficient to require that $\hat{x}(t+h) \to x(t+h)$ if $h \to 0$, as the directions of the vectors $(h, x(t + h) - x(t))^T$ and $(h, \hat{x}(t + h) - x(t))^T$ in the time-state space may remain different if $h \to 0$. This would imply that the slope of the numerically obtained approximation does not approach the slope of the exact solution, not even if $h \to 0$. Hence the crucial quantities are $(x(t+h) - x(t))/h$ and $(\hat{x}(t + h) - x(t))/h$. We therefore define the *local discretisation error* $\delta(\cdot)$ by

$$(6.7) \qquad \delta(x(t), h) := \frac{x(t+h) - x(t)}{h} - \frac{\hat{x}(t+h) - x(t)}{h} = \frac{x(t+h) - \hat{x}(t+h)}{h} \ .$$

A reasonable requirement for any numerical method is then

$$(6.8) \qquad \delta(x(t), h) \to 0 \qquad \text{if } h \to 0 \ .$$

We call this property *consistency*. For an explicit method consistency implies

$$\Phi(t, x(t), h) \to f(t, x) \qquad \text{if } h \to 0 \ .$$

A general definition of local discretisation errors comes from (6.5), viz.

(6.9) $\delta(x(t), h) := [x(t+h) - x(t) - h\,\Phi(t, x(t), x(t+h), h)]/h$.

In other words, $h\,\delta(\cdot)$ is *the residual that is found when the exact solution is substituted in the Δ-equation.* For an explicit one-step method (6.9) and (6.7) coincide. For an implicit method its interpretation is more complicated.

Example I.11.

(i) For the *Euler forward method,* where $\Phi(\cdot) = f(t, x(t))$, we can find the local discretisation error $\delta(\cdot)$ by a Taylor expansion of x at t:

$$x(t+h) = x(t) + h\,\dot{x}(t) + \frac{h^2}{2}\,\ddot{x}(\tau) , \qquad \tau \in (t, t+h)$$

$$= x(t) + h\,f\Big(t, x(t)\Big) + \frac{h^2}{2}\,\ddot{x}(\tau) .$$

Hence $\delta(x(t), h) = \frac{h}{2}\,\ddot{x}(\tau)$.

(ii) If we use $f(s, x(s)) \doteq f(t+h, x(t+h))$ in (6.2), we find the *Euler backward method,* with $\Phi(x(t), h) := f(t+h, x(t+h))$. It is apparently an implicit method. Using a Taylor expansion now at $t+h$ we find

$$x(t) = x(t+h) - h\,\dot{x}(t+h) + \frac{h^2}{2}\,\ddot{x}(\zeta) , \qquad \zeta \in (t, t+h)$$

$$= x(t+h) - h\,f\Big(t+h, x(t+h)\Big) + \frac{h^2}{2}\,\ddot{x}(\zeta) .$$

Hence $\delta(x(t), h) = -\frac{h}{2}\,\ddot{x}(\zeta)$.

Exercises Chapter I

1. Classify the following ODE and give the stationary points, if any.

$$\ddot{x} = (t-1)\,x + x^2$$
$$\dot{x} = \tan x + \log t , \qquad t > 0$$
$$\dot{x} = x\,\sin(tx)$$
$$\ddot{x} = x - \dot{x} + \sin x$$
$$\dot{x} = e^t\,x .$$

2. Consider the ODE

$$\dot{x} = (x^2 - 1)\,t^p , \qquad p \geq 0, \ t \geq 0 .$$

a) Determine the stationary points.

b) Sketch the vector field.

c) Find out (roughly) how the solutions behave that have initial values $x(0) = -2$, 0, and 2 respectively.

3. How does the solution of Example I.6 look like when $t_0 \neq 0$?

4. Consider the system

$$\begin{cases} \dot{x}_1 = (x_2 + 1)\cos at \\ \dot{x}_2 = x_1 + x_2 \, . \end{cases}$$

a) Classify the system (distinguish between $a = 0$ and $a \neq 0$).

b) Write the system in standard form using vectors.

c) Indicate stationary points, if any.

d) Sketch the vector field in $\mathbb{R}^2$.

e) Find out how the solutions, starting at $(0,0)^T$, $(1,1)^T$, $(-1,-1)^T$, $(1,-1)^T$ and $(-1,1)^T$, behave.

5. Consider the phase flow

$$\Psi_t(x_1, x_2) = \begin{bmatrix} e^{-t}\, x_1 \\ e^{-2t}\, x_2 \end{bmatrix} \, .$$

a) Show that Ψ_t is a group of transformations of $\mathbb{R}^2$ into itself. Determine the infinitesimal generator.

b) Sketch the trajectory of an arbitrary point under this transformation if t varies.

6. Rewrite the second order system

$$\begin{cases} \ddot{x} = -y + \dot{x} \\ \ddot{y} = -x + \dot{y} \end{cases}$$

as a first order system. Find out whether there are stationary points.

7. Investigate the stationary solutions of the Δ-equation

$$x_{i+1} = x_i \sqrt{1 - x_i^2} \, .$$

8. a) Let $g(t)$ be twice continuously differentiable on (t_i, t_{i+1}). Suppose we only know $g(t_i)$ and $g(t_{i+1})$. Then $g(t)$ can be approximated using linear interpolation :

$$g(t) \doteq l(t) := \frac{t - t_i}{t_{i+1} - t_i} g(t_{i+1}) + \frac{t - t_{i+1}}{t_i - t_{i+1}} g(t_i) \ .$$

The error is given by

$$g(t) - l(t) = \tfrac{1}{2} (t - t_i)(t - t_{i+1}) \ddot{g}(\tau) \ ,$$

for some $\tau \in (t_i, t_{i+1})$. From this derive the trapezoidal rule

$$\int_{t_i}^{t_{i+1}} g(t) \, dt = \tfrac{1}{2} h \left(g(t_i) + g(t_{i+1}) \right) - \tfrac{1}{12} h^3 \, \ddot{g}(\tau) \ ,$$

with $h := t_{i+1} - t_i$.

   b) Construct a one-step method based on the trapezoidal rule for the scalar ODE $\dot{x} = f(t, x)$. Determine whether this method is implicit or explicit, and check the consistency.

II

Existence, Uniqueness, and Dependence on Parameters

In this chapter we consider and analyse a number of fundamental properties of initial value problems (IVP). In §1 we first show the uniqueness of the solution if a so-called Lipschitz continuity requirement is satisfied. To this end the celebrated Lemma of Gronwall is employed. Then we prove the existence of a solution based on the construction of converging Picard iterates. The convergence is typically a local one. Therefore we show in §3 that we can continue a solution beyond such small intervals and indeed indicate maximum existence intervals. The dependence of a solution on the initial value and the vector field is the subject of the last two sections. In §4 we derive upper bounds for the variations in the solutions if the system is perturbed. These upper bounds are used in §5 to show that continuous perturbations give rise to continuous variations in the solutions. Similarly, we show that perturbations which are differentiable with respect to some parameter result in variations in the solution which are also differentiable with respect to this parameter.

1. Lipschitz continuity and uniqueness

In the following we shall frequently make use of norms. In Appendix B general properties and examples of norms are given. Unless explicitly stated otherwise we shall use the Euclidean norm, defined by

$$(1.1) \qquad \|\mathbf{x}\| := \sqrt{\mathbf{x}^T \mathbf{x}} = \sqrt{\sum_{i=1}^{n} x_i^2}$$

for a vector $\mathbf{x} := (x_1, ..., x_n)^T$ in $\mathbb{R}^n$.

If the components of $\mathbf{x}$ depend on $t \in I$, the norm of $\mathbf{x}(t)$ on I is defined as

$$(1.2) \qquad \|\mathbf{x}\|_I := \sup_{t \in I} \|\mathbf{x}(t)\| \ .$$

Note that in this case the interval over which the supremum is taken is indicated by the index I. In a similar way we introduce the norm of a vector field $\mathbf{f}(t, \mathbf{x})$ with $(t, \mathbf{x}) \in I \times \Omega$ as

$$(1.3) \qquad \|\mathbf{f}\|_{I \times \Omega} := \sup_{\substack{t \in I \\ \mathbf{x} \in \Omega}} \|\mathbf{f}(t, \mathbf{x})\| .$$

When using these norms we have to establish whether the suprema indeed exist. For example, if $\mathbf{x}(t)$ is continuous in $t \in I$ and I is closed and bounded, then the right-hand side in (1.2) indeed exists.

For matrices $\mathbf{A} \in I\!\!R^{n \times n}$ we can associate a matrix norm with any norm defined in $I\!\!R^n$ as follows:

$$\|\mathbf{A}\| := \max_{\mathbf{x} \neq 0} \frac{\|\mathbf{A}\mathbf{x}\|}{\|\mathbf{x}\|} = \max_{\|\mathbf{x}\|=1} \|\mathbf{A}\mathbf{x}\| .$$

For instance, with the Euclidean norm (1.1) we associate the (Euclidean) matrix norm

$$(1.4) \qquad \|\mathbf{A}\| = \sqrt{\max_{\|\mathbf{x}\|=1} (\mathbf{x}^T \mathbf{A}^T \mathbf{A} \mathbf{x})} = \max_i |\lambda_i|^{\frac{1}{2}} ,$$

with λ_i the eigenvalues of $\mathbf{A}^T \mathbf{A}$.

Now consider the initial value problem (IVP)

$$(1.5) \qquad \begin{cases} \dot{\mathbf{x}} = \mathbf{f}(t, \mathbf{x}) , \\ \mathbf{x}(t_0) = \mathbf{x}_0 , \quad t_0 \in I\!\!R . \end{cases}$$

Anticipating the existence of the solution, which will be considered in §2, we investigate the uniqueness of solutions. The following example demonstrates that the IVP (1.5) may have more than one solution if one requires only continuity of $\mathbf{f}$.

Example II.1.
Consider the scalar field

$$f(x) = \begin{cases} 2\sqrt{x} & x \geq 0 \\ 0 & x < 0 . \end{cases}$$

The corresponding IVP with $x(0) = 0$ has at least two solutions. Apparently, $x_1(t) \equiv 0$ is a solution, but also x_2 defined by $x_2(t) = 0$ for $t < 0$, and $x_2(t) = t^2$ for $t \geq 0$ satisfies the IVP.

Below we shall show that uniqueness of the solution of (1.5) is guaranteed if the vector field is required to be Lipschitz continuous. This property is stronger than continuity, but weaker than differentiability.

Definition 1.6 (Lipschitz continuity). *The vector field* $\mathbf{f}(t, \mathbf{x})$ *is Lipschitz continuous on* $I \times \Omega$ *if a constant* L *exists such that for all* $\mathbf{x}, \mathbf{y} \in \Omega$ *and all* $t \in I$

(1.6a) $\|\mathbf{f}(t, \mathbf{y}) - \mathbf{f}(t, \mathbf{x})\| \leq L \|\mathbf{y} - \mathbf{x}\|$.

If $\mathbf{f}$ is Lipschitz continuous on $I \times \Omega$ we denote this as $\mathbf{f} \in \text{Lip}(I \times \Omega)$. This property expresses that the function $\mathbf{f}$ can be bounded by a linear function on Ω for all $t \in I$. In practice, it may be tedious to establish whether a given vector field has this property. For convex Ω we may formulate a useful criterion. A set $\Omega \in I\!\!R^n$ is convex if the line segment between two arbitrary points in Ω is completely contained in Ω.

Property 1.7. *If* $\mathbf{f}(t, \mathbf{x})$ *is defined on* $I \times \Omega$, *with* $\Omega \subset I\!\!R^n$ *convex, and* $\mathbf{f}$ *is continuously differentiable with respect to* $\mathbf{x} \in \Omega$, *and, finally, the Jacobian matrix* $\mathbf{J}$ *is bounded on* $I \times \Omega$, *i.e.*

(1.7a) $L := \|\mathbf{J}(t, \mathbf{x})\|_{I \times \Omega} < \infty$,

then $\mathbf{f} \in \text{Lip}(I \times \Omega)$ *with Lipschitz constant* L.

Proof: The line segment joining two arbitrary points $\mathbf{x}, \mathbf{y} \in \Omega$ is given by

$$\mathbf{x} + s(\mathbf{y} - \mathbf{x}) , \qquad 0 \leq s \leq 1 .$$

For any $t \in I$ we may write

$$\mathbf{f}(t, \mathbf{x}) - \mathbf{f}(t, \mathbf{y}) = \int_0^1 \frac{d}{ds} \mathbf{f}\Big(t, \mathbf{x} + s(\mathbf{y} - \mathbf{x})\Big) ds =$$

$$= \int_0^1 \mathbf{J}\Big(t, \mathbf{x} + s(\mathbf{y} - \mathbf{x})\Big) (\mathbf{x} - \mathbf{y}) \, ds .$$

Taking norms at both sides we obtain the inequality:

$$\|\mathbf{f}(t, \mathbf{x}) - \mathbf{f}(t, \mathbf{y})\| \leq \int_0^1 \left\| \mathbf{J}\Big(t, \mathbf{x} + s(\mathbf{y} - \mathbf{x})\Big) \right\| \|\mathbf{x} - \mathbf{y}\| ds$$

$$\leq L \|\mathbf{x} - \mathbf{y}\|$$

with L given by (1.7a). Here we use the multiplicativity property $\|\mathbf{A}\mathbf{x}\| \leq \|\mathbf{A}\| \|\mathbf{x}\|$ mentioned in Appendix B. $\square$

Example II.2.
Consider the following scalar fields:

(i) $f(x) = \sin x$. Then $f \in \mathrm{Lip}(I\!\!R)$ with $L = \max\limits_{x \in I\!\!R} |\cos(x)| = 1$.

(ii) $f(t, x) = t \sin x$. Then $f \in \mathrm{Lip}(I \times I\!\!R)$ with I any bounded $\Omega \subset I\!\!R$. A Lipschitz constant is given by $L = \sup\limits_{t \in I} |t|$.

(iii) $f(x) = |x|$. Then $f \in \mathrm{Lip}(I\!\!R)$ with $L = 1$.

(iv) $f(x) = x^2$. For any bounded $\Omega \subset I\!\!R$, $f \in \mathrm{Lip}(\Omega)$ with $L = \sup\limits_{x \in \Omega} |2x|$.

(v) $f(t, x) = x/t$, $t > 0$. This f is Lipschitz continuous on $I \times I\!\!R$ for intervals I of the form $I = [t_0, \infty)$, $t_0 > 0$, with $L = 1/t_0$. This f is not Lipschitz continuous if I includes the origin $t = 0$.

(vi) $f(x) = \sqrt{x}$. Then $f \in \mathrm{Lip}(\Omega)$ with $\Omega = (x_0, \infty)$, $x_0 > 0$, with $L = 1/(2\sqrt{x_0})$.

(vii) $f(x) = x \ln x$. This f is Lipschitz continuous on any bounded subset of the form $\Omega = (a, b)$ with $0 < a < b < \infty$, and $L = \sup\limits_{x \in \Omega} \ln x + 1$.

In several derivations throughout this chapter we shall use a famous lemma named after Gronwall. It is known in several forms. We need it in the following formulation.

Lemma 1.8 (Gronwall). *Let $x(t)$ satisfy for $t \geq t_0$ the linear, scalar IVP*

(1.8a) $$\begin{cases} \dot{x} = a(t)x + b(t) \,, \\ x(t_0) = x_0 \,, \end{cases}$$

with $a(t)$, $b(t)$ continuous functions. If y satisfies for $t \geq t_0$ the inequalities

(1.8b) $$\begin{cases} \dot{y} \leq a(t)y + b(t) \,, \\ y(t_0) \leq x_0 \,, \end{cases}$$

then

(1.8c) $$y(t) \leq x(t) \,, \qquad t \geq t_0 \,.$$

Proof: The difference $y(t) - x(t)$ satisfies the inequalities

(*) $$\begin{cases} \dot{y} - \dot{x} \leq a(t)(y - x) \,, \\ y(t_0) - x(t_0) \leq 0 \,. \end{cases}$$

Multiplying both sides of these inequalities with the positive function $\exp(-\bar{a}(t))$, with $\bar{a}(t)$ defined by

$$\bar{a}(t) := \int_{t_0}^{t} a(s)\, ds \; ,$$

and defining $z(t)$ by

$$z(t) := \Big(y(t) - x(t)\Big) \exp\Big(-\bar{a}(t)\Big) \; ,$$

we may rewrite $(*)$ in the form

$$\begin{cases} \dot{z} \leq 0 \; , \\ z(t_0) \leq 0 \; . \end{cases}$$

We conclude that $z(t) \leq 0$, so $y(t) \leq x(t)$, for $t \geq t_0$. $\qquad\qquad\square$

We note that the Lemma of Gronwall assumes the existence of a solution of (1.8a). In Theorem IV.1.3 of Chapter IV the existence of such a solution for all $t \geq t_0$ is shown by construction. The result is given by

$$(1.9) \qquad x(t) = e^{\bar{a}(t)} \left(x_0 + \int_{t_0}^{t} e^{-\bar{a}(s)} b(s)\, ds \right) .$$

The lemma does not assume uniqueness of this solution. In this section and in Chapter IV it is shown that the lemma has this uniqueness as a consequence.

Theorem 1.10 (Uniqueness). *The IVP (1.5) with $\mathbf{f} \in \mathrm{Lip}(I \times \Omega)$ for some domain $I \times \Omega$ containing $(t_0, \mathbf{x}_0)$ has at most one solution.*

Proof: Suppose that both $\mathbf{x}_1$ and $\mathbf{x}_2$ are solutions of (1.5). The difference $\mathbf{y} := \mathbf{x}_1 - \mathbf{x}_2$ then satisfies the IVP

$$\begin{cases} \dot{\mathbf{y}} = \mathbf{f}(t, \mathbf{x}_1) - \mathbf{f}(t, \mathbf{x}_2) \; , \\ \mathbf{y}(t_0) = \mathbf{0} \; . \end{cases}$$

Multiplying both sides of the ODE by $\mathbf{y}(t)^T$ we find for the left-hand side

$$(*) \qquad \mathbf{y}(t)^T \dot{\mathbf{y}}(t) = \frac{1}{2} \frac{d}{dt}\Big(\mathbf{y}(t)^T \mathbf{y}(t)\Big) = \frac{1}{2} \dot{z}(t) \; ,$$

where we introduce the notation

$$z(t) := \mathbf{y}(t)^T \mathbf{y}(t) = \|\mathbf{y}(t)\|^2 \; .$$

From Appendix B we obtain for the right-hand side the inequality

$$\left| \mathbf{y}(t)^T \big(\mathbf{f}(t, \mathbf{x}_1) - \mathbf{f}(t, \mathbf{x}_2)\big) \right| \leq \|\mathbf{y}(t)\| \, \|\mathbf{f}(t, \mathbf{x}_1) - \mathbf{f}(t, \mathbf{x}_2)\| \leq L\, z(t)$$

with L the Lipschitz constant of $\mathbf{f}$. Combining these results we find for the scalar function $z(t)$:

$$\begin{cases} \dot{z} \leq 2Lz \ , \\ z(t_0) = 0 \ . \end{cases}$$

Application of Lemma 1.8 directly yields $z(t) \leq 0$. Since $z(t) \geq 0$ we conclude that $z(t) = 0$. This proves that $\mathbf{x}_1$ and $\mathbf{x}_2$ are identical for $t \geq 0$. The proof for $t \leq 0$ is the same after the transformation of $t \to -t$.   □

2. Local existence

To prove the existence of a solution of (1.5) it is sufficient to require that the vector field $\mathbf{f}(t, \mathbf{x})$ is continuous in a domain $I \times \Omega$ containing $(t_0, \mathbf{x}_0)$. The celebrated Cauchy-Peano theorem is given in many textbooks and we refer to, e.g., [19, 76, 13, 39, 5, 31]. The proof is elementary, but lengthy. Here, we prefer to present a much shorter derivation. It assumes that $\mathbf{f}$ is Lipschitz continuous, and proves at the same time existence and uniqueness. An essential ingredient is Picard iteration, a procedure of successive substitutions. It is based on Property I.2.6 of Chapter I, which states that a solution $\mathbf{x}(t)$ of (1.5) satisfies the integral equation

$$(2.1) \qquad \mathbf{x}(t) = \mathbf{x}_0 + \int_{t_0}^{t} \mathbf{f}\Big(s, \mathbf{x}(s)\Big)\, ds \ .$$

Every continuous function satisfying (2.1) is automatically differentiable, so the proof of the existence theorem can be formulated in terms of continuous functions. The following recursion yields a series $\mathbf{x}^i(t)$, $i = 0, 1, 2, \ldots$, of continuous functions:

$$(2.2) \qquad \begin{cases} \mathbf{x}^0(t) = \mathbf{x}_0 \ , \\ \\ \mathbf{x}^{i+1}(t) = \mathbf{x}_0 + \displaystyle\int_{t_0}^{t} \mathbf{f}\Big(s, \mathbf{x}^i(s)\Big)\, ds \ . \end{cases}$$

This so-called *Picard mapping* P maps continuous functions on continuous functions. By construction, the $\mathbf{x}^i(t)$ are even differentiable.

It is clear from (2.2) that all $\mathbf{x}^i$ satisfy the initial condition $\mathbf{x}^i(t_0) = \mathbf{x}_0$. The proof to be given below shows that the iterates $\mathbf{x}^i$ converge to a unique fixed point of the Picard mapping provided that $\mathbf{f}$ is Lipschitz continuous. But first we give some examples illustrating how Picard iteration works in practice.

Example II.3 (Picard iteration).
We consider a number of scalar IVP and calculate the Picard iterates explicitly.

(i) $\dot{x} = x$, $x(0) = 1$ with solution $x(t) = e^t$, $t \in \mathbb{R}$.

$$x^0(t) = 1 ,$$

$$x^1(t) = 1 + t ,$$

$$\vdots$$

$$x^i = 1 + t + ... + \frac{1}{i!} t^i .$$

Picard iteration thus yields the power series of e^t around $t = 0$.

(ii) $\dot{x} = x^2$, $x(0) = 1$ with solution $x(t) = \dfrac{1}{1-t}$, $t \in (-\infty, 1)$.

$$x^0(t) = 1 ,$$

$$x^1(t) = 1 + t ,$$

$$\vdots$$

$$x^i(t) = 1 + t + ... + t^i + O(t^{i+1}) .$$

We recognise the series expansion of $\dfrac{1}{1-t}$ around $t = 0$.

(iii) $\dot{x} = 2\sqrt{x}$, $x \geq 0$ and $\dot{x} = 0$, $x < 0$, $x(0) = 0$. See also Example II.1. This IVP has at least two solutions. Picard iteration yields $x^i(t) \equiv 0$ for all i, i.e. the zero solution.

(iv) $\dot{x} = \sqrt{x} + 1$, $x \geq 0$, and $\dot{x} = 1$, $x < 0$, with initial condition $x(0) = 0$. In this case Picard iteration already becomes very cumbersome after a few steps:

$$x^0(t) = 0 ,$$

$$x^1(t) = t ,$$

$$x^2(t) = t + \tfrac{2}{3} t^{3/2} ,$$

$$x^3(t) = \tfrac{36}{35} \left(-1 + (t + \tfrac{2}{3} t^{3/2})^{1/2} (1 - \tfrac{1}{3} t^{1/2} + \tfrac{1}{6} t + \tfrac{5}{9} t^{3/2}) \right) .$$

The following theorem shows that Lipschitz continuity of $\mathbf{f}$ guarantees the existence of a unique solution of (1.5) on a certain interval around t_0. It is a local result. The length of the interval is not of much importance, because in the next section it will be shown that the solution can be continued beyond this interval in general, thus leading to a global result.

Theorem 2.3 (Local existence). *The IVP (1.5) with* $\mathbf{f} \in \text{Lip}(I \times \Omega)$ *for* *some domain $I \times \Omega$ containing $(t_0, \mathbf{x}_0)$ in its interior has a unique solution on a certain interval*

$$(2.3\text{a}) \qquad I_\alpha = [t_0 - \alpha, t_0 + \alpha] \,, \qquad \alpha > 0 \,.$$

Proof: The proof is somewhat technical and based on the contractive mapping principle, see Appendix F. We show that an $\alpha > 0$ exists, such that the Picard mapping (2.2) is a contracting mapping on a closed subset of the set $C(I_\alpha)$ of continuous functions on the interval I_α. The latter set is a metric space, and the distance between two functions is measured with respect to the supremum norm (1.2). Note that $C(I_\alpha)$ is also complete, i.e. every Cauchy sequence has a limit in $C(I_\alpha)$. (A sequence of functions $\mathbf{x}^i$ is called a Cauchy sequence if $\|\mathbf{x}^i - \mathbf{x}^j\| \to 0$ for $i, j \to \infty$.) A complete, metric space is also called a Banach space. In such a space the contracting mapping principle applies, which guarantees that the Picard mapping (2.2) has a unique fixed point and thus that IVP (1.5) has a unique solution around t_0.

In the following, two constants are used. The first is the supremum M_f of the vector field $\mathbf{f}$ on $I \times \Omega$:

$$M_f = \|\mathbf{f}\|_{I \times \Omega} \,.$$

M_f exists because $f \in \text{Lip}(I \times \Omega)$. The second constant is the radius $\bar{\beta}$ of the largest sphere around $\mathbf{x}_0$ which is still completely contained in Ω. If we denote a sphere of radius β as

$$\Omega_\beta = \{\mathbf{x} \,|\, \|\mathbf{x} - \mathbf{x}_0\| < \beta\} \,,$$

then $\bar{\beta}$ is defined by

$$\bar{\beta} = \sup_\beta \{\beta \,|\, \Omega_\beta \subseteq \Omega\} \,.$$

Let us choose an α with $0 \leq \alpha \leq 1$. Using $\bar{\beta}$ we may introduce a closed subset $\bar{C}(I_\alpha)$ of $C(I_\alpha)$ as the set of all continuous functions on I_α that have a distance equal to or smaller than the product $\alpha\bar{\beta}$ to the constant function $\mathbf{x}^0(t) \equiv \mathbf{x}_0$. Thus, $\bar{C}(I_\alpha)$ contains all continuous functions which are contained in a tube of radius $\alpha\bar{\beta}$ around $\mathbf{x}_0$, i.e.

$$\bar{C}(I_\alpha) = \{\mathbf{x}(t) \in C(I_\alpha) \,|\, \|\mathbf{x} - \mathbf{x}_0\| \leq \alpha\bar{\beta}\} \,.$$

We take $\alpha \leq 1$ to ensure that the elements of $\bar{C}(I_\alpha)$ lie in $I_\alpha \times \Omega$. The smaller α, the shorter and narrower the tube is. The situation is shown in Fig. II.1.

Next, we show that Picard iteration maps $\bar{C}(I_\alpha)$ into itself and is a contracting mapping if α is small enough. For the distance between the i-th Picard iterate $\mathbf{x}^i$ and $\mathbf{x}_0$ we find

$$\|\mathbf{x}^{i+1} - \mathbf{x}_0\|_{I_\alpha} = \left\| \int_{t_0}^{t} \mathbf{f}\left(s, \mathbf{x}^i(s)\right) ds \right\|_{I_\alpha} \leq \alpha \, M_f \,.$$

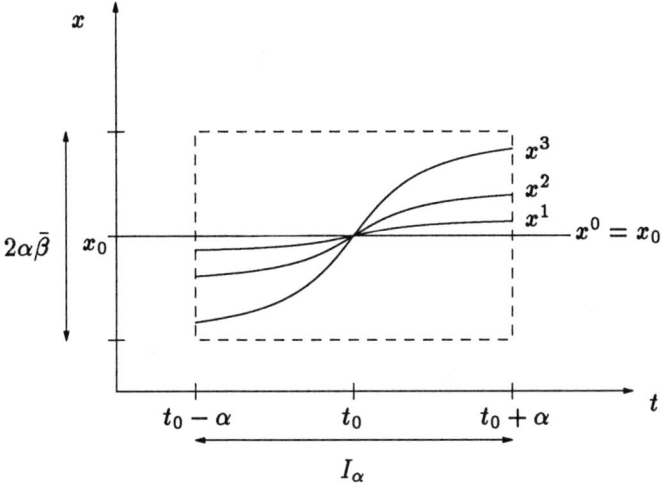

Figure II.1

So, if we set $\alpha < \bar{\beta}/M_f$ we have $\mathbf{x}^i \in \bar{C}(I_\alpha)$ for all i. For the distance between two successive Picard iterates we deduce

$$\|\mathbf{x}^{i+1} - \mathbf{x}^i\|_{I_\alpha} = \left\| \int_{t_0}^{t} \left[\mathbf{f}\big(s, \mathbf{x}^i(s)\big) - \mathbf{f}\big(s, \mathbf{x}^{i-1}(s)\big) \right] ds \right\|_{I_\alpha} \leq$$

$$\leq \alpha \sup_{t \in I_\alpha} \left\| \mathbf{f}\big(t, \mathbf{x}^i(t)\big) - \mathbf{f}\big(t, \mathbf{x}^{i-1}(t)\big) \right\| \leq$$

$$\leq \alpha L \|\mathbf{x}^i - \mathbf{x}^{i-1}\|_{I_\alpha}$$

with L the Lipschitz constant of $\mathbf{f}$ on $I \times \Omega$. Choosing $\alpha < 1/L$ ensures that the Picard mapping is a contraction. If all conditions on $\alpha > 0$ mentioned above are fulfilled, the mapping has a unique fixed point in $\bar{C}(I_\alpha)$, which satisfies (2.1) and is thus a solution of (1.5).

For later convenience we summarise the four conditions on $\alpha > 0$:

(i) $I_\alpha \subseteq I$

(ii) $\alpha < 1$

(iii) $\alpha < 1/L$

(iv) $\alpha \leq \bar{\beta}/M_f$.   □

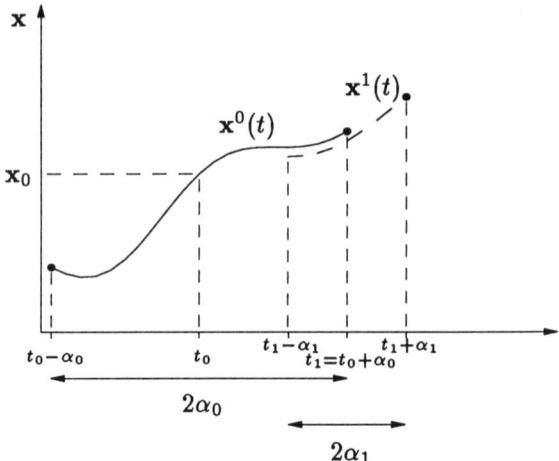

Figure II.2

3. Continuation of local solutions

From Theorem 2.3 we know that IVP (1.5) has a solution on a certain non-empty interval around t_0. Here, we shall discuss and prove that, in general, the solution can be continued uniquely over a much larger interval and may exist globally, i.e. for all $t \in \mathbb{R}$. The idea of continuation is rather straightforward. We shall first outline the procedure before formulating and proving the relevant theorem.

Let us investigate continuation for the case where $t > t_0$. The case $t < t_0$ is completely similar. Theorem 2.3 ensures that a unique solution $\mathbf{x}(t)$ exists on $I_{\alpha_0} := [t_0 - \alpha_0, t_0 + \alpha_0]$, with $\alpha_0 > 0$. Therefore, $\mathbf{x}(t_1)$ with $t_1 = t_0 + \alpha_0$ exists and $\mathbf{x}(t_1) \in \Omega$. If $(t_1, \mathbf{x}(t_1))$ is an interior point of $I \times \Omega$, we can apply the local existence theorem around this point, thereby obtaining a new interval $I_{\alpha_1} := [t_1 - \alpha_1, t_1 + \alpha_1]$ on which a unique solution of (1.5) exists. On $I_{\alpha_0} \cap I_{\alpha_1}$ this solution coincides with the solution in I_{α_0} and on $I_{\alpha_1} r I_{\alpha_0}$ it is its continuation. See Fig. II.2. Repeating this procedure while taking the α_i as large as possible, we get a series of overlapping intervals I_{α_i}, $i = 0, 1, \ldots$, and the solution exists uniquely on their union. If I and Ω are open and bounded, the I_{α_i} will become smaller and smaller, because $\mathbf{x}(t)$ will approach the boundary of $I \times \Omega$. If I and Ω are closed and bounded the procedure will terminate for certain $i = n$, because either t_n is at the boundary of I, or $\mathbf{x}(t_n)$ is at the boundary of Ω. We conclude that in all cases a maximal existence interval can be found. It may be very hard to estimate the size of this interval in advance without solving the problem explicitly. We illustrate this point with an example:

Example II.4.
Consider the scalar IVP

$$\begin{cases} \dot{x} = x^p \ , \\ x(0) = 1 \ . \end{cases}$$

This scalar field is autonomous, and the interval I is therefore not relevant. For any integer $p \geq 0$ we have $\Omega = \mathbb{R}$. If p is a negative integer we have to exclude the origin: $\Omega = \mathbb{R} \,r\, \{0\}$. If p is non-integer, negative x values are not allowed and then $\Omega = [0, \infty)$ for $p > 0$, and $\Omega = (0, \infty)$ for $p < 0$. The solutions are directly found from separation of variables leading to the implicit equation

$$\int\limits_{1}^{x} y^{-p}\, dy = t \ .$$

Hence

$$x(t) = \begin{cases} [1 + (1-p)t]^{\frac{1}{1-p}} \ , & p \neq 1 \\ e^t & , & p = 1 \ . \end{cases}$$

The maximum existence interval $I_{\max}$ depends on p in a remarkable way, as we can see from the following cases:

For $p > 1$, $I_{\max} = (-\infty,\ 1/(p-1))$.
For $p = 1$, $I_{\max} = \mathbb{R}$.
For $p < 1$ and $p \neq (n-1)/n$ for some integer $n \neq 0$, $I_{\max} = (1/(p-1)\,,\infty)$.
For $p = (n-1)/n$ for some positive integer n, $I_{\max} = \mathbb{R}$.
For $p = (n-1)/n$ for some negative integer n, $I_{\max} = (-\infty, |n|)$.

The following theorem formalises the continuation insights outlined above.

Theorem 3.1 (Continuation of local solutions). *Let* $\mathbf{f} \in \mathrm{Lip}(I \times \Omega)$, *$I$ and Ω closed and bounded, and $(t_0, \mathbf{x}_0)$ be an interior point of $I \times \Omega$; then the solution $\mathbf{x}$ of IVP (1.5) can be continued to the boundary of $I \times \Omega$.*

Proof: Because of symmetry we only consider continuation for $t > t_0$. For $t < t_0$ a similar argument applies. Let $I := [t_0, T]$ for some T. Applying Theorem 2.3 repeatedly we obtain a series of overlapping intervals $I_i = [t_0, t_i]$, with $t_{i+1} > t_i$, on which the solution uniquely exists. Let us assume that the t_i converge to some $\bar{t} < T$ and that the value $\bar{\mathbf{x}} := \mathbf{x}(\bar{t})$ is not on the boundary of Ω. So, $(\bar{t}, \bar{\mathbf{x}})$ is an interior point of $I \times \Omega$. If the solution cannot be continued beyond $\bar{t}$, this would imply that not all of the conditions (i)–(iv) listed at the end of the proof of Theorem 2.3 can be satisfied, when Theorem 2.3 is applied with $(\bar{t}, \bar{x})$ as initial values. Condition (i) can be satisfied, because $\bar{t}$ is an interior point of I. Conditions (ii) and (iii) can always be satisfied. So, we conclude that condition (iv) cannot be satisfied. This is only the case if $\bar{\beta} = 0$, which means that the distance of $\bar{x}$ to the boundary of Ω is vanishing. This is in contradiction of $\bar{x}$ being an interior point of Ω. We conclude that $(\bar{t}, \bar{x})$ is at the boundary of $I \times \Omega$, either because $\bar{t} = T$ or because $\bar{x}$ is on the boundary of Ω.   □

Theorem 3.1 has important consequences with respect to the question whether the solution of a given IVP exists globally, i.e. for all $t \in \mathbb{R}$. Sometimes the existence is already called global if it holds for all $t \geq t_0$. The following corollaries draw conclusions about this point if some more properties are known about the general behaviour of the solution. For global existence we must have $I = \mathbb{R}$ (or $I = [t_0, \infty)$), of course. For convenience we shall also assume $\Omega = \mathbb{R}^n$.

Corollary 3.2. *The solution* $\mathbf{x}$ *of IVP (1.5) with* $\mathbf{f} \in \text{Lip}(\mathbb{R} \times \mathbb{R}^n)$ *exists globally, i.e. for all* $t \in \mathbb{R}$, *if the following is known in advance:*

$$\{\mathbf{x}(t) \text{ exists for some } t\} \Rightarrow \{\|\mathbf{x}(t) - \mathbf{x}_0\| \leq M \text{ for some constant } M > 0\}.$$

Proof: Take some constants $T > 0$ and $\varepsilon > 0$. The domain $I_T \times \Omega$ with $I_T := [t_0 - T, t_0 + T]$ and $\Omega := \{\mathbf{x} \mid \|\mathbf{x} - \mathbf{x}_0\| \leq M + \varepsilon\}$ is closed and bounded. Theorem 3.1 guarantees that the solution $\mathbf{x}$ thus reaches the boundary of $I_T \times \Omega$. Because it cannot reach the boundary of Ω, it will reach the boundary of I_T. This implies that $\mathbf{x}(t_0 \pm T)$ exists. The value of T can be taken arbitrarily large, so that $\mathbf{x}(t)$ apparently exists for all $t \in \mathbb{R}$. The situation is sketched in Fig. II.3.   □

Corollary 3.3. *The solution* $\mathbf{x}$ *of IVP (1.5) exists globally if the following is known in advance:*

$$\{\mathbf{x}(t) \text{ exists for some } t\} \Rightarrow \{\|\mathbf{x}(t) - \mathbf{x}_0\| \leq g(t), \text{ with } g(t) > 0 \text{ a continuous function on } \mathbb{R}\}.$$

Proof: The proof is similar to the proof of Corollary 3.2. Here, Ω is defined as

$$\Omega := \{\mathbf{x} \mid \|\mathbf{x} - \mathbf{x}_0\| \leq M_T + \varepsilon\}$$

where $M_T = \max_{t \in I_T} g(t)$.   □

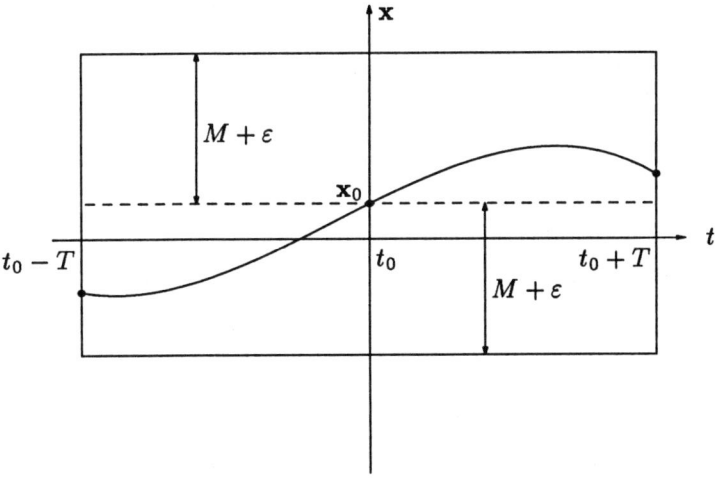

Figure II.3

An important consequence of Corollary 3.3 is that boundedness of the vector field $\mathbf{f}$ implies the global existence of the solution. We formulate this as a separate corollary:

Corollary 3.4. *The solution $\mathbf{x}$ of IVP (1.5) exists globally if the vector field $\mathbf{f}$ is bounded, i.e. if $\|\mathbf{f}\|_{I\!\!R \times I\!\!R^n} = M < \infty$.*

Proof: From (2.1) we have

$$\|\mathbf{x}(t) - \mathbf{x}_0\| = \left\| \int_{t_0}^{t} \mathbf{f}\Big(s, \mathbf{x}(s)\Big)\, ds \right\| \leq M(t - t_0) \ .$$

So, we can take $g(t) := M(t - t_0)$ and apply Corollary 3.3.   □

4. Dependence on initial value and vector field

The solution $\mathbf{x}$ of the IVP

$$(4.1) \qquad \begin{cases} \dot{\mathbf{x}} = \mathbf{f}(t, \mathbf{x}) \ , \\[2mm] \mathbf{x}(t_0) = \mathbf{x}_0 \end{cases}$$

clearly depends on $\mathbf{x}_0$, $\mathbf{f}$, and t_0. Perturbations of t_0 are of little practical importance and not studied here. In this section we investigate how $\mathbf{x}(t)$ varies with variations in x_0 and f. We derive upper bounds for the distance between the solutions of the perturbed and the unperturbed problem. The derivations are valid only for time intervals $[t_0, t_0 + T]$ with $T < \infty$. Stability of perturbed systems for $T \to \infty$ is studied separately in Chapter V. The finite interval results of this section are of great importance for the analysis of the stability of numerical methods, which are always subject to (discretisation) errors and necessarily deal with finite time intervals only.

Let us compare the solution x of IVP (4.1) with the solution y of the perturbed version

$$(4.2) \qquad \begin{cases} \dot{y} = f(t, y) + r(t, y) \;, \\ y(t_0) = x_0 + z_0 \;. \end{cases}$$

Assume that $f \in \mathrm{Lip}(I \times \mathbb{R}^n)$ with Lipschitz constant L, and that the solutions x and y uniquely exist on a time interval $[t_0, T] \subseteq I$.

We are interested in an apriori upper bound on the distance of $x(t)$ and $y(t)$. The difference $z(t) = y(t) - x(t)$ is the solution of the IVP

$$(4.3) \qquad \begin{cases} \dot{z} = f(t, y) - f(t, x) + r(t, y) \;, \\ z(t_0) = z_0 \;. \end{cases}$$

Following (2.1) we may write

$$(4.4) \qquad z(t) = z_0 + \int_{t_0}^{t} \left[f\Big(s, y(s)\Big) - f\Big(s, x(s)\Big) + r\Big(s, y(s)\Big) \right] ds \;.$$

From this we obtain for fixed t the inequality

$$(4.5) \qquad \|z(t)\| \leq \|z_0\| + \int_{t_0}^{t} \left[L\, \|z(s)\| + \left\| r\Big(s, y(s)\Big) \right\| \right] ds \;=:\; g(t) \;.$$

The right-hand side of this inequality satisfies

$$(4.6) \qquad \begin{cases} \dot{g} \leq L\, g + \left\| r\Big(t, y(t)\Big) \right\| \;, \\ g(t_0) = \|z_0\| \;. \end{cases}$$

Lemma 1.8 in combination with (1.9) then immediately provides the proof of the following theorem:

Theorem 4.7. *If x and y are solutions of (4.1) and its perturbed version (4.2) respectively on $J := [t_0, t_0 + T]$ and $f \in \mathrm{Lip}\,[J \times \mathbb{R}^n]$ with Lipschitz constant L, then for $t \in J$*

$$(4.7a) \qquad \|\mathbf{y}(t) - \mathbf{x}(t)\| \leq e^{L(t-t_0)} \|\mathbf{y}_0 - \mathbf{x}_0\| + \int_{t_0}^{t} \left\| \mathbf{r}\big(s, \mathbf{y}(s)\big) \right\| e^{L(t-s)} \, ds \ .$$

If $\mathbf{r}$ is bounded, i.e. $\|\mathbf{r}\|_{J \times \mathbb{R}^n} < M$ for some constant M, then

$$(4.7b) \qquad \|\mathbf{y}(t) - \mathbf{x}(t)\| \leq e^{L(t-t_0)} \|\mathbf{y}_0 - \mathbf{x}_0\| + \frac{M}{L} \left(e^{L(t-t_0)} - 1 \right) . \qquad \square$$

We conclude that two solutions, which are close at $t = t_0$, may diverge exponentially however small the perturbations are. The following example shows that it strongly depends on the specific form of $\mathbf{f}$ whether the upper bounds in (4.7) are realistic.

Example II.5.
We consider the trivial problem $\dot{x} = ax$ with initial value $x(0) = 1$ and $\dot{y} = ay$ with perturbed initial value $y(0) = 1 + \varepsilon$. The vector field is not perturbed, so $r(t, x) \equiv 0$. The Lipschitz constant L equals $|a|$. The exact solutions are $x(t) = e^{at}$ and $y(t) = (1 + \varepsilon) e^{at}$ and thus

$$|y(t) - x(t)| = |\varepsilon| e^{at} \ .$$

From (4.7) we find the estimate

$$|y(t) - x(t)| \leq |\varepsilon| e^{|a|t} \ .$$

We see that (4.7) is very accurate for $a > 0$, but very inaccurate for $a < 0$. The two cases are illustrated in Fig. II.4.

Although Theorem (4.7) may yield highly inaccurate results, inequality (4.7a) is still very useful and will be applied at many places in this book. A special application is the derivation of a maximum time interval during which $\mathbf{x}(t)$ remains in a given neighbourhood of the initial value $\mathbf{x}_0$. To investigate this, we take a ball $\Omega_R = \{\mathbf{x} \,|\, \|\mathbf{x} - \mathbf{x}_0\| \leq R\}$ around $\mathbf{x}_0$ and ask for which T $\mathbf{x}(t) \in \Omega_R$ for all $t \in [t_0, T]$. To answer this question we take $\mathbf{r} \equiv -\mathbf{f}$ and $\mathbf{z}_0 = 0$ in (4.2). The corresponding solution is then $\mathbf{y}(t) \equiv \mathbf{x}_0$ and we obtain a special case of Theorem 4.7, which we formulate as follows:

Corollary 4.8. *If $\mathbf{x}$ is a solution of (4.1) existing on $J \times \mathbb{R}^n$ with $J :=$ $[t_0, t_0 + T]$, then*

$$(4.8a) \qquad \|\mathbf{x}(t) - \mathbf{x}_0\| \leq \int_{t_0}^{t} \|\mathbf{f}(s, \mathbf{x}_0)\| e^{L(t-s)} \, ds \ .$$

For autonomous vector fields this reduces to

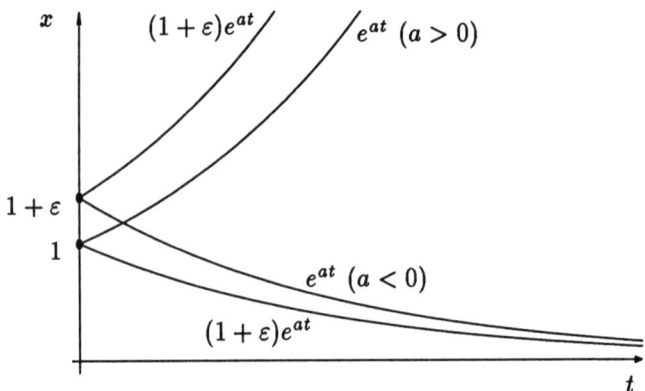

Figure II.4

(4.8b) $\|\mathbf{x}(t) - \mathbf{x}_0\| \le \|\mathbf{f}(\mathbf{x}_0)\| \dfrac{1}{L} \left(e^{L(t-t_0)} - 1 \right)$.   □

In inequalities (4.8a) and (4.8b) the Lipschitz constant L and the vector field at $\mathbf{x}_0$ apparently provide enough information to obtain upper bounds. From (4.8a) we find that $\mathbf{x}(t) \in \Omega_R$ as long as $t \in [t_0, T)$ with $T = T(R)$, the solution of the implicit relation

(4.9) $\displaystyle\int_{t_0}^{T} \|\mathbf{f}(s, \mathbf{x}_0)\| \, e^{L(T-s)} \, ds = R$.

For autonomous $\mathbf{f}$ this reduces to an explicit equation:

(4.10) $T(R) = t_0 + \dfrac{1}{\bar{L}} \ln \left(\dfrac{\bar{L}\,R}{\|\mathbf{f}(\mathbf{x}_0)\|} + 1 \right)$.

In (4.9) and (4.10) we may take a Lipschitz constant $\bar{L}$ corresponding to $\mathbf{f} \in \mathrm{Lip}(J \times \Omega_R)$, because we know that $\mathbf{x}(t) \in \Omega_R$ in advance. The value $\bar{L}$ may be smaller than the value L used in Theorem 4.7. If in (4.9) and (4.10) a fixed value L is used, T is a monotonically increasing function of R, but otherwise this may not be the case, as is illustrated in the following examples.

Example II.6.
(i) We consider first the autonomous IVP

$$\begin{cases} \dot{x} = x^2 \, , \\ x(0) = 1 \end{cases}$$

with solution $x(t) = (1-t)^{-1}$, which exists for $t \in (-\infty, 1)$. Following Property (1.7) the Lipschitz constant L on $\Omega = \{x \,|\, |x-1| \leq R\}$ is given by $\bar{L} = 2(1+R)$. From (4.10) we find that $x(t) \in \Omega$ as long as $t \leq T$ given by

$$T = \frac{1}{2(1+R)} \ln(2R(1+R)+1) \;.$$

Because the solution is known, we also know the exact value, T^* say, of this upper bound:

$$T^* = \frac{R}{R+1} \;.$$

Both T and T^* are plotted in Fig. II.5 as functions of R. For small R they agree quite well, but for $R \geq 1$ we see that T is a rather crude estimate of T^*.

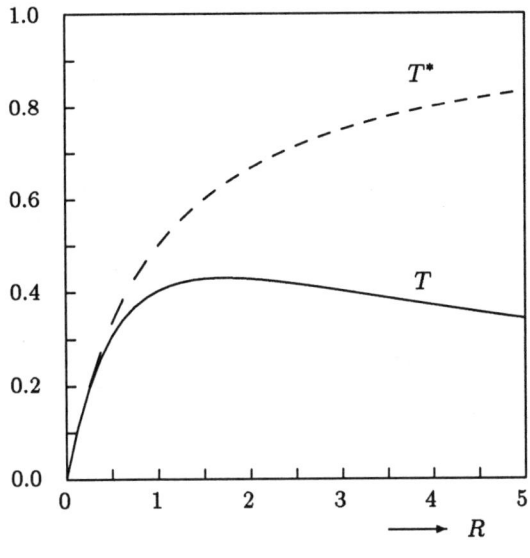

Figure II.5

(ii) Let us next consider the non-autonomous IVP

$$\begin{cases} \dot{x} = tx \\ x(0) = 1 \;. \end{cases}$$

Property (1.7) yields that a Lipschitz constant on $I = [0, T]$ is given by $L = T$. Because $f(t, x_0) = t$, we can evaluate the integral in (4.9), which leads to an implicit expression for T as a function of R:

$$-1 + \frac{1}{T^2} (e^{T^2} - 1) = R .$$

Solving this equation numerically we obtain the curve given in Fig. II.6. Separation of variables in the original problem yields

$$\int_1^x \frac{1}{y} \, dy = \int_0^t s \, ds ,$$

and thus $x(t) = \exp(\frac{1}{2} t^2)$. So, the exact upper bound is

$$T^* = \sqrt{2 \ln(R + 1)} ,$$

which is also plotted in Fig. II.6. We see that T and T^* agree quite well for small R, but deviate for increasing R.

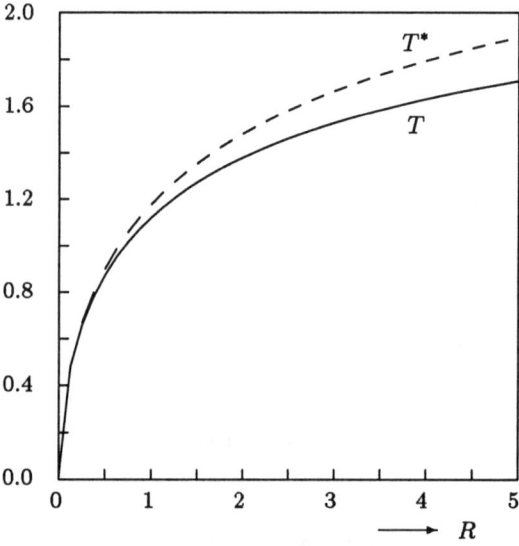

Figure II.6

5. Regular perturbations; linearisation

In §4 we studied the influence of perturbations of $\mathbf{f}$ and $\mathbf{x}_0$ in a general setting. Here, we assume that the perturbations can be parametrised by a quantity $\varepsilon \in I^\varepsilon$ with I^ε a neighbourhood of the origin in $\mathbb{R}$. To indicate the dependence of this parameter, we shall use the notations $\mathbf{f}^\varepsilon(t, \mathbf{x})$ and $\mathbf{x}_0^\varepsilon$. The case $\varepsilon = 0$ is the unperturbed problem, so in the following we identify $\mathbf{f}^0$ with $\mathbf{f}$ and $\mathbf{x}_0^0$ with $\mathbf{x}_0$.

We assume that $\mathbf{f}^\varepsilon$ and $\mathbf{x}_0^\varepsilon$ depend continuously on ε and that this dependence is uniform for all $(t, \mathbf{x}) \in I \times \Omega$. Perturbations with these properties are called *regular*, whereas perturbations not satisfying this condition are called *singular*. Singular perturbations may give rise to a discontinuous dependence of the solution on ε. They are investigated in Chapter VIII.

The existence of a local solution $\mathbf{x}^\varepsilon(t)$ of the IVP

$$(5.1) \qquad \begin{cases} \dot{\mathbf{x}}^\varepsilon = \mathbf{f}^\varepsilon(t, \mathbf{x}^\varepsilon) \, , \\ \mathbf{x}^\varepsilon(t_0) = \mathbf{x}_0^\varepsilon \end{cases}$$

is guaranteed if $\mathbf{f}^\varepsilon(t, \mathbf{x})$ depends continuously on its parameters $(\varepsilon, t, \mathbf{x})$ for some neighbourhood $I^\varepsilon \times I \times \Omega$ of $(0, t_0, \mathbf{x}_0^\varepsilon)$. The proof is completely similar to the theorems referred to in §2. Uniqueness of the solution is assured if $\mathbf{f}$ is Lipschitz continuous on $I \times \Omega$ for all $\varepsilon \in I^\varepsilon$, and the proof of Theorem (1.10) also applies to (5.1). In addition, the existence and uniqueness of a local solution of (5.1) also follow directly from Theorem 2.3. Here we shall show that $\mathbf{x}^\varepsilon(t)$ depends continuously or differentiably on ε, if this also holds for $\mathbf{f}^\varepsilon$.

Theorem 5.2 (Continuous perturbations). *Let the vector field $\mathbf{f}^\varepsilon(t, \mathbf{x})$ be continuous on $I^\varepsilon \times I \times \Omega$ containing $(0, t_0, \mathbf{x}_0^\varepsilon)$, and let the initial value be continuous on I^ε. Let $\mathbf{f}^\varepsilon(t, \mathbf{x}) \in \mathrm{Lip}(I^\varepsilon \times I \times \Omega)$. If the solution $\mathbf{x}^\varepsilon$ exists for all $\varepsilon \in I^\varepsilon$ on an interval $J = [t_0, t_0 + T] \subset I$, then $\mathbf{x}^\varepsilon$ is continuous in ε uniformly on J.*

Proof: We show continuity at $\varepsilon = 0$, but the derivation holds for arbitrary $\varepsilon \in I^\varepsilon$. The continuity of $\mathbf{f}^\varepsilon$ with respect to ε can be expressed as

$$\mathbf{f}^\varepsilon(t, \mathbf{x}) = \mathbf{f}(t, \mathbf{x}) + \mathbf{r}^\varepsilon(t, \mathbf{x})$$

with $\|\mathbf{r}^\varepsilon(t, \mathbf{x})\| \to 0$ if $\varepsilon \to 0$, uniformly on $I \times \Omega$. In terms of the norm introduced in (1.3):

$$\|\mathbf{r}^\varepsilon\|_{I \times \Omega} \to 0 \, , \qquad \text{if } \varepsilon \to 0 \, .$$

The continuity of $\mathbf{x}_0^\varepsilon$ similarly implies

$$\mathbf{x}_0^\varepsilon = \mathbf{x}_0 + \mathbf{z}_0^\varepsilon \, ,$$

with

$$\|\mathbf{z}_0^\varepsilon\| \to 0 \ , \qquad \text{if } \varepsilon \to 0 \ .$$

We note that $\mathbf{r}^\varepsilon$ and $\mathbf{z}_0^\varepsilon$ play the same role as $\mathbf{r}$ and $\mathbf{z}_0$ in (4.2). So, inequality (4.7b) applies, which results in

$$\|\mathbf{x}^\varepsilon(t) - \mathbf{x}(t)\| \le e^{L(t-t_0)} \|\mathbf{z}_0^\varepsilon\| + \|\mathbf{r}^\varepsilon\|_{I \times \Omega} \frac{e^{L(t-t_0)} - 1}{L} \ .$$

Here L is the Lipschitz constant of $\mathbf{f}^\varepsilon$ on $I^\varepsilon \times I \times \Omega$. This inequality holds for all $t \in J = [t_0, t_0 + T]$. Taking the supremum on both sides, we obtain an upper bound for the distance between $\mathbf{x}^\varepsilon$ and $\mathbf{x}$ on J:

$$\|\mathbf{x}^\varepsilon - \mathbf{x}\|_J \le e^{LT} \|\mathbf{z}_0^\varepsilon\| + \|\mathbf{r}^\varepsilon\|_{I \times \Omega} \frac{e^{LT} - 1}{L} \ .$$

So, $\mathbf{x}^\varepsilon$ approaches $\mathbf{x}$ uniformly on J if $\varepsilon \to 0$.   □

Example II.7.
Consider the scalar IVP

$$\dot{x} = x + r^\varepsilon(x) , \qquad x(0) = 1$$

with

$$r^\varepsilon = \begin{cases} 0 & \text{if } \varepsilon < 0 \ , \\ \varepsilon & \text{if } \varepsilon \ge 0 \ . \end{cases}$$

This perturbation is continuous with respect to ε, but not differentiable in $\varepsilon = 0$. Clearly the same holds for the solution

$$x^\varepsilon(t) = \begin{cases} e^t & , \quad \varepsilon < 0 \\ (1 + \varepsilon)e^t - \varepsilon & , \quad \varepsilon \ge 0 \ . \end{cases}$$

Theorem 5.3 (Differentiable perturbations). *Let $\mathbf{f}^\varepsilon(t, \mathbf{x})$ in (5.1) be differentiable with respect to ε and $\mathbf{x}$ on $I^\varepsilon \times I \times \Omega$ containing the points $(0, t_0, x_0^\varepsilon)$ and let $\mathbf{x}_0^\varepsilon$ in (5.1) be differentiable with respect to ε on I^ε. Let $\mathbf{f}^\varepsilon(t, \mathbf{x}) \in \mathrm{Lip}(I^\varepsilon \times I \times \Omega)$. If the solution $\mathbf{x}^\varepsilon(t)$ of IVP (5.1) exists for all $\varepsilon \in I^\varepsilon$ on an interval $J := [t_0, t_0 + T] \subset I$, then $\mathbf{x}^\varepsilon$ is differentiable with respect to ε for all $t \in J$.*

The derivative $\mathbf{z}(t) := \dfrac{\partial}{\partial \varepsilon} \mathbf{x}^\varepsilon(t)$ in $\varepsilon = 0$ satisfies the IVP

(5.3a)
$$\begin{cases} \dot{\mathbf{z}} = \mathbf{J}(t, \mathbf{x})\mathbf{z} + \dfrac{\partial \mathbf{f}^\varepsilon}{\partial \varepsilon}(t, \mathbf{x}) \Big|_{\varepsilon=0} \ , \\[2mm] \mathbf{z}(t_0) = \dfrac{d}{d\varepsilon} \mathbf{x}_0^\varepsilon \Big|_{\varepsilon=0} \ , \end{cases}$$

with $\mathbf{J}$ *the Jacobian matrix of the unperturbed vector field* $\mathbf{f}(t, \mathbf{x})$.

Proof: The proof will largely follow the proof of Theorem 5.2. Without loss of generality we can focus on the case $\varepsilon = 0$, because the same arguments apply for all $\varepsilon \in I^\varepsilon$ after a trivial adjustment.

We first show the existence of the derivative of $\mathbf{x}^\varepsilon(t)$ with respect to ε. The differentiability of $\mathbf{f}^\varepsilon$ with respect to ε implies that

$$(*) \qquad \mathbf{f}^\varepsilon(t, \mathbf{x}) = \mathbf{f}(t, \mathbf{x}) + \varepsilon\, \mathbf{r}(t, \mathbf{x}) + o(\varepsilon) \,,$$

where

$$\mathbf{r}(t, \mathbf{x}) = \frac{\partial}{\partial \varepsilon} \mathbf{f}^\varepsilon(t, \mathbf{x}) \Big|_{\varepsilon=0} \,.$$

The order symbol $o(\cdot)$ is defined in Appendix E. The differentiability of $\mathbf{x}_0^\varepsilon$ with respect to ε implies that

$$\mathbf{x}_0^\varepsilon = \mathbf{x}_0 + \varepsilon\, \mathbf{z}_0 + o(\varepsilon)$$

with

$$\mathbf{z}_0 = \frac{d}{d\varepsilon} \mathbf{x}_0^\varepsilon \Big|_{\varepsilon=0} \,.$$

We note that $\mathbf{r}$ and $\mathbf{z}_0$ play here the same rôle as in (4.2). Application of (4.7a) gives

$$\frac{\|\mathbf{x}^\varepsilon(t) - \mathbf{x}(t)\|}{\varepsilon} \leq e^{L(t-t_0)} \|\mathbf{z}_0\| + \int_{t_0}^{t} \left\| \mathbf{r}\big(s, \mathbf{x}(s)\big) \right\| e^{L(t-s)} ds \,.$$

The right-hand side of this inequality is independent of ε. So, on the left-hand side we can take the limit $\varepsilon \to 0$. The derivative $\mathbf{z}(t)$ of $\mathbf{x}^\varepsilon(t)$ thus exists for all $t \in J$.

To show that $\mathbf{z}(t)$ satisfies IVP (5.3a) we expand $\mathbf{f}^\varepsilon(t, \mathbf{x}^\varepsilon)$ in powers of ε. The differentiability of $\mathbf{x}^\varepsilon$ with respect to ε gives

$$(**) \qquad \mathbf{x}^\varepsilon(t) = \mathbf{x}(t) + \varepsilon\, \mathbf{z}(t) + o(\varepsilon)$$

where

$$\mathbf{z}(t) = \frac{\partial}{\partial \varepsilon} \mathbf{x}^\varepsilon(t) \Big|_{\varepsilon=0} \,.$$

Substituting this into $(*)$ we find

$$\mathbf{f}^\varepsilon(t, \mathbf{x}^\varepsilon) = \mathbf{f}(t, \mathbf{x}) + \varepsilon\, \mathbf{J}(t, \mathbf{x})\, \mathbf{z}(t) + \varepsilon\, \mathbf{r}(t, \mathbf{x}) + o(\varepsilon) \,.$$

From $(**)$ we have

$$\varepsilon\, \dot{\mathbf{z}} = \dot{\mathbf{x}}^\varepsilon - \dot{\mathbf{x}} + o(\varepsilon) \,.$$

So,

$$\varepsilon\dot{\mathbf{z}} = \varepsilon\,\mathbf{J}(t,\mathbf{x})\mathbf{z} + \varepsilon\,\mathbf{r}(t,\mathbf{x}) + o(\varepsilon) \ .$$

Dividing by $\varepsilon > 0$ and taking the limit $\varepsilon \to 0$ yields the ODE in (5.3a), while the initial value directly follows from the definition of $\mathbf{z}_0$ above.   □

The following example is meant to demonstrate Theorem 5.3. In this example the solutions $\mathbf{x}$ and $\mathbf{x}^\varepsilon$ of the unperturbed and perturbed system, respectively, are explicitly known, so that relation (5.3a) can be checked. In practice Theorem 5.3 is used in situations where only the unperturbed system can be solved. The IVP (5.3a) is then used to find a first approximation to $\mathbf{x}^\varepsilon(t)$. This IVP is linear and therefore easy to handle in many cases. In Chapter IV it is shown that the solutions of all linear systems have a general structure in common. IVP (5.3a) is usually called the *linearisation* of the original IVP (5.1). Chapter V, §4 discusses how linearisation can be used to determine the stability character of special solutions, e.g. stationary points.

Example II.8.
Consider the scalar problem

$$\begin{cases} \dot{x} = x^{1-\varepsilon} \ , \\ x(0) = 1 + \varepsilon \ . \end{cases}$$

The unperturbed problem with $\varepsilon = 0$ has the trivial solution

$$x(t) = e^t \ .$$

From separation of variables, cf. Example II.4, we find that for $\varepsilon > 0$ the solution is given by

$$x^\varepsilon(t) = [\varepsilon t + (1 + \varepsilon)^\varepsilon]^{1/\varepsilon} \ ,$$

which, for all $\varepsilon > 0$, exists for $t \in J := (-(1 + \varepsilon)^\varepsilon/\varepsilon, \infty)$. The perturbation is continuous, so that Theorem 5.2 implies that $x^\varepsilon(t) \to x(t)$, if $\varepsilon \to 0$, for all $t \in J$. This shows that two seemingly different representations of e^t are fully consistent. On the one hand we have that e^t is defined as the solution of the IVP $\dot{x} = x$, $x(0) = 1$, while from the form above and a continuity argument we have

$$e^t = \lim_{\alpha\to\infty} \left(\frac{t}{\alpha} + \left(1 + \frac{1}{\alpha} \right)^{\frac{1}{\alpha}} \right)^\alpha \ ,$$

where we have set $\alpha = 1/\varepsilon$ for convenience.

Now we turn to the evaluation of the linearised equation (5.3a). The Jacobian 'matrix', evaluated for $\varepsilon = 0$, is given by

$$J(x) = 1 .$$

The derivative of the initial value with respect to ε is

$$\frac{d\,x_0^\varepsilon}{d\varepsilon} = 1 .$$

The derivative of the scalar field reads

$$\left.\frac{\partial f^\varepsilon}{\partial\varepsilon}(x)\right|_{\varepsilon=0} = -x\ln x .$$

Substituting these findings and the unperturbed solution $x(t) = e^t$ in (5.3a) yields the IVP

$$\begin{cases} \dot{z} = z - t e^t , \\ z(0) = 1 . \end{cases}$$

This IVP has the solution

$$z(t) = (-\tfrac{1}{2} t^2 + 1) e^t .$$

We leave it to the reader to check that z is indeed a solution of the IVP, and that it is the derivative of x^ε with respect to ε, evaluated at $\varepsilon = 0$.

Exercises Chapter II

1. Determine whether the following scalar vector fields are Lipschitz continuous on the domains indicated.

    a) $f(x) = \tan x , \quad \Omega = (-\pi/2, \pi/2) .$
    b) $f(t, x) = \sin(tx) , \quad \Omega = \mathbb{R}, I = \mathbb{R} .$
    c) $f(x) = e^x , \quad \Omega = (-\infty, 0) .$
    d) $f(x) = e^x , \quad \Omega = \mathbb{R} .$
    e) $f(x) = (x^2 + 1)/x , \quad \Omega = (0, \infty) .$
    f) $f(t, x) = (\sin(tx))/x , \quad \Omega = \mathbb{R}, I = \mathbb{R} .$

2. Consider the scalar IVP

    $$\begin{cases} \dot{x} = \dfrac{x}{t + a} , & a > 0 \\ x(0) = x_0 . \end{cases}$$

    a) Prove that a unique solution exists for $t > -a$.
       Hint: Show that the vector field is $\text{Lip}(I \times \mathbb{R})$ for any $I = [b, c]$ with
       $a < b < c$, and apply Theorems 1.10, 2.3, and 3.4.

    b) Find in the same way as in Example II.6 an upper bound $T(R)$ for the
       period the solution remains in the interval $[x_0 - R, x_0 + R]$.

    c) Compare the result of b) with the behaviour of the true solution
       $x(t) = x_0(t + a)/a$.

3.   Consider the IVP

$$\begin{cases} \dot{x} = -\dfrac{x}{t + a}\,, & a > 0 \\ x(0) = x_0 \,. \end{cases}$$

    a) Prove, as in Exercise 2.a), the existence of a unique solution.

    b) The point $x_0 = 0$ is a stationary point of the vector field. It attracts
       all solutions with starting point $x_0 \neq 0$. Find an estimate $T(x_0, \varepsilon)$ for
       the period one has to wait before $|x(t)| < \varepsilon$ for given $\varepsilon > 0$.

    c) Compare the result of b) with the behaviour of the true solution
       $x(t) = a\,x_0/(t + a)$.

4.   a) Show that, for $t \geq t_0$, $\|\mathbf{x}(t)\| \leq \|\mathbf{x}_0\|$ for the standard IVP (1.5) if the
      inner product $(\mathbf{f}(t, \mathbf{x}), \mathbf{x}) \leq 0$ for all $\mathbf{x}$ with $\|\mathbf{x}\| \leq \|\mathbf{x}_0\|$.
      Hint: Utilise $(\mathbf{f}, \mathbf{x}) = (\dot{\mathbf{x}}, \mathbf{x}) = \dfrac{1}{2}\dfrac{d}{dt}\|\mathbf{x}\|^2$ using the Euclidean norm.

    b) What does this criterion imply if the vector field is homogeneous and
      linear, thus $\mathbf{f}(t, \mathbf{x}) = \mathbf{A}(t)\,\mathbf{x}$?

5.   For IVP (1.5) let Ω be a domain with $\mathbf{x}_0 \in \Omega$ and such that on the
    boundary of Ω $\mathbf{f}(t, \mathbf{x})$ points into Ω for all $t \geq t_0$. Show that then $\mathbf{x}(t) \in \Omega$
    for $t \geq t_0$.

6.   Consider the scalar version of IVP (1.5) with $f(t, x) \in \text{Lip}(\mathbb{R} \times \mathbb{R})$. Let
    x be the solution and x_1 and x_2 continuous functions on $\mathbb{R}$ such that for
    $t \geq t_0$

$$\begin{cases} \dot{x}_1 \leq f(t, x_1) \\ x_1(t_0) \leq x_0 \end{cases} \quad \text{and} \quad \begin{cases} \dot{x}_2 \geq f(t, x_2) \\ x_2(t_0) \geq x_0 \,. \end{cases}$$

    Show then that the following holds for $t \geq t_0$:

$$x_1(t) \leq x(t) \leq x_2(t)\,.$$

Hint: The same approach as in the proof of Theorem 1.10 could be used.

7.   Show that the solutions of the IVP

$$\begin{cases} \dot{x}_1 = x_2 - x_1 \\ \dot{x}_2 = -x_1 \end{cases}$$

are bounded.
Hint: Use the result of Exercise 4.

8.   Consider the vector field

$$f(t, x) = t \sin x \ .$$

a)   Show that all solutions are bounded for $t \to \infty$.
     Hint: Use the fact that the stationary points are $n\pi$, $n = 0, \pm1, \pm2, \dots$.
b)   Show that $0 < x_0 < x(t) \le \pi$ if $0 < x_0 < \pi$.
     Hint: Use the result of Exercise 5.

9.   Show that $x_0 \le x(t) < 1$ for $t \ge 0$, and $0 < x(t) \le x_0$ for $t \le 0$, if $x(t)$ is the solution of the IVP

$$\begin{cases} \dot{x} = x - x^2 \\ x(0) = x_0 \end{cases}$$

with $0 < x_0 < 1$.
Hint: Use the result of Exercise 5.

10. Find, using the result of Exercise 5, for which initial values x_0 the following differential equations have bounded solutions:

a) $\dot{x} = x^2 - 1$.

b) $\dot{x} = \dfrac{\sin x}{x}$.

c) $\dot{x} = -\log x$.

11. Linearise the following vector fields:

a) $\dot{x} = \tan x + e^{\sin x} - 1$ around $x = 0$.
b) $\ddot{x} = \dot{x} \log x + x^2$ around $(x, \dot{x}) = (1, 0)$.

III

Numerical Analysis of One-Step Methods

In this chapter the discretisations which were briefly introduced in the first chapter will be discussed and analysed in more detail. First we consider the construction of one-step methods, in particular Runge-Kutta methods, in §1. Next we investigate the local discretisation error and the consistency (order) of a method in §2. In §3 the notion of convergence (of the numerical approximation to the exact solution) and a theorem, with sufficient conditions to achieve this, are treated. Since the result of the latter theorem is rather crude, §4 is devoted to giving a more precise estimate of the magnitude of the global discretisation error (viz. by deriving the leading term in the asymptotic expansion). In §5 two methods are given to estimate the – in practice more important – local error. In §6 this idea is implemented in a practical recommendation for an adaptive integrator. An analysis of the actual errors is also given, for various tolerance criteria, and a number of examples illustrate its effectiveness.

1. Runge-Kutta methods

One-step methods are not only mathematically quite natural (as we shall see in Chapter VII on multistep methods); they are also attractive because they do not complicate the computations when the step size is changed. An important class of one-step methods are the Runge-Kutta methods. Consider the equation (cf. (I.6.3))

$$(1.1) \qquad x(T) = x(t) + \int_t^T f\Big(\tau, x(\tau)\Big) d\tau \ .$$

By taking $t = t_i$ and $T = t_{i+1}$ we may approximate the integral in (1.1) by a quadrature formula, i.e. the discrete sum (cf. Appendix A)

$$(1.2a) \qquad x(t_{i+1}) \doteq x(t_i) + h \sum_{j=1}^{m} \beta_j \, f\Big(t_{ij}, x(t_{ij})\Big) \, ,$$

where $h := t_{i+1} - t_i$ and $t_{ij} := t_i + \rho_j \, h$, $0 \le \rho_j \le 1$ (so the t_{ij} are nodes on $[t_i, t_{i+1}]$). In order to find the (unknown) $x(t_{ij})$ we apply another quadrature formula to a relation like (1.1), but now with $t := t_i$ and $T := t_{ij}$, using (a subset of) the nodes t_{ij}, $j = 1, ..., m$. This results in

$$(1.2b) \qquad x(t_{ij}) \doteq x(t_i) + h \sum_{l=1}^{m} \gamma_{jl} \, f\Big(t_{il}, x(t_{il})\Big) \, .$$

Combining (1.2a), (1.2b) we then find a so-called *Runge-Kutta (RK) formula* with *m stages*:

$$(1.3a) \qquad x(t_{i+1}) \doteq x(t_i) + h \sum_{j=1}^{m} \beta_j \, k_j \, ,$$

with

$$(1.3b) \qquad k_j := f\Big(t_i + \rho_j \, h, x(t_i) + h \sum_{l=1}^{m} \gamma_{jl} \, k_l\Big) \, .$$

A compact notation for such an RK formula is the *Butcher matrix*

$$(1.4a) \qquad
\begin{array}{c|ccc}
\rho_1 & \gamma_{11} & \cdots & \gamma_{1m} \\
\vdots & & & \\
\rho_m & \gamma_{m1} & & \gamma_{mm} \\
\hline
& \beta_1 & \cdots & \beta_m
\end{array}
\ ,$$

denoted for short by

$$(1.4b) \qquad
\begin{array}{c|c}
\rho & \Gamma \\
\hline
& \beta^T
\end{array}
\ .$$

Note that (1.3) is *explicit* (i.e. an approximation of $x(t_{i+1})$ can be found by direct evaluation, given $x(t_i)$) if $\gamma_{jl} = 0$, $l \ge j$. If this is not the case we call it *implicit*; in particular if $\gamma_{jl} = 0$ for $l \ge j + 1$, we call it *diagonally implicit*. Note that (1.3) is in the form (I.6.5).

Example III.1.

(i)   Approximate f by a zeroth order interpolation polynomial at t_i:

$$x(t_{i+1}) \doteq x(t_i) + h \, f\Big(t_i, x(t_i)\Big) \, ;$$

this is the *Euler forward method*, which is clearly explicit.

(ii) Approximate f by a zeroth order interpolation polynomial at t_{i+1}:

$$x(t_{i+1}) \doteq x(t_i) + h f\Big(t_{i+1}, x(t_{i+1})\Big) \ ;$$

this is the *Euler backward method*, which is implicit.

(iii) Approximate f by a linear interpolation polynomial at t_i and t_{i+1} (the resulting quadrature rule is a so-called second order 'Newton-Cotes' formula):

$$x(t_{i+1}) \doteq x(t_i) + \tfrac{1}{2} h \left[f\Big(t_i, x(t_i)\Big) + f\Big(t_{i+1}, x(t_{i+1})\Big) \right] \ ;$$

this is the (implicit) *trapezoidal rule*.

(iv) If we combine (iii) with (i), in order to 'eliminate' $x(t_{i+1})$ in the former, we obtain

$$x(t_{i+1}) \doteq x(t_i) + \tfrac{1}{2} h \left[f\Big(t_i, x(t_i)\Big) + f\Big(t_{i+1}, x(t_i)\Big) + h f\Big(t_i, x(t_i)\Big) \right] \ ;$$

this is *Heun's method* (apparently explicit). The Butcher matrix is given by

$$
\begin{array}{c|cc}
0 & 0 & 0 \\
1 & 1 & 0 \\
\hline
 & \tfrac{1}{2} & \tfrac{1}{2}
\end{array} \ .
$$

(v) The *classical formula of Runge* (1895) *and Kutta* (1901) can be expressed as

$$x(t_{i+1}) \doteq x(t_i) + h\left[\tfrac{1}{6} k_1 + \tfrac{1}{3} k_2 + \tfrac{1}{3} k_3 + \tfrac{1}{6} k_4\right] \ ,$$

where

$$k_1 := f\Big(t_i, x(t_i)\Big)$$
$$k_2 := f(t_i + \tfrac{1}{2} h, x(t_i) + \tfrac{1}{2} h k_1)$$
$$k_3 := f(t_i + \tfrac{1}{2} h, x(t_i) + \tfrac{1}{2} h k_2)$$
$$k_4 := f(t_i + h, x(t_i) + h k_3) \ .$$

(vi) One can construct families of RK formulae, where m (the number of stages) equals 1, 2, 3 etc. and where all k_j (cf. (1.3b)) are the same (as far as they occur). In particular pairs (with m and $m+1$ stages respectively) play an important rôle as they provide for a straightforward method to estimate discretisation errors. Our working example in this book is the Runge-Kutta Fehlberg family, in particular RKF4 and RKF5 (see Table III.1).

Table III.1

0	0						
$\frac{1}{4}$	$\frac{1}{4}$	0					
$\frac{3}{8}$	$\frac{3}{32}$	$\frac{9}{32}$	0				
$\frac{12}{13}$	$\frac{1932}{2197}$	$\frac{-7200}{2197}$	$\frac{7296}{2197}$	0			
1	$\frac{439}{216}$	-8	$\frac{3680}{513}$	$\frac{-845}{4104}$	0		
$\frac{1}{2}$	$\frac{-8}{27}$	2	$\frac{-3544}{2565}$	$\frac{1859}{4104}$	$\frac{-11}{40}$	0	extra in RKF5
	$\frac{25}{216}$	0	$\frac{1408}{2565}$	$\frac{2197}{4104}$	$\frac{-1}{5}$		$\{\beta_i\}$ in RKF4
	$\frac{16}{135}$	0	$\frac{6656}{12825}$	$\frac{28561}{56430}$	$\frac{-9}{50}$	$\frac{2}{55}$	$\{\beta_i\}$ in RKF5

2. Local discretisation errors, consistency

In order to distinguish the approximating solution of the Δ-equation at time t_i from the exact solution of the ODE, we shall denote the latter as x_i^h. This notation implies a constant step size h, i.e. $ih = t_i$. Later it will be necessary to consider variable steps; if the i-th step size is then denoted by h_i (and thus $t_{i+1} = \sum_{j=0}^{i} h_j$) then the approximate solution is denoted by $x_i^{h_i}$.

A one-step method, such as an RK method, is then typically written as

$$(2.1) \qquad x_{i+1}^h = x_i^h + h\,\Phi(t_i, x_i^h, x_{i+1}^h, h) \; .$$

We recall from (I.6.7) that the *local discretisation error* δ is defined as the residual divided by h, when we substitute the exact solution in a method; hence

$$(2.2) \qquad \delta(x(t_{i+1}), h) := \Big[x(t_{i+1}) - x(t_i) - h\,\Phi\Big(t_i, x(t_i), x(t_{i+1}), h\Big) \Big]/h \; .$$

As we saw in §I.6 *consistency* means that the local error approaches zero when $h \to 0$. In general we call p the *consistency order* if $\delta(x(t_{i+1}), h) = O(h^p)$, for p as large as possible. If the solution is sufficiently smooth we can find the local error from Taylor expansions (e.g. around t_i) of the expressions in (2.1). By identifying the coefficients of appropriate terms in the expansion

$$(2.3) \qquad x(t_{i+1}) = x(t_i) + h\,\dot{x}(t_i) + \frac{h^2}{2}\,\ddot{x}(t_i) + \frac{h^3}{3!}\,\dddot{x}\,(t_i) + \ldots + O(h^{p+1}) \; ,$$

we can say that the order equals p if the coefficients of the zeroth up to the p-th term match with (2.3).

For simplicity we shall assume f to be autonomous, which avoids partial derivatives with respect to t. We then find

$$(2.4) \quad \begin{cases} \dot{x} = f \\[2mm] \ddot{x} = \dfrac{\partial f}{\partial x}\, \dot{x} = \dfrac{\partial f}{\partial x}\, f \\[3mm] \dddot{x} = \left(\dfrac{\partial f}{\partial x}\right)^2 f + \dfrac{\partial^2 f}{\partial x^2}\, f^2 \ . \end{cases}$$

Example III.2.

(i) For Euler backward we find

$$h\,\delta(x(t_{i+1}), h) = x(t_{i+1}) - x(t_i) - h\,\dot{x}(t_{i+1})$$

$$= x(t_i) + h\,\dot{x}(t_i) + \frac{h^2}{2}\,\ddot{x}(t_i) - x(t_i) - h\,\dot{x}(t_i)$$

$$- h^2\,\ddot{x}(t_i) + O(h) =$$

$$= -\frac{h^2}{2}\,\ddot{x}(t_i) + O(h^3) \ .$$

By expanding around t_{i+1} we directly find

$$h\,\delta(x(t_{i+1}), h) = -\frac{h^2}{2}\,\ddot{x}(\vartheta_i) \ , \qquad \vartheta_i \in (t_i, t_{i+1}) \ .$$

Clearly Euler backward has consistency order 1.

(ii) For the trapezoidal rule we obtain (cf. (2.3))

$$\dot{x}(t_{i+1}) = \dot{x}(t_i) + h\,\ddot{x}(t_i) + O(h^2) \ .$$

Hence we find from Example III.1(iii):

$$x(t_{i+1}) = x(t_i) + \frac{h}{2}\,\dot{x}(t_i) + \frac{h}{2}\,\dot{x}(t_i) + \frac{h^2}{2}\,\ddot{x}(t_i) + O(h^3) \ ,$$

i.e. the terms up to the second order one match the Taylor expansion.

(iii) Consider the RK method of which the Butcher matrix reads

$$\begin{array}{c|ccc} 0 & 0 & 0 & 0 \\ \frac{1}{2} & \frac{1}{2} & 0 & 0 \\ 1 & -1 & 2 & 0 \\ \hline & \frac{1}{6} & \frac{2}{3} & \frac{1}{6} \end{array} \ .$$

We obtain then (all quantities (evaluated) at time t_i and exact x)

$$k_1 = f \ ,$$

$$k_2 = f + \tfrac{1}{2} h \frac{\partial f}{\partial x}\, f + \tfrac{1}{8} h^2 \frac{\partial^2 f}{\partial x^2}\, f^2 + O(h^3) \ ,$$

$$k_3 = f\left(x - hf + 2hf + h^2 \frac{\partial f}{\partial x}\, f + O(h^3)\right) =$$

$$= f + h \frac{\partial f}{\partial x} f + h^2 \frac{\partial^2 f}{\partial x^2} f + \tfrac{1}{2} h^2 \frac{\partial^2 f}{\partial x^2} f^2 + O(h^3) \; .$$

Therefore

$$x(t_{i+1}) = x(t_i) + h \sum_{j=1}^{3} \beta_j \, k_j =$$

$$= x(t_i) + h \, f + \tfrac{1}{2} h^2 \frac{\partial f}{\partial x} f + \tfrac{1}{6} h^3 \left[\left(\frac{\partial f}{\partial x} \right)^2 f + \frac{\partial^2 f}{\partial x^2} f^2 \right] +$$

$$+ \; O(h^4) \; .$$

From (2.3) and (2.4) we therefore conclude that the order is 3.

(iv) With a little more (tedious) work than for the previous cases, one
      can show that RKF4 and RKF5 (see Table III.1) have order 4 and
      5 respectively; this explains the last character in their name.

For $m = 1, 2, 3$ and 4 one can construct explicit RK methods of order 1,
2, 3 and 4 respectively. For $m \geq 5$ the order is always *smaller* than m. Since
computing an extra stage is a considerable investment, the classical Runge-
Kutta formula (see Example III.1(v)) is in a way optimal. Note that RKF4
and RKF5 have $m = 5$ and $m = 6$ respectively. Implicit RK methods, based
on Hermite interpolation, i.e. Gaussian quadrature (see Appendix A), can be
of order $2m$.

From the construction of RK formulae as quadrature rules consistency implies some simple relationships between their coefficients.

Property 2.5. *For a consistent RK formula we have*

(2.5a) $$\sum_{j=1}^{m} \beta_j = 1$$

(2.5b) $$\rho_j = \sum_{l=1}^{m} \gamma_{jl} \; , \qquad j = 1, ..., m \; .$$

Proof: If a quadrature formula is exact for at least the constant function, then
the sum of weights must be equal to the length of the integration interval, i.e.
$\sum h \, \beta_j = h$, which proves (2.5a).

From (1.2b) we note that a similar argument can be used to find that
$\sum_{l=1}^{m} h \, \gamma_{jl} = t_{ij} - t_i = \rho_j \, h$, which proves (2.5b).   □

3. Convergence of one-step methods

Thus far we simply assumed that the exact $x(t_i)$ was known for computing x_{i+1}^h. In practice, of course, we do not have the *exact* $x(t_i)$, but only its *approximation* x_i^h $(i > 1)$. In this section we shall investigate the *global discretisation error*

$$(3.1) \qquad e_i^h := x(t_i) - x_i^h \ .$$

It will appear that this global error is the cumulative (propagated) effect of all local errors made during the integration up to t_i. In order to understand this effect it is helpful to investigate the 'error propagation' for an explicit method. At time t_i the computed value x_i^h can be viewed as the initial value of a solution, $x_i(t)$ say, with $x_i(t_i) = x_i^h$. The local error then equals $\left[x_i(t_{i+1}) - x_{i+1}(t_{i+1})\right]/h$, i.e. $h\,\delta(x_i(t_{i+1}), h)$ is precisely the 'jump' at t_{i+1} going from one integral curve to the next (see Fig. III.1). Note that for an implicit method we can only say that

$$h\,\delta(x_i(t_{i+1}), h) = x_i(t_{i+1}) - x_{i+1}(t_{i+1})$$

$$- h\left[\Phi(t_i, x_i^h, x_i(t_{i+1}), h) - \Phi(t_i, x_i^h, x_i(t_{i+1}), h)\right] \ .$$

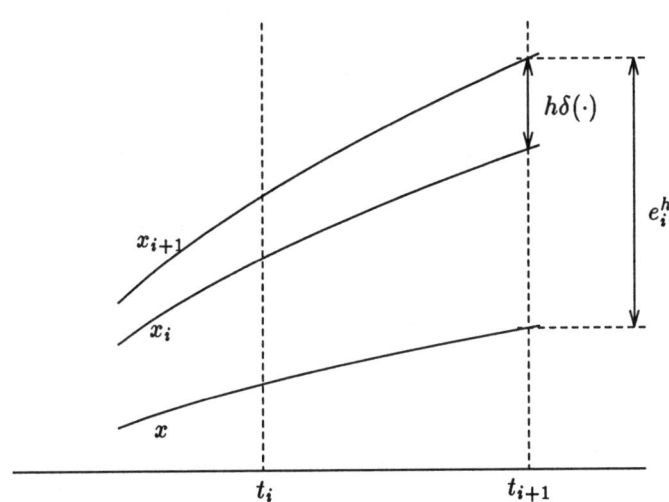

Figure III.1

If Φ is continuously differentiable with respect to its third variable, and $\partial\Phi/\partial z$ denotes that derivative, then

$$h\,\delta(x_i(t_{i+1}), h)\Big/\Big[1 - h\,\frac{\partial\Phi}{\partial z}(t_i, x_i^h, \hat{z}, h)\Big] =$$

$$= x_{i+1}(t_{i+1}) - x_i(t_{i+1})$$

(where $\hat{z}$ is somewhere 'between' $x_i(t_{i+1})$ and $x_{i+1}(t_{i+1})$) denotes a similar 'jump'; if the consistency order is p, then the latter is $O(h^p)$, as can be verified easily.

We can now describe the global effect of discretisation errors by providing the ODE with a perturbation 'source' term r, say (cf. (II.4.2)), such that the solution y of an appropriate IVP coincides with $\{x_i^h\}_{i\geq 0}$ at the grid points $t_0, t_1, \dots$:

$$(3.2) \quad \begin{cases} \dfrac{dy}{dt} = f(t, y) + r(t, y)\,, \qquad r(t, y) = O(h^p)\,, \\[2mm] y(t_0) = x(t_0) = x_0^h\,. \end{cases}$$

Such a solution y can be constructed as follows: consider the solution $x_i(t)$ as well as $x_{i+1}(t)$ on the interval (t_i, t_{i+1}) (x_{i+1} exists there under very mild conditions). Then y can be defined by interpolation. The simplest way is to try linear interpolation:

$$(3.3a) \quad y(t) := \frac{t - t_i}{h}\, x_{i+1}(t) + \frac{t_{i+1} - t}{h}\, x_i(t)\,.$$

From the definition of $x_i(t)$ and $x_{i+1}(t)$ it follows that $y(t_i) = x_i(t_i) = x_i^h$ and $y(t_{i+1}) = x_{i+1}(t_{i+1}) = x_{i+1}^h$. From smoothness and consistency it simply follows that $y(t) - x_i(t) = O(h^{p+1})$ and also $y(t) - x_{i+1}(t) = O(h^{p+1})$, $t_i < t < t_{i+1}$. We also obtain

$$(3.3b) \quad \frac{dy(t)}{dt} = \frac{x_{i+1}(t)}{h} + \frac{t - t_i}{h}\,\frac{dx_{i+1}(t)}{dt} - \frac{x_i(t)}{h} + \frac{t_{i+1} - t}{h}\,\frac{dx_i(t)}{dt} =$$

$$= \frac{1}{h}\,[f(t, x_{i+1})(t - t_i) + f(t, x_i)(t_{i+1} - t)] + O(h^p) =$$

$$= f(t, y) + O(h^p)\,.$$

The last equality follows from expanding $f(t, x_{i+1})$ and $f(t, x_i)$ around $f(t, y)$.

Unfortunately such a solution y is not in C^1 (i.e. the space of continuously differentiable functions). However, by employing cubic Hermite interpolation (requiring that also $\dot{y}(t_i) = \dot{x}_i(t_i)$ and $\dot{y}(t_{i+1}) = \dot{x}_{i+1}(t_{i+1})$) one can obtain a solution y satisfying (3.2). For details we refer to Exercise 12.

Now, if x is sufficiently smooth and f is Lipschitz continuous with Lipschitz constant L, we may apply Theorem II.4.7. Since $y_0 = x_0$, it follows from that theorem that $y(t_i) - x(t_i) = O(h^p)$. Another proof follows below.

Theorem 3.4. *Let f be locally Lipschitz continuous on a domain Ω with constant L. Let T ($< \infty$) be such that y as defined in (3.2) remains in Ω. Then* *there exists a constant C such that $|e_i^h| \leq C \max\limits_{j \leq i} |\delta(x_j(t_{j+1}), h)|$, $ih \leq T$. Consequently $e_i^h = O(h^p)$ if p is the consistency order of the method.*

Proof: We only prove the result for the explicit case. Since

$$x_{i+1}^h = x_i^h - x_i(t_{i+1}) + h\,\Phi(t_i, x_i^h, h) + x_i(t_{i+1}) =$$
$$= -h\,\delta(x_i(t_{i+1}), h) + x_i(t_{i+1})$$

we find

$$e_{i+1}^h = x(t_{i+1}) - x_i(t_{i+1}) + h\,\delta(x_i(t_{i+1}), h) \ .$$

Since $x(\tau)$ and $x_i(\tau)$ satisfy the same ODE, we derive from (II.4.7a):

$$|e_{i+1}^h| \leq e^{hL} |e_i^h| + h\,|\delta(\cdot)|.$$

This gives

$$|e_i^h| \leq \left\{ \sum_{l=0}^{i-1} (e^{hL})^l \right\} h \max |\delta(\cdot)| \leq \frac{e^{ihL} - 1}{e^{hL} - 1} h \max |\delta(\cdot)| \ .$$

Now we use $e^{hL} - 1 \geq h\,L$ and take $C := (e^{TL} - 1)/L$. $\qquad \square$

We shall call a method *convergent* if there exists a domain $\Omega \in \mathbb{R}$, such that for $t = i\,h$

$$\lim_{\substack{h \to 0 \\ i \to \infty}} e_i^h = 0 \ .$$

If $e_i^h = O(h^q)$, q is called the *convergence order*. Such a domain coincides roughly speaking with the existence domain of the problem (cf. the regular perturbation in §II.5).

Corollary 3.5. *A one-step method is consistent of order p iff it is convergent of order p.*

Remark 3.6. If the RK method is implicit, we have to divide δ in the proof of Theorem 3.4 by $(1 - h\,\partial\Phi/\partial z)$, where z denotes the third variable, at some suitable intermediate value. This results in an only slightly different definition of the constant C in that theorem.

Remark 3.7. The proof of Theorem 3.4 is based on relating $\{x_i^h\}$ to the *continuous solution* $y(t)$ of the perturbed problem (3.2); y is a mapping from some interval $I \subset \mathbb{R}$ into $\mathbb{R}$, as x is. Conversely, it is also possible to project the solution x on the space of sequences $\{x(t_0), x(t_1), ...\}$, equate the solution with $\{x_i^h\}$, and estimate the errors in this space. Note that a convergence proof then requires a demonstration first that $\Phi(t, x, z, h)$ is Lipschitz continuous in the second and third variables.

Corollary 3.8. *Under the assumptions of Theorem 3.4 we have*

$$\delta(x_i(t_{i+1}), h) = \delta(x(t_{i+1}), h) + O(h^{p+1}) .$$

This corollary says that we may identify, up to $O(h^{p+1})$, the local discretisation errors made during the real process of time stepping, with the local errors we would find for the exact solution. This is often useful for theoretical analyses.

4. Asymptotic expansion of the global error

The global error can be linked to the local error even more explicitly than was done above. We proceed as follows. Assuming that $x \in C^p$ (the space of p times continuously differentiable functions) we find for the local discretisation of a method with consistency order $\geq p$ (see also Corollary 3.8)

$$(4.1) \qquad \delta(x(t_i), h) = c\left(t_{i-1}, x(t_{i-1})\right) h^p + O(h^{p+1}) ,$$

where $c \subset C^0(I \times \Omega)$ (i.e. a continuous function) depends on the method used as well. We shall call c the *local error function*. We then have the following more precise result for the global discretisation error:

Theorem 4.2. *Let f be sufficiently smooth; then there exists a so-called global error function $d \in C^1(I \times \Omega)$ with*

$$(4.2a) \qquad \begin{cases} \dfrac{d}{dt} d\left(t, x(t)\right) = \dfrac{\partial f}{\partial x}\left(t, x(t)\right) d\left(t, x(t)\right) + c(t, x) , \\[2mm] d(t_0) = 0 . \end{cases}$$

Moreover the global discretisation error is given by

$$(4.2b) \qquad e_i^h = d\left(t_i, x(t_i)\right) h^p + O(h^{p+1}) .$$

Proof: Clearly we have

$$e_{i+1}^h - e_i^h = x(t_{i+1}) - [x_i(t_{i+1}) - h\,\delta(x(t_{i+1}), h)] - [x(t_i) - x_i(t_i)]$$

$$= \int\limits_{t_i}^{t_{i+1}} \left[f\left(\tau, x(\tau)\right) - f\left(\tau, x_i(\tau)\right)\right] d\tau + c\left(t_i, x(t_i)\right) h^{p+1} + O(h^{p+2})$$

$$= \int_{t_i}^{t_{i+1}} \left[\frac{\partial f}{\partial x} \Big(\tau, x(\tau) \Big) e_i(\tau) + O\Big(e_i^2(\tau) \Big) \right] d\tau + c(\cdot) \, h^{p+1} + O(h^{p+2}) \, ,$$

where we defined $e_i(t) := x(t) - x_i(t)$ (so $e(t_i) \equiv e_i^h$). By introducing the 'magnified' quantity $\hat{e}_i(t) := h^p \, e_i(t)$, we obtain, using the fact that

$$\int_{t_i}^{t_{i+1}} e_i^2(\tau) d\tau = O(h^{2p+1}) = O(h^{p+2}) \text{ for } p > 1,$$

$$\hat{e}_i(t_{i+1}) = \hat{e}_i(t_i) + \int_{t_i}^{t_{i+1}} \frac{\partial f}{\partial x} \Big(\tau, x(\tau) \Big) \hat{e}_i(\tau) \, d\tau + c \Big(t_i, x(t_i) \Big) h + O(h^2)$$

$$= \hat{e}_i(t_i) + h \left[\frac{\partial f}{\partial x} \Big(t_i, x(t_i) \Big) \hat{e}_i(\tau) + c \Big(t_i, x(t_i) \Big) + O(h) \right] .$$

Here we used a first order approximation

$$\frac{\partial f}{\partial x} \Big(\tau, x(\tau) \Big) \hat{e}_i(\tau) \doteq \frac{\partial f}{\partial x} \Big(t_i, x(t_i) \Big) \hat{e}_i(t_i)$$

in the integral (cf. Appendix A)). The expression thus found, however, can be viewed as an Euler forward approximation step of the ODE in (4.2a) with an extra source term $O(h)$. Hence, on account of Theorem 3.4, we find $\hat{e}_i(t_i) - d \Big(t_i, x(t_i) \Big) = O(h)$, from which (4.2b) immediately follows. □

The previous theorem actually specifies the first term in the so-called *asymptotic expansion* of the global error

(4.3) $\qquad e_i^h = d_i^0 \, h^p + d_i^1 \, h^{p+1} + d_i^2 \, h^{p+2} + \dots \, ,$

which can be shown for sufficiently smooth x (see [72]). We shall not prove this more general theoretical result. It can be used to compute more accurate approximations as follows: if we have an approximation of $x(t)$ from using the method with step h_1, and also one with step h_2 (for example $h_2 = h_1/2$), then a suitable linear combination of these approximations will effectively eliminate the $O(h^p)$ term in the asymptotic expansion of this combination. Thus, neglecting the effect of rounding errors and assuming sufficient smoothness, this leads to an $O(h^{p+1})$ approximation for $x(t)$. This process can be repeated to eliminate even higher order terms, thus leading to a so-called *Richardson extrapolation* method. The first step goes, for example, as follows. If we have approximations x_i^h and $x_{2i}^{\frac{1}{2}h}$, then

(4.4a) $\qquad x_i^h = x(t) - d \Big(t, x(t) \Big) h^p + O(h^{p+1})$

(4.4b) $\qquad x_{2i}^{\frac{1}{2}h} = x(t) - (\tfrac{1}{2})^p \, d \Big(t, x(t) \Big) h^p + O(h^{p+1}) \, .$

Neglecting the $O(h^{p+1})$ term we see that $x_{2i}^{\frac{1}{2}h}$ has an error which is a factor $(\frac{1}{2})^p$ smaller than that of x_i^h. Define a new approximant by

(4.5) $\tilde{x}_i^h := [x_{2i}^{\frac{1}{2}h} - (\frac{1}{2}p)\,x_i^h]/[1 - (\frac{1}{2})^p]$.

Clearly the $O(h^p)$ term is missing in the error expansion of $\tilde{x}_i^h$, in other words $\tilde{x}_i^h = x(t) + O(h^{p+1})$.

Another result of Theorem 4.2 is that we can employ (4.4a,b) to estimate the error in x_i^h (or $x_{2i}^{\frac{1}{2}h}$). Indeed, we have

(4.6a) $h^p\, d\left(t, x(t)\right) = [x_{2i}^{\frac{1}{2}h} - x_i^h]/[1 - (\frac{1}{2})^p] + O(h^{p+1})$,

whence

(4.6b) $e_i^h = [x_i^h - x_{2i}^{\frac{1}{2}h}]/[1 - (\frac{1}{2})^p] + O(h^{p+1})$.

The expression

$$[x_i^h - x_{2i}^{\frac{1}{2}h}]/[1 - (\frac{1}{2})^p]$$

is called the *extrapolated global error estimate*. Given an absolute error threshold, the *tolerance*, say TOL, one may use the latter expression to conversely predict the step h using $|e_i^h| \approx$ TOL.

Example III.3.
Consider the initial value problem

$$\begin{cases} \dot{x} = -x + 2x^2\,e^{-t} , \\ x(0) = \frac{1}{3} . \end{cases}$$

As can be verified easily, the solution is given by $x(t) = [2e^t + e^{-t}]^{-1}$.

In Table III.2 we have given the exact global error at some points $\in [0,1]$ for various step sizes using Euler forward. In Table III.3 the errors are given when one Richardson extrapolation step as above is carried out. Finally, in Table III.4 we have given the error estimates as in (4.6b); the latter show a remarkable accuracy when compared to the actual errors in **Table III.2**.

5. Practical error estimation

If we would like to use the global error estimate as outlined in the previous section, we end up with a very costly method, since we need to compute numerical approximations of $x(t)$ at all grid points each time we want to

Table III.2 $h = 0.125$.

t	x_i^h	error	$x_{2i}^{h/2}$	error	$x_{4i}^{h/4}$	error
0.125	3.1944E-01	-1.86E-03	3.1850E-01	-9.17E-04	3.180E-01	-4.55E-04
0.250	3.0203E-01	-3.24E-03	3.0037E-01	-1.58E-03	2.996E-01	-7.78E-04
0.375	2.8204E-01	-4.06E-03	2.7994E-01	-1.95E-03	2.789E-01	-9.57E-04
0.500	2.6045E-01	-4.30E-03	2.5820E-01	-2.05E-03	2.571E-01	-1.00E-04
0.625	2.3818E-01	-4.08E-03	2.3602E-01	-1.93E-03	2.350E-01	-9.34E-04
0.750	2.1600E-01	-3.52E-03	2.1412E-01	-1.64E-03	2.133E-01	-7.90E-04
0.875	1.9451E-01	-2.74E-03	1.9302E-01	-1.25E-03	1.924E-01	-5.98E-04
1.000	1.7414E-01	-1.85E-03	1.7311E-01	-8.24E-04	1.727E-01	-3.87E-04

Table III.3 $h = 0.125$.

t	x	error	$x_{2i}^{h/2}$	error
0.125	3.1755E-01	2.84E-05	3.1757E-01	7.98E-06
0.250	2.9871E-01	8.29E-05	2.9877E-01	2.20E-05
0.375	2.7784E-01	1.45E-04	2.7795E-01	3.74E-05
0.500	2.5595E-01	1.98E-04	2.5610E-01	4.99E-05
0.625	2.3386E-01	2.31E-04	2.3404E-01	5.73E-05
0.750	2.1224E-01	2.42E-04	2.1242E-01	5.92E-05
0.875	1.9154E-01	2.32E-04	1.9171E-01	5.63E-05
1.000	1.7208E-01	2.06E-04	1.7223E-01	4.99E-05

Table III.4 $h = 0.125$.

t	$2[x_{2i}^{h/2} - x_i^h]$	$2[x_{4i}^{h/4} - x_{2i}^{h/2}]$
0.125	-1.891E-03	-9.251E-04
0.250	-3.322E-03	-1.600E-03
0.375	-4.191E-03	-1.988E-03
0.500	-4.497E-03	-2.100E-03
0.625	-4.313E-03	-1.982E-03
0.750	-3.760E-03	-1.697E-03
0.875	-2.969E-03	-1.309E-03
1.000	-2.060E-03	-8.737E-04

change the step size. This problem is only aggravated by the fact that we have assumed a uniform step size, which does not seem wise if a solution tends to become smoother as it evolves (cf. Example III.3 for $t \geq 0.5$). Hence we really want to have a nonuniform step to be found from computed results which should be as local as possible. As we already saw in (4.2a) the global error quite closely follows the behaviour of the solution, which in turn essentially determines the local error (apart from the factor h^p). All this then leads to the seemingly contradictory idea of estimating an appropriate step size from local errors. In this section we first investigate practical ways for doing this. In the subsequent section we validate these ideas.

There are two methods for estimating the *local error*:

(i) *local extrapolation*
(ii) *embedding*.

First we look at extrapolation. Suppose we compute, besides x_i^h,

$$(5.1) \qquad \bar{x}_{i+1}^h := x_i^h + \frac{h}{2}\,\Phi\left(t_i, x_i^h, \frac{h}{2}\right) +$$

$$+ \frac{h}{2}\,\Phi\left(t_i + \frac{h}{2}, x_i^h + \frac{h}{2}\,\Phi\left(t_i, x_i^h, \frac{h}{2}\right), \frac{h}{2}\right) .$$

As in §2 let $x_i(t)$ be defined by

$$\begin{cases} \dfrac{d}{dt}\,x_i(t) = f\left(t, x_i(t)\right) , \\[2mm] x_i(t_i) = x_i^h . \end{cases}$$

We obviously have for the local discretisation error

$$(5.2) \qquad \delta(x_i(t_{i+1}), h) = c\left(t_i, x_i(t_i)\right)h^p + O(h^{p+1}) .$$

For the local error arising from (5.1), say $\bar{\delta}(\cdot)$, we find using Taylor expansions

$$(5.3) \qquad \bar{\delta}(x_i(t_{i+1}), h) = \frac{1}{h}\left[c\left(t_i, x_i(t_i)\right)\left(\frac{h}{2}\right)^{p+1} \right.$$

$$\left. + c\left(t_i + \tfrac{1}{2} h, x_i(t + \tfrac{1}{2} h) + O(h^{p+1})\right)\left(\frac{h}{2}\right)^{p+1} \right] + O(h^{p+1})$$

$$= c\left(t_i, x_i(t_i)\right)\left(\frac{h}{2}\right)^{p} + O(h^{p+1})$$

$$= (\tfrac{1}{2})^p\, \delta(x_i(t_{i+1}), h) + O(h^{p+1}) .$$

Since $x_i(t_{i+1}) = x_{i+1}^h + h\,\delta(\cdot) = \bar{x}_{i+1}^h + h\,\bar{\delta}(\cdot)$, we thus obtain

$$(5.4) \qquad \delta(x_i(t_{i+1}), h) = \frac{1}{h}\,[\bar{x}_{i+1}^h - x_{i+1}^h]/[1 - (\tfrac{1}{2})^p] + O(h^{p+1}) ,$$

from which the local error may be estimated. Note that this expression is similar to (4.6b) (where two 'global' solutions were compared).

The other technique, called *embedding*, employs another, higher order, method as well. Consider, for example, the following pair:

(5.5a) $\quad x_{i+1}^h = x_i^h + h\,\Phi_1(t_i, x_i^h, h)$

(5.5b) $\quad \bar{x}_{i+1}^h = x_i^h + h\,\Phi_2(t_i, x_i^h, h)$,

and let for $q \geq p + 1$

(5.6a) $\quad h\,\delta(x_i(t_{i+1}), h) := x_i(t_{i+1}) - x_{i+1}^h$, $\qquad \delta(\cdot) = O(h^p)$,

(5.6b) $\quad h\,\bar{\delta}(x_i(t_{i+1}), h) := x_i(t_{i+1}) - \bar{x}_{i+1}^h$, $\qquad \delta(\cdot) = O(h^q)$.

Then $\bar{x}_{i+1}^h - x_{i+1}^h = -h\big(\bar{\delta}(\cdot) - \delta(\cdot)\big) = h\,\delta(\cdot) + O(h^{p+2})$, so

(5.7) $\quad \delta(x_i(t_{i+1}), h) = [\bar{x}_{i+1}^h - x_{i+1}^h]/h + O(h^{p+1})$.

In order to economise on such a higher order method one can construct 'families' of RK formulae where the higher order members employ the same function evaluations as the lower order ones at relevant points. A very well-known family has been constructed by Fehlberg [26]. The fourth-fifth order pair has been given in Table III.1 (note that the former needs 5 and the latter 6 levels). We thus see that the local error is estimated by

$$\delta(\cdot) = \sum_{j=1}^{6} (\bar{\beta}_j - \beta_j)\,k_j \ ,$$

where the β_j ($\beta_6 := 0$) and the $\bar{\beta}_j$ are the weights of RKF4 and RKF5 respectively.

6. Step size control

In this section we shall again consider explicit methods only. Given an (absolute) tolerance TOL, there are two criteria according to which the local error estimates may be employed for determining the step sizes. Our previous notation, where we assumed a constant step size h, has to be adjusted here. Therefore let now $h_i := t_{i+1} - t_i$. Then we introduce (cf. [68])

(i) *Error Per Step* (EPS): determine h_i such that

$$|h_i\,\delta(x_i(t_{i+1}), h_i)| = \text{TOL}.$$

(ii) *Error Per Unit Step* (EPUS): determine h_i such that

$$|\delta(x_i(t_{i+1}), h_i)| = \text{TOL}.$$

When using EPS one simply considers the error going from one integral curve to the next in the time stepping process (cf. §3). With EPUS one anticipates the global order (when $O(1/h)$ steps have been done). One proceeds as follows:

(i) Let EST_i be an estimate for $\left|h_i\left(\delta(x_i(t_{i+1}), h_i)\right)\right|$.

    (a) If $\frac{1}{4} \leq \text{EST}_i/\text{TOL} \leq 4$, we accept h_i and continue our integration on the next interval with $h_{i+1} = h_i$.

    (b) If $\frac{1}{4} > \text{EST}_i/\text{TL}$ or $4 < \dfrac{\text{EST}_i}{\text{TOL}}$, we determine a new h_i, h_{new} say,

$$h_{\text{new}} := 0.9\, h_i \left(\frac{\text{TOL}}{\text{EST}_i}\right)^{\frac{1}{p+1}},$$

and perform the integration again on the present interval.

In (b) the step size choice is based on the $(p+1)$ st order behaviour of $h_i\,\delta(\cdot)$ (so EST_i) i.e. we seek h_{new} such that $\left(h_{\text{new}}/h_i\right)^{p+1} \doteq \text{TOL}/\text{EST}_i$. The factor 0.9 is a safety factor, which makes the next step size choice somewhat more 'conservative'. The numbers $\frac{1}{4}$ and 4 in (a) and (b) are fairly arbitrary. We might as well have chosen $1/\alpha$ and α, $\alpha > 0$, instead. Clearly, the closer α is to 1 the more often the step size selector will become active and henceforth may perform an (often not really necessary) integration repeat.

(ii) For EPUS we can do the same as for EPS. Let EST_i be an estimate for $|\delta(x_i(t_{i+1}), h_i)|$ now. Only if $\frac{1}{4} > \text{EST}_i/\text{TOL}$ or $4 < \text{EST}_i/\text{TOL}$, we determine h_{new} from

$$h_{\text{new}} := 0.9\, h_i \left(\frac{\text{TOL}}{\text{EST}_i}\right)^{\frac{1}{p}}.$$

There remain two more practical problems, viz. how to start (i.e. to choose h_0) and where to place output points. There are various ways to find a reasonable estimator for h_0. The most crude one is to take h_0 as the length of the integration interval. Often, however, one also has an input parameter h_0, to be assigned by the user. This last possibility is very useful when a large number of output points are required (e.g. when the solution has to be visualised). Clearly a crude h_0 will require a fair amount of overhead in the latter situation.

In order to appreciate this step size mechanism we consider its application to the model problem:

(6.1) $\dfrac{dx}{dt} = \lambda x$, $x(0) = 1$;   so $x(t) = e^{\lambda t}$.

If the local error estimate is fairly accurate we have ideally

$$\text{EST}_i = |\zeta| \, (h_i)^{p+1} \, |\lambda|^p \, e^{\lambda t_i} \, ,$$

where ζ is some constant (more specifically, $\zeta \lambda^{p+1} e^{\lambda t_i} \doteq c\big(t_i, x(t_i)\big)$, cf. (4.1)).

- **With EPS we obtain for (6.1)**

$$(6.2) \qquad h_{i+1} = \frac{1}{|\lambda|} \left| \frac{\text{TOL}}{\zeta} \right|^{\frac{1}{p+1}} \exp\left(-\frac{\lambda t_i}{p+1}\right) .$$

Now consider the interval $[0, T]$. The next theorem then gives an estimate of the number of necessary steps with tolerance TOL.

Theorem 6.3. *Let N be the number of steps necessary to determine the solution of (6.1) on $[0, T]$ with tolerance TOL and EPS control. Then*

$$N \doteq \left| \frac{\zeta}{\text{TOL}} \right|^{\frac{1}{p+1}} (p+1) \left| \exp\left(\frac{\lambda T}{p+1}\right) - 1 \right| + 1 \, .$$

Proof: Let $\nu := \dfrac{\lambda}{p+1}$. Let $t_0, t_0 + h_0, ..., t_0 + \sum_{j=1}^{N-1} h_j$ be the grid points. Then we find

$$h_{i+1} = h_0 \exp\left(-\nu \sum_{j=0}^{i} h_j\right) , \qquad i \geq 0 \, .$$

The function $s(\xi)$, defined by

$$s(\xi) := h_0 \exp\left[-\nu \int_1^{\xi} s(\tau) d\tau \right] , \qquad 1 \leq \xi \leq N \, ,$$

is a reasonable estimator for h_{i-1} (at least for $\lambda > 0$, cf. Example III.5). Differentiating s gives a so-called *Ricatti ODE*:

$$\frac{d}{d\xi} s(\xi) = -\nu s^2(\xi) \Rightarrow s(\xi) = \frac{1}{\nu} \, \frac{1}{\xi + c} \, ,$$

where c follows from $s(1) = h_0$. Hence we find

$$s(N) = \frac{h_0}{\nu h_0 (N-1) + 1} \doteq h_{N-1} \, .$$

Realising that $\sum_{j=0}^{N-1} h_j = T$, we thus obtain

$$(N-1) = \frac{1}{h_0 \nu} \left\{ \exp\left(\nu T\right) - 1 \right\} \, .$$

Note that s is increasing for $\lambda < 0$ with a possible singular point. However, this corresponds to the case $h_{N-1} = \infty \Rightarrow T = \infty$. The estimate now follows directly. $\square$

Corollary 6.4. *Let* $\left| \dfrac{\lambda T}{p+1} \right|$ *be not 'small':*

(a) $\lambda < 0 \Rightarrow N \doteq \left| \dfrac{\zeta}{\text{TOL}} \right|^{\frac{1}{p+1}} (p+1)$

(b) $\lambda > 0 \Rightarrow N \doteq \left| \dfrac{\zeta}{\text{TOL}} \right|^{\frac{1}{p+1}} (p+1) \exp\!\left(\dfrac{\lambda T}{p+1} \right).$

In other words, for $\lambda < 0$ the step size eventually becomes almost independent of T and arbitrarily large; for $\lambda > 0$ the step size becomes arbitrarily small. However, for extreme values our assumption about the reliability of the estimates EST_i is no longer valid; hence $s(i)$ in the proof of Theorem 6.3 will no longer be a good estimator of h_{i-1}. In particular for λ large and negative, we have so-called stiffness, which will be dealt with in Chapter VIII.

– With EPUS we obtain for (6.1)

(6.5) $h_{i+1} = \left| \dfrac{1}{\lambda} \right|^{\frac{p+1}{p}} \left| \dfrac{\text{TOL}}{\zeta} \right|^{\frac{1}{p}} \exp\!\left(-\dfrac{\lambda t_i}{p} \right).$

The results for EPUS are not much different from those for EPS. We obtain estimates similar to those in Theorem 6.3 and Corollary 6.4 if we replace the factor $\left| \dfrac{\zeta}{\text{TOL}} \right|^{\frac{1}{p+1}} (p+1)$ by $\left| \dfrac{\zeta\lambda}{\text{TOL}} \right|^{\frac{1}{p}} p$ there.

Of practical importance is the question how large the global discretisation error becomes if we use these step size controls. Ideally we should have

$$e_{i+1} \doteq \exp(\lambda t_i)e_i + \text{EST}_i\ ,$$

where e_i is short for the global error $e_i^{h_i} := x(t_i) - x_i^{h_i}$. Apparently we have

(6.6) $e_i \doteq \displaystyle\sum_{l=1}^{i-1} \exp\!\left(\lambda(t_i - t_l) \right) \text{EST}_l\ .$

From this we find the following:

Theorem 6.7.
(a) *With EPS control the global error is estimated by*

$$|e_i| \doteq \frac{p+1}{p} |\zeta|^{\frac{1}{p+1}} (\text{TOL})^{\frac{p}{p+1}} \left| \exp(\lambda t_i) - \exp\!\left(\frac{\lambda t_i}{p+1} \right) \right|.$$

(b) *With EPUS control the global error is estimated by*

$$|e_i| \doteq \frac{1}{|\lambda|} \, \mathrm{TOL} \, |\exp(\lambda t_i) - 1| \, .$$

Proof:

(a) $\mathrm{EST}_l \doteq \mathrm{TOL} \Rightarrow e_i \doteq e^{\lambda t_i} \, \mathrm{TOL} \displaystyle\sum_{l=1}^{i-1} e^{-\lambda t_l}.$

Let $g(t_l) := \exp(-\lambda t_l - \ln h_l)$, then

$$\int_0^{t_i} g(t) \, dt \doteq \sum_{l=0}^{i-1} e^{-\lambda t_l} \, .$$

Using $\ln h_l \doteq c - \dfrac{\lambda t_l}{p+1}$, with $c := \ln\left(\dfrac{1}{|\lambda|} \left| \dfrac{\mathrm{TOL}}{\zeta} \right|^{\frac{1}{p+1}}\right)$ we thus find

$$\sum_{l=0}^{i-1} e^{-\lambda t_l} \doteq e^{-c} \int_0^{t_i} \exp\left(\frac{-p}{p+1} \lambda t\right) dt \, ,$$

from which the estimate follows.

(b) $\mathrm{EST}_l \doteq h_l \, \mathrm{TOL} \, ,$ whence $e_i \doteq e^{\lambda t_i} \, \mathrm{TOL} \displaystyle\int_0^{t_i} e^{-\lambda t} \, dt \, .$ $\square$

Remark 6.8. Qualitatively EPS and EPUS give the same result, although only EPUS gives errors proportional to TOL.

Remark 6.9. For negative λ the actual absolute error can be much smaller than TOL, and for positive λ this error can be much larger. In the latter case, however, we still expect the relative tolerance to be proportional to TOL.

The error control may also be performed in a *relative* way:

(iii) *Relative Error Per Step* (REPS): find h_i such that

$$|h_i \, \delta(x_i(t_{i+1}), h_i)| / |x_i^{h_i}| \doteq \mathrm{TOL} \, .$$

(iv) *Relative Error Per Unit Step* (REPUS): find h_i such that

$$|\delta(x_i(t_{i+1}), h_i)| / |x_i^{h_i}| \doteq \mathrm{TOL} \, .$$

– **With REPS we obtain for (6.1)**

$$h_{i+1} = \frac{1}{|\lambda|} \left| \frac{\text{TOL}}{\zeta} \right|^{\frac{1}{p+1}} .$$

– **With REPUS we obtain for (6.1)**

$$h_{i+1} = \left| \frac{1}{\lambda} \right|^{\frac{p+1}{p}} \left| \frac{\text{TOL}}{\zeta} \right|^{\frac{1}{p}} .$$

Note that in both cases the step size is constant, say $h_i = h$, so that

$$N \doteq \frac{T}{h} .$$

This results in the following estimates:

Theorem 6.10.
(a) *With REPS control the global error is estimated by*

$$|e_i| \doteq i\, e^{\lambda t_i}\, TOL, \quad \text{and so the relative error by } \left| \frac{e_i}{x_i} \right| \doteq i\, \text{TOL} .$$

(b) *With REPUS control, the global error is estimated by*

$$|e_i| \doteq t_i\, e^{\lambda t_i}\, TOL, \quad \text{and so the relative error by } \left| \frac{e_i}{x_i} \right| \doteq t_i\, \text{TOL} .$$

Proof:

(a) $$e_i \doteq \sum_{l=0}^{i-1} e^{\lambda(t_i - t_l)}\, e^{\lambda t_l}\, \text{TOL} = i\, e^{\lambda t_i}\, \text{TOL} .$$

(b) $$e_i \doteq \sum_{l=0}^{i-1} e^{\lambda t_i}\, h\, \text{TOL} . \qquad\qquad \square$$

From the foregoing it appears that for $|\lambda|$ 'large' a step size based on relative error control is more meaningful than one based on absolute error control; for $|\lambda|$ 'small' this is less relevant. Finally, for $\lambda < 0$ but $|\lambda|$ not 'large' EPS or EPUS is the most reasonable choice (note that h is then monotonically increasing).

In practice the error control is a mixture of absolute and relative tolerances, ABSTOL and RELTOL respectively. An obvious choice is then to require

$$\text{EST}_i \doteq \text{ABSTOL} + \text{RELTOL} \cdot |x_i| .$$

The Runge-Kutta Fehlberg implementation RKF45 is based on EPS/REPS. Although the error of the fourth order method is controlled, the code uses the higher order approximation. Hence effectively $p = 5$. For ζ one may take a

Table III.5

TOL $\backslash$ G	10^3	10^2	10	10^{-1}	10^{-2}	10^{-3}	10^{-4}
10^{-3}	16(16.4)	8(8.3)	3(3.2)	4(3.0)	6(4.3)	7(5.1)	8(5.6)
10^{-4}	25(26.0)	13(13.2)	5(5.1)	5(4.2)	8(5.2)	10(7.5)	11(8.3)
10^{-5}	41(41.2)	21(20.9)	8(8.1)	7(6.1)	11(9.4)	14(11.3)	16(12.6)
10^{-6}	67(63.4)	34(33.0)	13(12.8)	11(8.4)	17(14.2)	21(17.4)	24(19.5)

constant $\approx 10^{-3}$. At each step the remainder of the integration interval is compared to two times the actual step length; if the latter is larger then the remainder interval is just halved so that the last two steps are equal.

This is also used in the PASCAL procedure Fehlberg by which the following examples have been computed.

Example III.4.
From Theorem 6.3 it appears that the number of steps N is proportional to $\left|\exp\left(\dfrac{\lambda T}{p+1}\right) - 1\right|$; hence λT is a relevant parameter to test the actual value of N against the *growth* $G := \exp(\lambda T)$.

In Table III.5 we give, for various tolerances TOL (= ABSTOL) and values of G, the actually used number of grid points N and, between parentheses, their estimates as found in Theorem 6.3.

Below we illustrate step sizes and errors through three test problems.

Example III.5.

(i) $\dot{x} = 2x$, $x(0) = 1$; $T = 5$.

   In Table III.6 we have displayed time, step size, absolute and relative errors, approximates and finally the exact solution values for EPS, with ABSTOL $= 10^{-4}$ and RELTOL $= 10^{-10}$, which effectively makes this EPS control. Note that the relative error soon becomes constant, as predicted by Theorem 6.7.

   In Table III.7 we have an analogous result for REPS control, with RELTOL $= 10^{-4}$ (and ABSTOL $= 0$, which is allowed). Note that h_i now becomes constant after a few steps.

(ii) $\dot{x} = -2x$, $x(0) = 1$; $T = 5$.

   In Table III.8 we have displayed the results as in case (i), for EPS (ABSTOL $= 10^{-4}$). The absolute error is almost constant, as predicted by Theorem 6.7.

   In Table III.9 we have the results for REPUS (RELTOL $= 10^{-4}$). Note that h_i becomes asymptotically constant; moreover the relative error becomes proportional to iTOL, see Theorem 6.10.

Table III.6

i	t_i	h_i	e_i^h	$e_i^h/x(t_i)$	x_i^h	$x(t_i)$
0	0.00E-00					1.0000E+00
1	1.38E-01	1.38E-01	4.26E-07	3.23E-07	1.3178E+00	1.3178E+00
2	4.14E-01	2.76E-01	3.91E-05	1.71E-05	2.2892E+00	2.2892E+00
3	6.54E-01	2.39E-01	9.10E-05	2.46E-05	3.6952E+00	3.6953E+00
4	8.92E-01	2.38E-01	1.90E-04	3.19E-05	5.9499E+00	5.9501E+00
5	1.11E+00	2.16E-01	3.32E-04	3.62E-05	9.1716E+00	9.1719E+00
6	1.30E+00	1.96E-01	5.24E-04	3.86E-05	1.3572E+01	1.3572E+01
7	1.48E+00	1.79E-01	7.78E-04	4.01E-05	1.9415E+01	1.9416E+01
8	1.65E+00	1.65E-01	1.11E-03	4.10E-05	2.7009E+01	2.7010E+01
9	1.80E+00	1.53E-01	1.53E-03	4.16E-05	3.6698E+01	3.6700E+01
10	1.95E+00	1.43E-01	2.05E-03	4.20E-05	4.8868E+01	4.8870E+01
11	2.08E+00	1.35E-01	2.70E-03	4.22E-05	6.3945E+01	6.3948E+01
12	2.21E+00	1.27E-01	3.50E-03	4.24E-05	8.2401E+01	8.2404E+01
13	2.33E+00	1.20E-01	4.46E-03	4.26E-05	1.0475E+02	1.0476E+02
14	2.44E+00	1.14E-01	5.62E-03	4.27E-05	1.3156E+02	1.3156E+02
15	2.55E+00	1.09E-01	6.99E-03	4.28E-05	1.6344E+02	1.6344E+02
16	2.65E+00	1.04E-01	8.61E-03	4.28E-05	2.0105E+02	2.0106E+02
17	2.75E+00	9.91E-02	1.05E-02	4.29E-05	2.4511E+02	2.4512E+02
18	2.85E+00	9.50E-02	1.27E-02	4.29E-05	2.9639E+02	2.9641E+02
19	2.94E+00	9.12E-02	1.53E-02	4.29E-05	3.5573E+02	3.5575E+02
20	3.03E+00	8.78E-02	1.82E-02	4.30E-05	4.2401E+02	4.2403E+02
21	3.11E+00	8.46E-02	2.16E-02	4.30E-05	5.0218E+02	5.0220E+02
22	3.19E+00	8.16E-02	2.54E-02	4.39E-05	5.9124E+02	5.9127E+02
23	3.27E+00	7.89E-02	2.98E-02	4.30E-05	6.9228E+02	6.9231E+02
24	3.35E+00	7.63E-02	3.47E-02	4.30E-05	8.0644E+02	8.0648E+02
25	3.42E+00	7.39E-02	4.02E-02	4.30E-05	9.3493E+02	9.3497E+02
26	3.49E+00	7.17E-02	4.64E-02	4.30E-05	1.0790E+03	1.0791E+03
27	3.56E+00	6.96E-02	5.34E-02	4.30E-05	1.2401E+03	1.2401E+03
28	3.63E+00	6.76E-02	6.11E-02	4.30E-05	1.4195E+03	1.4196E+03
29	3.70E+00	6.57E-02	6.97E-02	4.31E-05	1.6189E+03	1.6190E+03
30	3.76E+00	6.39E-02	7.92E-02	4.31E-05	1.8397E+03	1.8400E+03
31	3.82E+00	6.23E-02	8.97E-02	4.31E-05	2.0836E+03	2.0837E+03
32	3.88E+00	6.07E-02	1.01E-01	4.31E-05	2.3525E+03	2.3526E+03
33	3.94E+00	5.92E-02	1.14E-01	4.31E-05	2.6480E+03	2.6481E+03
34	4.00E+00	5.77E-02	1.28E-01	4.31E-05	2.9722E+03	2.9723E+03
35	4.06E+00	5.64E-02	1.43E-01	4.31E-05	3.3269E+03	3.3271E+03
36	4.11E+00	5.51E-02	1.60E-01	4.31E-05	3.7144E+03	3.7146E+03
37	4.16E+00	5.39E-02	1.78E-01	4.31E-05	4.1368E+03	4.1370E+03
38	4.22E+00	5.27E-02	1.98E-01	4.31E-05	4.5964E+03	4.5966E+03
39	4.27E+00	5.15E-02	2.20E-01	4.31E-05	5.0955E+03	5.0957E+03
40	4.32E+00	5.05E-02	2.43E-01	4.31E-05	5.6366E+03	5.6369E+03
41	4.37E+00	4.94E-02	2.68E-01	4.31E-05	6.2224E+03	6.2226E+03
42	4.42E+00	4.84E-02	2.95E-01	4.31E-05	6.8554E+03	6.8557E+03
43	4.46E+00	4.75E-02	3.25E-01	4.31E-05	7.5385E+03	7.5388E+03
44	4.51E+00	4.66E-02	3.56E-01	4.31E-05	8.2745E+03	8.2749E+03
45	4.56E+00	4.57E-02	3.91E-01	4.31E-05	9.0665E+03	9.0669E+03
46	4.60E+00	4.49E-02	4.27E-01	4.31E-05	9.9176E+03	9.9180E+03
47	4.65E+00	4.41E-02	4.67E-01	4.31E-05	1.0831E+04	1.0831E+04

Table III.7

i	t_i	h_i	e_i^h	$e_i^h/x(t_i)$	x_i^h	$x(t_i)$
0	0.00E-00					1.0000E+00
1	1.38E-01	1.38E-01	4.26E-07	3.23E-07	1.3178E+00	1.3178E+00
2	4.22E-01	2.84E-01	4.67E-05	2.01E-05	2.3273E+00	2.3274E+00
3	7.25E-01	3.02E-01	2.04E-04	4.78E-05	4.2613E+00	4.2615E+00
4	1.03E+00	3.05E-01	6.03E-04	7.69E-05	7.8408E+00	7.8414E+00
5	1.34E+00	3.05E-01	1.53E-03	1.06E-04	1.4437E+01	1.4439E+01
6	1.64E+00	3.05E-01	3.60E-03	1.35E-04	2.6585E+01	2.6588E+01
7	1.95E+00	3.05E-01	8.06E-03	1.65E-04	4.8955E+01	4.8963E+01
8	2.25E+00	3.05E-01	1.75E-02	1.93E-04	9.0149E+01	9.0166E+01
9	2.56E+00	3.05E-01	3.70E-02	2.23E-04	1.6601E+02	1.6604E+02
10	2.86E+00	3.05E-01	7.71E-02	2.52E-04	3.0569E+02	3.0577E+02
11	3.17E+00	3.05E-01	1.59E-01	2.82E-04	5.6292E+02	5.6308E+02
12	3.47E+00	3.05E-01	3.22E-01	3.11E-04	1.0366E+03	1.0369E+03
13	3.78E+00	3.05E-01	6.49E-01	3.40E-04	1.9089E+03	1.9095E+03
14	4.08E+00	3.05E-01	1.30E+00	3.69E-04	3.5151E+03	3.5164E+03
15	4.39E+00	3.05E-01	2.58E+00	3.98E-04	6.4730E+03	6.4755E+03
16	4.69E+00	3.05E-01	5.10E+00	4.28E-04	1.1920E+04	1.1925E+04
17	4.85E+00	1.53E-01	6.94E+00	4.28E-04	1.6200E+04	1.6207E+04
18	5.00E+00	1.53E-01	9.45E+00	4.29E-04	2.2017E+04	2.2027E+04

(iii) $\dot{x} = -x + 2x^2 e^{-t}$, $x(0) = \frac{1}{3}$; $T = 20$.

The solution of this problem is given by $x(t) = [2e^t + e^{-t}]^{-1}$ (see Example III.3). For larger values of t one has $\dot{x} \doteq -x$, i.e. the ODE is effectively linear there (so fits our analysis, despite the nonlinearity). Table III.10 and Table III.11 have been built up similarly to the two previous cases. Note that for $t \geq 5$ this example is qualitatively similar to example (ii).

Table III.8

i	t_i	h_i	e_i^h	$e_i^h/x(t_i)$	x_i^h	$x(t_i)$
0	0.00E-00					1.0000E+00
1	0.00E+00	0.00E+00	0.00E+00	0.00E+00	1.0000E+00	1.0000E+00
2	1.38E-01	1.38E-01	3.78E-07	4.98E-07	7.5885E-01	7.5885E-01
3	4.03E-01	2.65E-01	1.38E-05	3.09E-05	4.4674E-01	4.4676E-01
4	6.78E-01	2.75E-01	1.80E-05	6.99E-05	2.5757E-01	2.5758E-01
5	9.84E-01	3.06E-01	2.05E-05	1.47E-04	1.3974E-01	1.3976E-01
6	1.32E+00	3.40E-01	2.12E-05	3.00E-04	7.0778E-02	7.0799E-02
7	1.71E+00	3.83E-01	2.08E-05	6.32E-04	3.2913E-02	3.2933E-02
8	2.14E+00	4.36E-01	1.96E-05	1.43E-03	1.3743E-02	1.3763E-02
9	2.65E+00	5.05E-01	1.78E-05	3.56E-03	4.9916E-03	5.0095E-03
10	3.25E+00	5.97E-01	1.59E-05	1.03E-02	1.5018E-03	1.5174E-03
11	3.97E+00	7.24E-01	1.29E-05	3.61E-02	3.4364E-04	3.5653E-04
12	4.49E+00	5.15E-01	4.90E-06	3.85E-02	1.2233E-04	1.2723E-04
13	5.00E+00	5.15E-01	1.85E-06	4.08E-02	4.3546E-05	4.5400E-05

Table III.9

i	t_i	h_i	e_i^h	$e_i^h/x(t_i)$	x_i^h	$x(t_i)$
0	0.00E-00					1.0000E+00
1	1.38E-01	1.38E-01	3.78E-07	4.98E-07	7.5885E-01	7.5885E-01
2	3.96E-01	2.58E-01	1.19E-05	2.63E-05	4.5279E-01	4.5280E-01
3	6.45E-01	2.49E-01	1.29E-05	4.69E-05	2.7502E-01	2.7503E-01
4	8.95E-01	2.50E-01	1.13E-05	6.78E-05	1.6684E-01	1.6685E-01
5	1.15E+00	2.50E-01	8.98E-06	8.87E-05	1.0122E-01	1.0123E-01
6	1.40E+00	2.50E-01	6.73E-06	1.10E-04	6.1409E-02	6.1416E-02
7	1.65E+00	2.50E-01	4.87E-06	1.31E-04	3.7256E-02	3.7261E-02
8	1.90E+00	2.50E-01	3.42E-06	1.52E-04	2.2603E-02	2.2606E-02
9	2.15E+00	2.50E-01	2.36E-06	1.72E-04	1.3713E-02	1.3715E-02
10	2.39E+00	2.50E-01	1.61E-06	1.93E-04	8.3196E-03	8.3212E-03
11	2.64E+00	2.50E-01	1.08E-06	2.14E-04	5.0474E-03	5.0485E-03
12	2.89E+00	2.50E-01	7.20E-07	2.35E-04	3.0622E-03	3.0629E-03
13	3.14E+00	2.50E-01	4.76E-07	2.56E-04	1.8578E-03	1.8583E-03
14	3.39E+00	2.50E-01	3.12E-07	2.77E-04	1.1271E-03	1.1274E-03
15	3.64E+00	2.50E-01	2.04E-07	2.98E-04	6.8381E-04	6.8401E-04
16	3.89E+00	2.50E-01	1.32E-07	3.19E-04	4.1486E-04	4.1499E-04
17	4.14E+00	2.50E-01	8.55E-08	3.40E-04	2.5169E-04	2.5178E-04
18	4.39E+00	2.50E-01	5.51E-08	3.61E-04	1.5270E-04	1.5276E-04

Table III.10

i	t_i	h_i	e_i^h	$e_i^h/x(t_i)$	x_i^h	$x(t_i)$
0	0.00E-00					3.3333E-01
1	0.00E+00	0.00E+00	0.00E+00	0.00E+00	3.3333E-01	3.3333E-01
2	2.46E-01	2.46E-01	4.39E-08	1.47E-07	2.9943E-01	2.9943E-01
3	1.06E+00	8.09E-01	-2.59E-05	-1.58E-04	1.6418E-01	1.6416E-01
4	1.86E+00	8.09E-01	9.41E-07	1.23E-05	7.6613E-02	7.6613E-02
5	2.74E+00	8.80E-01	2.53E-05	7.90E-04	3.2060E-02	3.2086E-02
6	3.62E+00	8.78E-01	2.15E-05	1.61E-03	1.3342E-02	1.3364E-02
7	4.64E+00	1.02E+00	1.87E-05	3.89E-03	4.7995E-03	4.8182E-03
8	5.84E+00	1.20E+00	1.58E-05	1.09E-02	1.4330E-03	1.4488E-03
9	7.30E+00	1.46E+00	1.29E-05	3.82E-02	3.2384E-04	3.3669E-04
10	9.14E+00	1.83E+00	9.62E-06	1.79E-01	4.4150E-05	5.3771E-05
11	1.16E+01	2.43E+00	5.74E-06	1.21E+00	-9.8830E-07	4.7466E-06
12	1.51E+01	3.53E+00	-6.46E-07	-4.64E+00	7.8584E-07	1.3946E-07
13	2.00E+01	4.91E+00	3.46E-06	3.36E+03	-3.4591E-06	1.0306E-09

Exercises Chapter III

1. Show that RKF4 and RKF5 are consistent (see Table III.1).

2. Determine the local discretisation error of the *modified Euler method*, of which the Butcher matrix reads

$$
\begin{array}{c|cc}
0 & 0 & 0 \\
\frac{1}{2} & \frac{1}{2} & 0 \\
\hline
& 0 & 1
\end{array} \;.
$$

3. The same question as in Exercise 2, but now for *Heun's second order method* with the Butcher matrix

$$
\begin{array}{c|cc}
0 & 0 & 0 \\
1 & 1 & 0 \\
\hline
& \frac{1}{2} & \frac{1}{2}
\end{array} \;.
$$

4. Show that *Heun's third order method* has consistency order 3. Its Butcher matrix reads

Table III.11

i	t_i	h_i	e_i^h	$e_i^h/x(t_i)$	x_i^h	$x(t_i)$
0	0.00E-00					3.3333E-01
1	1.97E-01	1.97E-01	1.37E-08	4.46E-08	3.0700E-01	3.0700E-01
2	8.13E-01	6.15E-01	-4.93E-06	-2.44E-05	2.0197E-01	2.0197E-01
3	1.43E+00	6.15E-01	-3.77E-06	-3.24E-05	1.1657E-01	1.1657E-01
4	1.95E+00	5.23E-01	-1.05E-06	-1.50E-05	7.0371E-02	7.0370E-02
5	2.47E+00	5.23E-01	4.07E-07	9.69E-06	4.1984E-02	4.1984E-02
6	2.99E+00	5.12E-01	8.36E-07	3.32E-05	2.5220E-02	2.5220E-02
7	3.49E+00	5.04E-01	8.37E-07	5.49E-05	1.5253E-02	1.5254E-02
8	3.99E+00	5.01E-01	7.03E-07	7.61E-05	9.2435E-03	9.2442E-03
9	4.49E+00	5.00E-01	5.44E-07	9.71E-05	5.6056E-03	5.6062E-03
10	4.99E+00	5.00E-01	4.01E-07	1.18E-04	3.4003E-03	3.4007E-03
11	5.49E+00	5.00E-01	2.87E-07	1.39E-04	2.0628E-03	2.0631E-03
12	5.99E+00	5.00E-01	2.00E-07	1.60E-04	1.2514E-03	1.2516E-03
13	6.49E+00	5.00E-01	1.37E-07	1.81E-04	7.5916E-04	7.5930E-04
14	6.99E+00	5.00E-01	9.29E-08	2.02E-04	4.6053E-04	4.6063E-04
15	7.49E+00	5.00E-01	6.22E-08	2.23E-04	2.7937E-04	2.7942E-04
16	7.99E+00	5.00E-01	4.13E-08	2.44E-04	1.6944E-04	1.6948E-04
17	8.49E+00	5.00E-01	2.72E-08	2.65E-04	1.0275E-04	1.0278E-04
18	8.99E+00	5.00E-01	1.78E-08	2.86E-04	6.2296E-05	6.2313E-05
19	9.49E+00	5.01E-01	1.16E-08	3.07E-04	3.7750E-05	3.7762E-05
20	9.99E+00	5.02E-01	7.51E-09	3.29E-04	2.2859E-05	2.2867E-05
21	1.05E+01	5.03E-01	4.84E-09	3.50E-04	1.3825E-05	1.3830E-05
22	1.10E+01	5.05E-01	3.11E-09	3.73E-04	8.3448E-06	8.3480E-06
23	1.15E+01	5.08E-01	1.99E-09	3.96E-04	5.0206E-06	5.0226E-06
24	1.20E+01	5.13E-01	1.26E-09	4.21E-04	3.0050E-06	3.0063E-06
25	1.25E+01	5.22E-01	8.00E-10	4.48E-04	1.7839E-06	1.7846E-06
26	1.31E+01	5.34E-01	5.02E-10	4.80E-04	1.0455E-06	1.0460E-06
27	1.36E+01	5.54E-01	3.13E-10	5.20E-04	6.0109E-07	6.0140E-07
28	1.42E+01	5.82E-01	1.94E-10	5.76E-04	3.3601E-07	3.3620E-07
29	1.48E+01	6.21E-01	1.19E-10	6.60E-04	1.8055E-07	1.8067E-07
30	1.55E+01	6.75E-01	7.41E-11	8.06E-04	9.1936E-08	9.2010E-08
31	1.63E+01	7.46E-01	4.74E-11	1.09E-03	4.3582E-08	4.3629E-08
32	1.71E+01	8.40E-01	3.21E-11	1.70E-03	1.8805E-08	1.8837E-08
33	1.81E+01	9.63E-01	2.33E-11	3.24E-03	7.1666E-09	7.1899E-09
34	1.90E+01	9.71E-01	1.33E-11	4.87E-03	2.7088E-09	2.7221E-09
35	2.00E+01	9.71E-01	6.69E-12	6.50E-03	1.0239E-09	1.0306E-09

$$
\begin{array}{c|ccc}
0 & 0 & 0 & 0 \\
\frac{1}{2} & \frac{1}{2} & 0 & 0 \\
\frac{2}{3} & 0 & \frac{2}{3} & 0 \\
\hline
 & \frac{1}{4} & 0 & \frac{3}{4}
\end{array} \quad .
$$

5. Find the requirements that the coefficients of an explicit two-level RK method should satisfy to make it consistent. Under which conditions is the order maximal?

6. Let f be Lipschitz continuous on $I \times \Omega \subset \mathbb{R}^2$, and therefore let us say $|f(t, x) - f(t, y)| \leq L|x - y|$. Consider the explicit method $x_{i+1}^h = x_i^h + h\,\Phi(t_i, x_i^h, h)$. Show that Φ is Lipschitz continuous in the second variable on $\hat{I} \times \hat{\Omega} \subset I \times \Omega$ for $h \leq \hat{h}$, $\hat{h}$ sufficiently small, say $|\Phi(t, x, h) - \Phi(t, y, h)| \leq \hat{L}|x - y|$.
   Hint: Determine a Lipschitz constant for k_1, k_2 etc. recursively first.

7. Give another convergence proof of Theorem 3.2 for an explicit RK method by considering the exact solution to satisfy the Δ-equation with the local error as additional perturbation term and using the Lipschitz continuity of Φ (cf. Exercise 6).

8. Let L be a Lipschitz constant for $f(t, x)$. Determine a Lipschitz constant for Heun's second order method (cf. Exercise 3).

9. Determine the functions $c(t, x)$ and $d(t, x)$ in (4.1) and (4.2a) respectively for Heun's second order method applied to the ODE $\dot{x} = \lambda x$.

10. The so-called Taylor method reads: given the ODE $\dot{x} = f(t, x)$, determine $\ddot{x}$, $\dddot{x}$ etc. explicitly. A k-level method is then given by

$$
x_{i+1} = x_i + \tfrac{1}{2} h\,\dot{x}_i + \dots + \frac{1}{k!}\, h^h\, x_i^{(k)} ,
$$

where $x_i^{(j)} \doteq x^{(j)}(t_i)$ (derived from explicit differentiation).

   a) Determine the local discretisation error.

   b) Let the partial derivatives of f on a certain domain be bounded by a constant M. Determine the Lipschitz constant of a three-level Taylor method.

   c) Indicate why this method may be inefficient for more complicated problems (take, e.g., $f(t, x) = t^2 + x^2$).

11. The following (implicit) method is named after *Obrechkoff*:

$$x_{i+1}^h = x_i^h + \frac{h}{2} \left(\dot{x}_{i+1}^h + \dot{x}_i^h \right) - \frac{h^2}{12} \left(\ddot{x}_{i+1}^h - \ddot{x}_i^h \right) .$$

Find the consistency order.

12. Consider a Hermite interpolation polynomial for a y on (t_i, t_{i+1}) (cf. Appendix A), i.e. let

$$y(t_j) = x(t_j), \quad j = i, i+1 ,$$
$$\dot{y}(t_j) = \dot{x}(t_j), \quad j = i, i+1 .$$

a)   Give an explicit expression for $y(t)$ in terms of $x_i(t)$ and $x_{i+1}(t)$ (cf. (3.3a))

b)   Let x_i and x_{i+1} be the solutions of the ODE

$$\frac{dx}{dt} = f(t, x) ,$$

with $x(t_i)$, $x(t_{i+1})$ given. Moreover, assume $x_i(t) - x_{i+1}(t) = O(h^{p+1})$ for given p and $t \in (t_i, t_{i+1})$. Show that y is a solution of the IVP

$$\begin{cases} \dfrac{dy}{dt} = f(t, y) + r , & r = O(h^p) , \\ y(t_i) = x_i(t_i) . \end{cases}$$

c)   If $x_i(t_i) = x_i^h$ for all i and we are using some one-step p-th order method as in §3, show that the function y constructed above is a globally differentiable function on any interval (t_0, t) for t small enough to ensure existence.

IV

Linear Systems

In this chapter we consider linear systems in detail. In §1 the global existence and uniqueness of the solutions are shown to follow directly from the general theorems in Chapter II. In §2 and §3 explicit expressions for the solutions are derived. These expressions are studied for the special cases of constant (§4) and periodic (§5) systems. The general theory is extensively applied in §6 to planar systems, i.e. systems of order 2. These problems are of great importance in many mathematical models, such as mechanical systems. The classification of planar, autonomous systems is the subject of §7. Finally, in §8 a theory of linear difference equations is given with some emphasis on the apparent similarities between difference and differential equations.

1. Introduction

The linear IVP

$$(1.1) \qquad \begin{cases} \dot{\mathbf{x}} = \mathbf{A}(t)\,\mathbf{x} + \mathbf{b}(t) \,, \\ \mathbf{x}(t_0) = \mathbf{x}_0 \,, \end{cases}$$

plays a fundamental rôle in the theory of ODE. Because the structure of its solutions is completely understood, the analysis of general, nonlinear ODE is often based on local reductions to linear systems. A typical example of this idea is the technique of linearisation introduced in §II.5 and used in the stability theory of §V.4. Linearisation is also an essential ingredient in most numerical algorithms to solve ODE.

Throughout this chapter we assume $\mathbf{A}(t)$ and $\mathbf{b}(t)$ to be continuous functions on $I\!\!R$. From this we can show that the IVP (1.1) has a unique solution for all $t \in I\!\!R$. We remark that this does not hold for nonlinear systems in general, as was shown by Example II.1. The proofs of existence and uniqueness use the following property.

Property 1.2. *The linear vector field* $\mathbf{A}(t)\,\mathbf{x} + \mathbf{b}(t)$, *where* $\mathbf{A}$ *and* $\mathbf{b}$ *are* *continuous functions for $t \in \mathbb{R}$, is Lipschitz continuous on $I \times \mathbb{R}^n$ for every bounded and closed (and thus compact) interval $I \subset \mathbb{R}$.*

Proof: Because $\mathbf{A}(t)$ is continuous on I, this also holds for $\|\mathbf{A}(t)\|$ (for any matrix norm mentioned in Appendix B). A continuous function attains a maximum value on a compact interval, so

$$L := \max_{t \in I} \|\mathbf{A}(t)\|$$

exists. It is a Lipschitz constant for the linear vector field on $I \times \mathbb{R}^n$. $\qquad \square$

Theorem 1.3 (Global existence). *The linear IVP (1.1) has a solution for all $t \in \mathbb{R}$.*

Proof: We shall treat the cases $n = 1$ and $n > 1$ separately.

The scalar case $n = 1$. A solution of the scalar version of (1.1),

$$(1.3\mathrm{a}) \qquad \begin{cases} \dot{x} = a(t)\,x + b(t)\ , \\ x(t_0) = x_0\ , \end{cases}$$

is found directly from elementary calculus. Let us define

$$\bar{a}(t) := \int_{t_0}^{t} a(s)\,ds\ ,$$

which exists and is differentiable for all $t \in \mathbb{R}$, because $a(t)$ is continuous on $\mathbb{R}$. Multiplying the scalar ODE above by the integrating factor $\exp(-\bar{a}(t))$, which is positive for all $t \in \mathbb{R}$, and integrating, we obtain the explicit form

$$(1.3\mathrm{b}) \qquad x(t) = e^{\bar{a}(t)} \left(x_0 + \int_{t_0}^{t} e^{-\bar{a}(s)}\, b(s)\,ds \right)$$

which exists for all $t \in \mathbb{R}$.

The general case $n > 1$. The local existence theorem II.2.3, in combination with Property 1.2 above, guarantees that (1.1) has a solution on some time interval around t_0. To prove global existence for $t \in \mathbb{R}$ we invoke the continuation theorem II.3.1, and especially its corollary II.3.3. To apply this corollary, we have to show that if $\mathbf{x}$ is a solution of (1.1) for $t \in \mathbb{R}$, then $\|\mathbf{x}(t) - \mathbf{x}_0\| \le g(t)$ for all $t \in \mathbb{R}$ with $g(t)$ a continuous, scalar function on $\mathbb{R}$. According to Property I.2.6, possible solutions of (1.1) can be written in the form

$$\mathbf{x}(t) = \mathbf{x}_0 + \int_{t_0}^{t} \{ \mathbf{A}(s)\,\mathbf{x}(s) + \mathbf{b}(s) \}\, ds\ .$$

Taking norms we obtain the inequality

$$\|\mathbf{x}(t)\| \le \|\mathbf{x}_0\| + \int_{t_0}^{t} \|\mathbf{A}(s)\,\mathbf{x}(s) + \mathbf{b}(s)\|\,ds =: g(t) \ .$$

We may identify the right-hand side of this inequality with the function $g(t)$ we are looking for. This function is differentiable on $I\!\!R$, because the integrand is a continuous function on $I\!\!R$, and it satisfies

$$\begin{cases} \dot{g} = \|\mathbf{A}(t)\,\mathbf{x}(t) + \mathbf{b}(t)\| \le \|\mathbf{A}(t)\|\,g + \|\mathbf{b}(t)\| \ , \\ g(t_0) = \|\mathbf{x}_0\| \ . \end{cases}$$

The solution of this IVP with $\le$ replaced by $=$ exists as shown in the 'scalar' part of this proof. From the Lemma of Gronwall (II.1.8) we conclude that $g(t)$ exists for all $t \ge t_0$, and this proves the existence of $\mathbf{x}(t)$ for $t \ge t_0$. For $t \le t_0$ a similar proof applies after the substitution $t \to -t$. $\square$

Theorem 1.4 (Uniqueness). *The solution of the linear IVP (1.1) is unique.*

Proof: To prove uniqueness we could directly rely on Theorem II.1.10 in combination with Property 1.2 above. However, we wish to present a direct and elementary approach which explicitly uses the linearity of the system. Assume that IVP (1.1) has two solutions $\mathbf{x}^1(t)$ and $\mathbf{x}^2(t)$. Their difference $\mathbf{y}(t) := \mathbf{x}^1(t) - \mathbf{x}^2(t)$ then satisfies the IVP

$$\begin{cases} \dot{\mathbf{y}} = \mathbf{A}(t)\,\mathbf{y} \ , \\ \mathbf{y}(t_0) = \mathbf{0} \ . \end{cases}$$

Because the origin is a stationary point of this system and the IVP has the origin as initial value, this IVP has the trivial solution $\mathbf{y}(t) \equiv 0$. To show that the zero solution is the only solution, we note that $\mathbf{y}(t)$ can formally be written as

$$\mathbf{y}(t) = \int_{t_0}^{t} \mathbf{A}(s)\,\mathbf{y}(s)\,ds \ .$$

Taking norms we find for each t the inequality

$$\|\mathbf{y}(t)\| \le \int_{t_0}^{t} \|\mathbf{A}(s)\|\,\|\mathbf{y}(s)\|\,ds =: h(t) \ .$$

As in the proof of Theorem 1.3 we now have for h

$$\begin{cases} \dot{h} \le \|\mathbf{A}(t)\|\,h \ , \\ h(t_0) = 0 \ . \end{cases}$$

From the Lemma of Gronwall (II.1.8) we conclude that $h(t) \equiv 0$. Because $\|\mathbf{y}(t)\| = 0$ for $t \geq t_0$, we have $\mathbf{x}^1 \equiv \mathbf{x}^2$.   □

2. The homogeneous case

The solutions of the homogeneous equation

(2.1) $\dot{\mathbf{x}} = \mathbf{A}(t)\,\mathbf{x}$

have a linear structure: if $\mathbf{x}^1$ and $\mathbf{x}^2$ satisfy (2.1), then so does the linear combination $c_1\,\mathbf{x}^1 + c_2\,\mathbf{x}^2$, with $c_1, c_2 \in I\!R$ (or C). This insight is called the *superposition principle*; this principle is the backbone of all theories of linear systems. Here we formalise this principle as follows:

Property 2.2. *The solutions of (2.1) form a linear vector space V over $I\!R$ (or C).*

Definition 2.3. *A set of solutions $\mathbf{x}^1, ..., \mathbf{x}^m \in V$ is called linearly independent on $I\!R$ if the condition*

(2.3a) $c_1\,\mathbf{x}^1(t) + ... + c_m\,\mathbf{x}^m(t) = 0 \,, \quad \forall t \in I\!R$

implies that $c_1 = ... = c_m = 0$. So, none of these functions can be written as a linear combination (with constant coefficients) of the other ones.

To get a full understanding of the structure of the solutions of (2.1), the following property is important :

Property 2.4. *If two elements $\mathbf{x}^1$ and $\mathbf{x}^2$ of V are linearly independent, then the vectors $\mathbf{x}^1(t)$ and $\mathbf{x}^2(t)$ are linearly independent in $I\!R^n$ for all $t \in I\!R$.*

Proof: If we assume that for some time t_0 these vectors are dependent, then $\mathbf{x}^1(t_0) = c\,\mathbf{x}^2(t_0)$ for some scalar c. Then we may use $\mathbf{x}^1(t_0)$ and $\mathbf{x}^2(t_0)$ as initial values at time t_0 for IVP (2.1). Because of the linearity, the corresponding unique solutions satisfy $\mathbf{x}^1(t) = c\,\mathbf{x}^2(t)$ for all $t \in I\!R$ and are thus linearly dependent in V.   □

An alternative formulation of this property may be useful to understand its consequences:

Property 2.5. *Let the initial values $\mathbf{x}^1(t_0)$ and $\mathbf{x}^2(t_0)$ be linearly independent vectors in $I\!R^n$. The corresponding solutions $\mathbf{x}^1$ and $\mathbf{x}^2$ are then linearly independent in V.*

We conclude that each set of n linearly independent initial values in $\mathbb{R}^n$ induces a set of linearly independent solutions in V. Property 2.5 directly implies:

Property 2.6. *The dimension of V is equal to the dimension of the space $\mathbb{R}^n$ of initial values, i.e. equals n.*

We can use these insights to construct the general solution of (2.1). Let us take n linearly independent initial values $x_0^1, ..., x_0^n$ and consider the solutions $x^1, ..., x^n$ of the IVP $(i = 1, ..., n)$

$$(2.7) \qquad \begin{cases} \dot{x}^i = A(t)\, x^i \,, \\ x^i(t_0) = x_0^i \,. \end{cases}$$

According to Property 2.5 the solutions x^i not only form a basis in V, but the vectors $x^i(t)$ form also a basis in $\mathbb{R}^n$ for each $t \in \mathbb{R}$. It is convenient to introduce the so-called *fundamental solution* $Y(t)$ by interpreting the $x^i(t)$ as columns of a matrix:

$$(2.8) \qquad Y(t) := \left(x^1(t) \,|\, ... \,|\, x^n(t) \right) \,.$$

The columns of $Y(t)$ are linearly independent for each t, thus Y is nonsingular for all $t \in \mathbb{R}$. Because the x^i are a basis in V, an arbitrary solution x of (2.1) can be written in terms of Y as

$$(2.9) \qquad x(t) = Y(t)\, c \,,$$

with c a constant vector, whose elements c_i are uniquely determined from the initial value x_0:

$$(2.10) \qquad x_0 = Y(t_0)\, c \equiv c_1 x^1(t_0) + ... + c_n x^n(t_0) \,.$$

An alternative way to express (2.9) is

$$(2.11) \qquad x(t) = Y(t, t_0)\, x_0$$

with the *evolution matrix function* $Y(t, s)$ defined by

$$(2.12) \qquad Y(t, s) := Y(t)\, Y^{-1}(s) \,.$$

Expression (2.11) provides the general solution of (2.1).

We conclude this section with some remarks concerning the fundamental solution Y, i.e. a matrix solution of (2.1). Choosing for the x_0^i the standard basis with orthogonal unit vectors, the so-called *principal fundamental solution* is obtained. It is the solution of the IVP

$$(2.13) \qquad \begin{cases} \dot{Y} = A(t)\, Y \,, \\ Y(t_0) = I \,, \end{cases}$$

with $\mathbf{I}$ the identity matrix in $I\!\!R^n$. The determinant of $\mathbf{Y}(t)$ is called the *Wronskian*. Its time dependence is determined by the diagonal elements of $\mathbf{A}(t)$ only, as is shown in the following theorem.

Theorem 2.14. *The time dependence of the Wronskian* $W(t) := det\,\mathbf{Y}(t)$ *of (2.1) is given by*

$$(2.14a) \quad W(t) = W(0)\exp\left[\int_{t_0}^{t} \mathrm{Tr}\,\mathbf{A}(s)\,ds\right]$$

with $\mathrm{Tr}\,\mathbf{A} := \mathrm{Trace}(\mathbf{A})$.

Proof: We shall show that $W(t)$ is a solution of

$$(*) \qquad \dot{W} = \mathrm{Tr}\Big(\mathbf{A}(t)\Big)\,W\ .$$

Because the columns of the fundamental solution $\mathbf{Y}(t)$ are solutions of (2.1), this matrix function is differentiable. A Taylor expansion yields

$$\mathbf{Y}(t+\varepsilon) = \mathbf{Y}(t) + \varepsilon\,\dot{\mathbf{Y}}(t) + o(\varepsilon) = \Big(1 + \varepsilon\,\mathbf{A}(t)\Big)\mathbf{Y}(t) + o(\varepsilon)\ .$$

For the order symbol $o(\varepsilon)$ we refer to Appendix E. Taking determinants at both sides and using that

$$\det\Big(\mathbf{I} + \varepsilon\,\mathbf{A}(t)\Big) = \prod_{i}\Big(1 + \varepsilon\,\mathbf{A}_{ii}(t)\Big) + o(\varepsilon^2) =$$

$$= 1 + \varepsilon\,\mathrm{Tr}\Big(\mathbf{A}(t)\Big) + o(\varepsilon^2)\,,$$

we find that

$$\frac{W(t+\varepsilon) - W(t)}{\varepsilon} = \mathrm{Tr}\Big(\mathbf{A}(t)\Big)\,W(t) + o(\varepsilon)\ .$$

From the limit $\varepsilon \downarrow 0$ we see that W satisfies ODE $(*)$, which has (2.14a) as its solution.   □

Theorem 2.14 has an important interpretation. The Wronskian $W(t)$ measures the volume in the phase space spanned by the solution vectors at time t. If at t_0 the basis vectors are linearly independent, we have $W(t_0) \neq 0$. According to Property 2.5 they remain independent for all $t > t_0$, and we then have $W(t) \neq 0$ for all t. This is also reflected by (2.14a). Systems with the property $W(t) = W(0)$ for all t are *volume preserving*. They are dealt with in §XI.3. On the other hand, if $\mathrm{Tr}(\mathbf{A}(t)) < 0$ for all $t \geq t_0$, the volume $W(t)$ will monotonically decrease, indicating that the solution vectors tend to coalesce as time evolves. The phase space thus tends to contract.

3. The inhomogeneous case

For the inhomogeneous IVP

(3.1) $\begin{cases} \dot{\mathbf{x}} = \mathbf{A}(t)\,\mathbf{x} + \mathbf{b}(t)\ , \\ \mathbf{x}(t_0) = \mathbf{x}_0\ . \end{cases}$

we can conveniently use the insights developed in §2. The solutions of (3.1) do not form a linear vector space and a relation like (2.9) does not hold. However, Property 2.5 implies that the columns of the fundamental matrix $\mathbf{Y}(t)$ of (2.1) are a basis in $I\!\!R^n$ for each $t \in I\!\!R$. So, instead of (2.9) we may write solutions of (3.1) as

(3.2) $\mathbf{x}(t) = \mathbf{Y}(t)\,\mathbf{c}(t)$.

The analogy between (2.9) and (3.2) is the reason that (3.2) is called the *variation of constants formula*. Substitution of (3.2) into (3.1) yields an IVP for the coefficient vector $\mathbf{c}(t)$:

(3.3) $\begin{cases} \dot{\mathbf{c}} = \mathbf{Y}^{-1}(t)\,\mathbf{b}\ , \\ \mathbf{c}(t_0) = \mathbf{Y}^{-1}(t_0)\,\mathbf{x}_0\ , \end{cases}$

with solution

(3.4) $\mathbf{c}(t) = \mathbf{Y}^{-1}(t_0)\,\mathbf{x}_0 + \displaystyle\int_{t_0}^{t} \mathbf{Y}^{-1}(s)\,\mathbf{b}(s)\,ds$.

The solution of (3.1) is therefore given by

(3.5) $\mathbf{x}(t) = \mathbf{Y}(t,t_0)\,\mathbf{x}_0 + \displaystyle\int_{t_0}^{t} \mathbf{Y}(t,s)\,\mathbf{b}(s)\,ds$.

We recognise in this expression a second superposition principle. The solution (3.5) of (3.1) turns out to be the sum of the solution of the homogeneous part of (3.1), and a so-called *particular solution*, given by the integral in (3.5). This particular solution is the solution of (3.1) with vanishing initial value, thus $\mathbf{x}_0 = 0$.

4. Constant coefficients

For constant matrices $\mathbf{A}$ the principal fundamental solution $\mathbf{Y}$, i.e. the solution of (2.13), is given by

(4.1) $\mathbf{Y}(t) = e^{\mathbf{A}t} := \displaystyle\sum_{n=0}^{\infty} \mathbf{A}^n\,\frac{t^n}{n!}$.

Following (2.11), the solutions of the *homogeneous* system (2.1) can be written as

(4.2) $\mathbf{x}(t) = e^{\mathbf{A}(t-t_0)} \mathbf{x}_0$.

The behaviour of $\mathbf{x}(t)$ is completely determined by the eigenvalues of $\mathbf{A}$. If $\mathbf{A}$ has n different eigenvectors $\mathbf{v}_i$ corresponding to (not necessarily different) eigenvalues λ_i, we may expand $\mathbf{x}_0$ in terms of the $\mathbf{v}_i$:

$$(4.3)   \mathbf{x}_0 = \sum_{i=1}^{n} c_i\, \mathbf{v}_i \ .$$

Substitution of (4.3) into (4.2) yields

$$(4.4)   \mathbf{x}(t) = \sum_{i=1}^{n} c_i\, e^{\lambda_i(t-t_0)} \mathbf{v}_i \ .$$

According to Property 2.5 expansion (4.4) indicates that the functions $\mathbf{v}_i\, e^{\lambda_i t}$, $i = 1, ..., n$, form a basis in the vector space V of solutions of $\dot{\mathbf{x}} = \mathbf{A}\,\mathbf{x}$. We note that the λ_i and $\mathbf{v}_i$ may be complex-valued. If an eigenvalue λ is complex with eigenvector $\mathbf{v}$, then also its complex conjugate $\bar{\lambda}$ is an eigenvalue with eigenvector $\bar{\mathbf{v}}$. If $\mathbf{A}$ and $\mathbf{x}_0$ are real-valued, we can always combine complex conjugate pairs to form (independent) real pairs. We may assume (4.4) to be in the latter form already.

If $\mathbf{A}$ has $p < n$ eigenvectors, the situation is more delicate. We shall use the Jordan form of $\mathbf{A}$ (see Appendix C). This implies that a nonsingular matrix $\mathbf{S}$ exists such that

(4.5) $e^{\mathbf{A}t} = \mathbf{S}\, e^{\mathbf{J}t}\, \mathbf{S}^{-1}$

with $e^{\mathbf{J}t}$ an $n \times n$ matrix of the form

$$(4.6)   e^{\mathbf{J}t} = \begin{bmatrix} e^{\lambda_1 t}\,\mathbf{K}_1 & & & \emptyset \\ & e^{\lambda_2 t}\,\mathbf{K}_2 & & \\ & & \ddots & \\ \emptyset & & & e^{\lambda_p t}\,\mathbf{K}_p \end{bmatrix} .$$

The Jordan blocks $\mathbf{K}_i$, $i = 1, ..., p$, of sizes $r_i \times r_i$, are given by

$$(4.7)   \mathbf{K}_i(t) = \begin{bmatrix} 1 & t^2 & \dfrac{t^2}{2!} & \cdots & \dfrac{t^{r_i-1}}{(r_i-1)!} \\ & 1 & t & & \\ & & 1 & \ddots & \\ & & & \ddots & t \\ \emptyset & & & & 1 \end{bmatrix} .$$

Substituting (4.6) and (4.7) into (4.2) we find

$$(4.8) \qquad \mathbf{x}(t) = \mathbf{S} \, e^{\mathbf{J}(t-t_0)} \, \mathbf{S}^{-1} \mathbf{x}_0 \ .$$

Let us next turn to the *inhomogeneous* case. Comparing expressions (2.11) and (3.5) we see that the solutions in the inhomogeneous case have an extra term, the so-called *particular solution* $\mathbf{x}_p(t)$, given by

$$(4.9) \qquad \mathbf{x}_p(t) := \int_{t_0}^{t} e^{\mathbf{A}(t-s)} \, \mathbf{b}(s) \, ds \ .$$

We shall analyse this integral for periodic $\mathbf{b}$, which is an important case in practice. If $\mathbf{b}(t)$ has period T, we can write it as the Fourier sum

$$(4.10) \qquad \mathbf{b}(t) = \sum_{l=-\infty}^{+\infty} \mathbf{b}_l \, e^{i\omega_l(t-t_0)}$$

with $\omega_l = 2\pi l/T$, and where i denotes the imaginary unit. It therefore suffices to investigate (4.9) with $\mathbf{b}$ of the form

$$(4.11) \qquad \mathbf{b}(t) = e^{i\omega(t-t_0)} \, \mathbf{b}$$

for some driving frequency ω and constant vector $\mathbf{b}$.

If $\mathbf{A}$ has n different eigenvectors $\mathbf{v}_j$ we may expand $\mathbf{b}$ in terms of $\mathbf{v}_j$:

$$(4.12) \qquad \mathbf{b}(t) = e^{i\omega(t-t_0)} \sum_{j=1}^{n} b_j \, \mathbf{v}_j \ .$$

Substituting this expansion into (4.9) we find directly

$$(4.13) \qquad \mathbf{x}_p(t) = \sum_{j=1}^{n} b_j \, I_j(t,\omega) \, \mathbf{v}_j \ ,$$

where the integrals $I_j(t,\omega)$ are defined as

$$(4.14) \qquad I_j(t,\omega) := e^{\lambda_j(t-t_0)} \int_{t_0}^{t} e^{(i\omega-\lambda_j)(s-t_0)} \, ds \ .$$

Evaluation of this integral yields

$$(4.15) \qquad I_j(t,\omega) = \begin{cases} \dfrac{1}{i\omega - \lambda_j} \left(e^{i\omega(t-t_0)} - e^{\lambda_j(t-t_0)} \right) & \text{if } \lambda_j \neq i\omega \\[2mm] (t-t_0) \, e^{i\omega(t-t_0)} & \text{if } \lambda_j = i\omega \ . \end{cases}$$

If $p < n$ the situation is more delicate, but still similar. We then substitute representation (4.5) into (4.9). This leads to an expression with an integration involved:

$$(4.16) \qquad \mathbf{x}(t) = \mathbf{S} \int_{t_0}^{t} e^{\mathbf{J}(t-t_0)+\mathrm{i}\omega(s-t_0)} \, ds \, \mathbf{S}^{-1} \mathbf{b} \, .$$

The time dependence of $\mathbf{x}_p(t)$ for $t \to \infty$ is thus determined by the time dependence of integrals $I_{jk}(t)$ given by

$$(4.17) \qquad I_{jk}(t,\omega) = e^{\lambda_j(t-t_0)} \int_{t_0}^{t} (s-t_0)^k \, e^{(\mathrm{i}\omega-\lambda_j)(s-t_0)} \, ds \, .$$

Evaluation of the integration yields

$$(4.18) \qquad I_{jk}(t,\omega) = \begin{cases} q\Big[e^{\mathrm{i}\omega(t-t_0)}(t^k - kqt^{k-1} + k(k-1)q^2 t^{k-2}\ldots) \\ \qquad\qquad -e^{\lambda_j(t-t_0)}(1 - kq + k(k-1)q^2\ldots)\Big] & \text{if } \lambda_j \neq \mathrm{i}\omega \\[2mm] e^{\mathrm{i}\omega(t-t_0)} \dfrac{1}{k+1}(t-t_0)^{k+1} & \text{if } \lambda_j = \mathrm{i}\omega \, , \end{cases}$$

where $q := (\mathrm{i}\omega - \lambda_j)^{-1}$.

5. Periodic coefficients

In §4 we derived explicit expressions for the solution of a linear IVP with $\mathbf{A}$ constant, based on the fact that the evolution takes the simple forms (4.2) and (4.9). Here, we consider the case of periodic $\mathbf{A}$:

$$(5.1) \qquad \mathbf{A}(t+T) = \mathbf{A}(t) \, .$$

Now, an expression for $\mathbf{Y}(t,s)$ similar to that in (4.2) can be found. Recall from (2.13) that the principal fundamental solution $\mathbf{Y}$ is the solution of the IVP

$$(5.2) \qquad \begin{cases} \dot{\mathbf{Y}} = \mathbf{A}(t)\,\mathbf{Y} \, , \\ \mathbf{Y}(t_0) = \mathbf{I} \, . \end{cases}$$

We first solve this IVP on $[t_0, t_0 + T]$. The solution at the end of this interval is denoted by

$$(5.3) \qquad \mathbf{C} := \mathbf{Y}(T) \, .$$

Next, we solve the IVP (5.2) on $[t_0 + T, t_0 + 2T]$, i.e. t_0 is replaced by $t_0 + T$. Defining $\mathbf{Z}(t) := \mathbf{Y}(T+t)$ and using the periodicity property (5.1) we find that $\mathbf{Z}$ is the solution of

$$(5.4) \qquad \begin{cases} \dot{\mathbf{Z}} = \mathbf{A}(t)\,\mathbf{Z} \ , \\ \mathbf{Z}(t_0) = \mathbf{C} \ . \end{cases}$$

So, $\mathbf{Y}$ and $\mathbf{Z}$ satisfy the same differential equation and their initial values are related by the constant, nonsingular matrix $\mathbf{C}$. Because of the linearity it follows that

$$(5.5) \qquad \mathbf{Z}(t) = \mathbf{Y}(t + T) = \mathbf{Y}(t)\,\mathbf{C} \ .$$

The periodicity of the system suggests writing t as

$$(5.6) \qquad t = nT + s \ ,$$

with n the smallest integer such that $t_0 \le s < t_0 + T$. Repeated application of (5.5) then yields

$$(5.7) \qquad \mathbf{Y}(t) = \mathbf{Y}(s)\,\mathbf{C}^n \ , \qquad t_0 \le s < t_0 + T \ .$$

For showing the analogy with (4.1), one often introduces a (possibly complex) matrix $\mathbf{B}$ such that

$$(5.8) \qquad \mathbf{C} := e^{\mathbf{B}T} \ .$$

See, e.g., [19]. In terms of $\mathbf{B}$, $\mathbf{Y}(t)$ in (5.7) can be rewritten as

$$(5.9) \qquad \mathbf{Y}(t) = \mathbf{Y}(s)\,e^{\mathbf{B}nT} = \mathbf{Y}(s)\,e^{-\mathbf{B}s}\,e^{\mathbf{B}(nT+s)} = \mathbf{F}(s)\,e^{\mathbf{B}t} \ ,$$

with the T-periodic matrix function $\mathbf{F}$ defined by

$$(5.10) \qquad \mathbf{F}(t) := \mathbf{Y}(t)\,e^{-\mathbf{B}t} \ .$$

After substitution of (5.9) into (2.11) we have actually shown the following theorem:

Theorem 5.11 (Floquet). *The solution of the IVP*

$$(5.11a) \qquad \begin{cases} \dot{\mathbf{x}} = \mathbf{A}(t)\,\mathbf{x} \ , \\ \mathbf{x}(t_0) = \mathbf{x}_0 \ , \end{cases}$$

with $\mathbf{A}$ T-periodic can be written as

$$(5.11b) \qquad \mathbf{x}(t) = \mathbf{F}(t)\,e^{\mathbf{B}(t-t_0)}\,\mathbf{x}_0 \ ,$$

with $\mathbf{F}$ a T-periodic and $\mathbf{B}$ a constant matrix.

Note that, in general, periodic systems have no periodic solutions.

The factor $\exp(\mathbf{B}(t-t_0))$ in (5.11b) can be analysed in the same way as the factor $\exp(\mathbf{A}(t-t_0))$ in (4.2). Expressions similar to (4.4) and (4.8) are then obtained with an extra T-periodic factor $\mathbf{F}$. In general, $\mathbf{x}(t)$ in (5.11b) can be written as

(5.12) $\mathbf{x}(t) = \mathbf{F}(t)\,\mathbf{S}\,e^{\mathbf{J}(t-t_0)}\,\mathbf{S}^{-1}\,\mathbf{x}_0$.

In the matrix exponent in (5.12) the eigenvalues of $\mathbf{J}$ and thus $\mathbf{B}$ occur. They are called the *characteristic exponents* and related to the eigenvalues μ_i of $\mathbf{C}$ by

(5.13) $\mu_i := e^{\lambda_i T}$.

The μ_i are usually called the *characteristic multipliers*. Actually, the λ_i are often defined through (5.13) instead. One should realise that this definition is not unique: if λ_i satisfies (5.13) for given μ_i, then so does $\lambda_i + (2\pi i n/T)$ for arbitrary integer n.

No general strategy is known to calculate the λ_i and $\mathbf{F}(t)$ analytically. In practice, they have to be determined numerically. This requires the solution of $\dot{\mathbf{x}} = \mathbf{A}\mathbf{x}$ on $[0, T]$ for the n standard basis vectors as initial values. An extensive example will be worked out in §XII.9.

Theorem 5.14. *The characteristic multipliers μ_i satisfy*

(5.14a) $$\prod_{i=1}^{p} \mu_i^{r_i} = \exp\left[\int_{t_0}^{t_0+T} \mathrm{Tr}\,\mathbf{A}(s)\,ds \right] .$$

The characteristic exponents λ_i satisfy

(5.14b) $$\sum_{i=1}^{p} r_i\,\lambda_i = \frac{1}{T} \int_{t_0}^{t_0+T} \mathrm{Tr}\,\mathbf{A}(s)\,ds .$$

Proof: This follows immediately from Theorem 2.14, because the Wronskian $W(T)$, evaluated after one period, is the determinant of $\mathbf{Y}(T) \equiv \mathbf{C}$. So,

$$W(T) = \det \mathbf{Y}(T) = \det \mathbf{C} = \prod_{i=1}^{p} \mu_i^{r_i} ,$$

with $p \leq n$ the number of different eigenvectors of $\mathbf{C}$ and r_i the dimension of the i-th Jordan block of $\mathbf{C}$.   □

Let us now turn to the *inhomogeneous system*

(5.15) $\dot{\mathbf{x}} = \mathbf{A}(t)\,\mathbf{x} + \mathbf{b}(t)$,

with $\mathbf{A}(t)$ T-periodic and $\mathbf{b}(t)$ of the harmonic form (4.11). Following (3.5), a particular solution $\mathbf{x}_p(t)$ of (5.15) is given by

(5.16) $$\mathbf{x}_p(t) = \int_{t_0}^{t} \mathbf{Y}(t, s)\,\mathbf{b}(s)\,ds .$$

The evolution function $\mathbf{Y}(t,s)$ for periodic systems has the form

(5.17) $\qquad \mathbf{Y}(t,s) = \mathbf{F}(t)\, e^{\mathbf{B}(t-s)}\, \mathbf{F}^{-1}(s)$,

as directly follows from (2.12) and (5.9). The nonsingular matrix function $\mathbf{F}(t)$ is T-periodic and $\mathbf{B}$ is a constant matrix. Substitution of (5.17) and (4.11) into (5.16) yields:

(5.18) $\qquad \mathbf{x}_p(t) = \mathbf{F}(t) \int_{t_0}^{t} e^{\mathbf{B}(t-s)}\, \mathbf{F}^{-1}(s)\, \mathbf{b}\, e^{i\omega(s-t_0)}\, ds$.

To evaluate the time integration we use the T-periodicity of $\mathbf{F}(t)$ and $\mathbf{F}^{-1}(t)$, and expand $\mathbf{F}^{-1}(t)$ in a Fourier series:

(5.19) $\qquad \mathbf{F}^{-1}(t) = \sum_{l=-\infty}^{+\infty} \mathbf{F}_l\, e^{i\omega_l(t-t_0)}$

with angular frequencies

(5.20) $\qquad \omega_l = \dfrac{2\pi l}{T}$.

Let $\mathbf{S}$ be the similarity matrix transforming $\mathbf{B}$ into its Jordan form $\mathbf{J}$. So,

(5.21) $\qquad e^{\mathbf{B}t} = \mathbf{S}\, e^{\mathbf{J}t}\, \mathbf{S}^{-1}$.

Substituting (5.19) and (5.20) into (5.18) we obtain the expression

(5.22) $\qquad \mathbf{x}_p(t) = \mathbf{F}(t)\, \mathbf{S} \sum_{l=-\infty}^{+\infty} \int_{t_0}^{t} e^{\mathbf{J}(t-s)+i(\omega+\omega_l)(t-t_0)}\, ds\, \mathbf{S}^{-1}\, \mathbf{F}_l\, \mathbf{b}$.

The matrix exponents should be read as in (4.6). From (5.22) we see that the time dependence of $\mathbf{x}_p(t)$ for $t \to \infty$ is essentially dependent on the evolution of the integrals

(5.23) $\qquad I_{jk}(t, \omega + \omega_l) = e^{\lambda_j(t-t_0)} \int_{t_0}^{t} (s-t_0)^k\, e^{(i(\omega+\omega_l)-\lambda_j)(s-t_0)}\, ds$.

The integrals in (5.23) and (4.17) are quite similar. In (4.17) the λ_j are the eigenvalues of the constant matrix $\mathbf{A}$, whereas in (5.23) the λ_j are the characteristic exponents of the T-periodic matrix $\mathbf{A}(t)$. In (4.17) only the driving angular frequency ω is present, whereas in (5.23) the sum $\omega + \omega_l$ with ω_l given by (5.20) occurs.

6. Planar systems

In many mathematical models we encounter second order equations of the form

(6.1) $a(t)\,\ddot{y} + b(t)\,\dot{y} + c(t)\,y = f(t)$,

where a, b, c and f are continuous functions of $t \in \mathbb{R}$. We shall work out the theory from the previous sections for this equation in detail. Without loss of generality we may take $a(t) \equiv 1$; when $a(t)$ vanishes for some t, the system is no longer second order. It is convenient to apply the *Liouville transformation*

(6.2) $x(t) = y(t)\,e^{\frac{1}{2}\bar{b}(t)}$,

with

(6.3) $\displaystyle \bar{b}(t) := \int_{t_0}^{t} b(s)\,ds$.

In the resulting equation for x no first derivative $\dot{x}$ is present. Because the term with $\dot{x}$ usually stems from a friction force, the equation is sometimes said to be brought into the 'frictionless' form

(6.4) $\ddot{x} + d(t)\,x = g(t)$,

with

(6.5) $\begin{cases} d(t) := c(t) - \frac{1}{4}b^2(t) - \frac{1}{2}\dot{b}(t) \\[2mm] g(t) := f(t)\,e^{\frac{1}{2}\bar{b}(t)} \ . \end{cases}$

The system is two-dimensional and the phase plane is spanned by the vectors $\mathbf{x} = (x, \dot{x})^T$. In standard form (6.4) reads as

(6.6) $\dot{\mathbf{x}} = \mathbf{A}(t)\,\mathbf{x} + \mathbf{b}(t)$,

with

(6.7) $\mathbf{A}(t) := \begin{bmatrix} 0 & 1 \\ -d(t) & 0 \end{bmatrix}$, $\mathbf{b}(t) := \begin{bmatrix} 0 \\ g(t) \end{bmatrix}$.

Because $\operatorname{Tr}\mathbf{A}(t) = 0$, we find from Theorem 2.14 that the Wronskian $W(t)$ is preserved:

(6.8) $W(t) = W(t_0)$, $\forall t \in \mathbb{R}$.

The flow in the phase plane induced by (6.4) is thus volume preserving.

Let us now deal with the *homogeneous case*, in which f in (6.1) vanishes. If $\mathbf{x}_1$ and $\mathbf{x}_2$ are two linearly independent solutions of

(6.9) $\dot{\mathbf{x}} = \mathbf{A}(t)\mathbf{x}$,

the corresponding Wronskian is given by

(6.10) $W(t) = \det(\mathbf{x}_1 \,|\, \mathbf{x}_2) = x_1(t)\,\dot{x}_2(t) - \dot{x}_1(t)\,x_2(t)$.

Since this expression is preserved, we can draw some interesting conclusions. If $x_1(t) \neq 0$ for some $t \geq t_0$, we may write

(6.11) $\dot{x}_2 = \frac{1}{x_1}\left(\dot{x}_1 x_2 + W(t_0)\right)$.

So, if a solution $x_1(t)$ of (6.9) is known, we can find a second, linearly independent solution x_2 by integrating (6.11). The latter equation has the scalar form (1.3a) with $a \equiv \dot{x}_1/x_1$ and $b = W(t_0)/x_1$. The solution is thus given by (1.3b) and reads as

(6.12) $x_2(t) = x_1(t)\left(\dfrac{x_2(t_0)}{x_1(t_0)} + W_0 \displaystyle\int_{t_0}^{t} \dfrac{1}{x_1^2(s)}\,ds\right)$.

The procedure followed here is called *reduction of order*. It is applicable as long as the integral in (6.12) exists, and if $x_1(t_0) \neq 0$. We illustrate this technique with an example.

Example IV.1.
Consider the equation

$$\ddot{x} - \frac{2}{(1+t)^2}\,x = 0 , \qquad t > -1 .$$

The solution with initial values $x_1(0) = 1$, $\dot{x}_1(0) = -1$ is given by

$$x_1(t) = \frac{1}{1+t} .$$

We look for a (linearly independent) solution x_2 with initial values $x_2(0) = 0$, $\dot{x}_2(0) = 1$. The Wronskian is given by

$$W(t) = W(t_0) = x_1(0)\,\dot{x}_2(0) - x_2(0)\,\dot{x}_1(0) = 1 .$$

Evaluation of (6.12) yields

$$x_2(t) = \frac{1}{1+t}\int_0^t (1+s)^2\,ds = \frac{t}{1+t}\left(1 + t + \frac{1}{3}t^2\right) .$$

Equations (6.8) and (6.10) lead to the following property:

Property 6.13. *Let x_1 and x_2 be two linearly independent solutions of (6.9).*

(i)   *If $x_1(t) = 0$ for some t, then $\dot{x}_1(t) \neq 0$ and $x_2(t) \neq 0$.*

(ii) *If $x_1(t_1) = x_1(t_2) = 0$, and $x_1(t) \neq 0$ for $t_1 < t < t_2$, then $x_2(t)$ has one and only one zero in the interval (t_1, t_2).*

The zeros of x_1 and x_2, if any, are thus separating each other on the t-axis. This implies that all solutions of (6.9) are oscillating, if one solution oscillates.

Example IV.2. (Damped, undriven harmonic oscillator)
The one-dimensional harmonic oscillator with mass $m > 0$, spring constant $k > 0$, and friction coefficient $c \geq 0$ is described by the equation

$$m\ddot{y} + c\dot{y} + ky = 0 .$$

Application of transformation (6.2) yields the equation

$$\ddot{x} + dx = 0$$

with

$$d = \frac{k}{m} - \left(\frac{c}{2m}\right)^2 .$$

The constant matrix $\mathbf{A}$ of this system is given by

$$\mathbf{A} = \begin{bmatrix} 0 & 1 \\ -d & 0 \end{bmatrix},$$

and has eigenvalues

$$\lambda_\pm = \pm i\sqrt{d}$$

and eigenvectors

$$\mathbf{v}_\pm = \begin{bmatrix} 1 \\ \lambda_\pm \end{bmatrix} .$$

So, for $d \neq 0$ two linearly independent eigenvectors exist. With the notation $\mathbf{x}(t) := (x(t), \dot{x}(t))^T$, we may write, following (4.4),

$$\mathbf{x}(t) = c_+ \, \mathbf{v}_+ \, e^{\lambda_+ t} + c_- \, \mathbf{v}_- \, e^{\lambda_- t}$$

where $t_0 = 0$ is taken for convenience. The coefficients $c_\pm$ follow from the expansion of $\mathbf{x}(0)$ in terms of $\mathbf{v}_+$ and $\mathbf{v}_-$. The solution $y(t)$ is given by

$$y(t) = c_+ e^{-(\frac{c}{2m} - \lambda_+)t} + c_- e^{-(\frac{c}{2m} - \lambda_-)t} .$$

This solution dies out for $t \to \infty$ if $c > 0$. For $d > 0$, the $\lambda_\pm$ are purely imaginary and $y(t)$ will oscillate if t increases. For $d < 0$, the $\lambda_\pm$ are real and $y(t)$ does not change sign for $t \to \infty$. If $d = 0$, i.e. if $k = c^2/4m$, the problem has only one eigenvalue $\lambda = 0$ with eigenvector $(1,0)^T$. In this case the solution is

$$x(t) = c_1 + c_2 t ,$$

which corresponds to

$$y(t) = (c_1 + c_2 t) e^{-\frac{c}{2m} t} ,$$

where the coefficients c_1 and c_2 are to be determined from $y(0)$ and $\dot{y}(0)$. The cases $d > 0$, $d = 0$, $d < 0$ are referred to as the *under-damped*, the *critically damped*, and the *over-damped* case respectively. Critical damping is often the ideal in practice. It corresponds to the highest value of the friction for which oscillations, and thus a considerable time delay, do not yet occur. In Fig. IV.1 solutions for the initial values $y(0) = 1$, $\dot{y}(0) = 0$ are given for various values of d. For $d > 0$ we also give the solution starting with $y(0) = 0$, $\dot{y} = 1$. According to Property 6.13, the zeros of the linearly independent, oscillating solutions lie alternatingly.

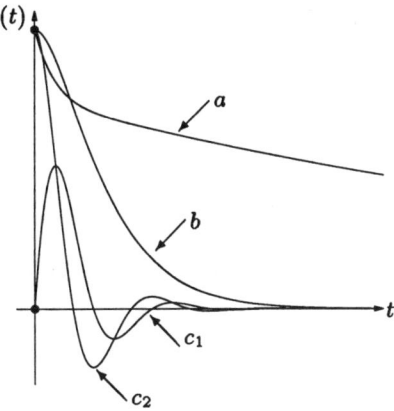

Figure IV.1 Solutions of Example IV.2 for the under-damped (a), critically damped (b), and over-damped (c_1, c_2) cases. The zeros of the two solutions c_1 and c_2 lie alternatingly on the t-axis.

Example IV.3.(Damped, driven harmonic oscillator)
Here we include an oscillating driving force into the harmonic oscillator
equation in the previous example. The Liouville transformation is
omitted, because it is less convenient in this case. As discussed in §4,
it suffices to study a harmonic driving force with one specific driving
frequency ω:

$$m\ddot{x} + c\dot{x} + kx = f_0\, e^{i\omega t} \ .$$

The matrix $\mathbf{A}$ of the corresponding two-dimensional system $\dot{\mathbf{x}} = \mathbf{A}\,\mathbf{x} + \mathbf{b}(t)$ with $\mathbf{x} := (x, \dot{x})^T$ is given by

$$\mathbf{A} := \begin{bmatrix} 0 & 1 \\ -\dfrac{k}{m} & -\dfrac{c}{m} \end{bmatrix} \ ,$$

and the vector $\mathbf{b}$ by

$$\mathbf{b}(t) = \begin{bmatrix} 0 \\ \dfrac{f_0}{m} \end{bmatrix} e^{i\omega t} =: \mathbf{b}_0\, e^{i\omega t} \ .$$

The eigenvalues of $\mathbf{A}$ are

$$\lambda_\pm = -\frac{c}{2m} \pm i\sqrt{d}$$

with corresponding eigenvectors

$$\mathbf{v}_\pm = \begin{bmatrix} 1 \\ \lambda_\pm \end{bmatrix} \ .$$

The homogeneous part of the solution has been dealt with in the previous
example. We have seen that it dies out if $c > 0$. The inhomogeneous
part x_p is given by (4.13). The coefficients in (4.13) are determined
from the expansion of the constant amplitude vector $\mathbf{b}_0$ in terms of the
eigenvectors :

$$\mathbf{b}_0 = b_+\,\mathbf{v}_+ + b_-\,\mathbf{v}_-$$

with

$$b_+ = -b_- = \frac{-i}{\sqrt{d}}\,\frac{f_0}{2m} \ .$$

Following (4.13), $x_p(t)$ is given by

$$x_p(t) = b_+ \left(I_+(t, \omega) - I_-(t, \omega) \right)$$

with the integrals $I_\pm$ given by (4.15). Here, we ignore the (exceptional) case of critical damping and assume $\lambda_+ \neq \lambda_-$. From (4.15) we find that for $t_0 = 0$

$$I_\pm(t, \omega) = \frac{e^{i\omega t} - e^{\lambda_\pm t}}{i\omega - \lambda_\pm} .$$

If $c > 0$, we have $\mathrm{Re}\,\lambda_\pm < 0$, and thus

$$I_\pm(t, \omega) \to \frac{e^{i\omega t}}{i\omega - \lambda_\pm} , \qquad t \to \infty .$$

After the transient phenomena have damped out, the particular solution is thus oscillatory with frequency ω, and given by

$$x_p(t) = b_+ \left(\frac{1}{i\omega - \lambda_+} - \frac{1}{i\omega - \lambda_-} \right) e^{i\omega t} .$$

Example IV.4. (Periodic coefficients)
Consider the equation

$$\ddot{y} + b(t)\,\dot{y} + c(t)\,y = 0$$

with $b(t)$ and $c(t)$ T-periodic and continuous on $\mathbb{R}$. Application of the Liouville transformation (6.2) yields the equation

$$\ddot{x} + d(t)\,x = 0$$

with $d(t)$ given by (6.5) and thus T-periodic. Because $\mathrm{Tr}\,\mathbf{A} = 0$, we conclude from Theorem 5.14 that the characteristic multipliers and exponents satisfy the relations

$$\mu_+ \mu_- = 1 ,$$
$$\lambda_+ + \lambda_- = 0 ,$$

respectively. The multipliers $\mu_\pm$ are the eigenvalues of the principal fundamental matrix $\mathbf{C} = \mathbf{Y}(T)$. Following the proof of Theorem 5.14 we have $\det \mathbf{C} = \mu_+ \, \mu_- = 1$. Denoting $p := 1/2\, \mathrm{Tr}\,\mathbf{C}$ we find that the $\mu_\pm$ satisfy

$$\det(\mathbf{C} - \mu\mathbf{I}) = \mu^2 - 2p\mu + 1 = 0$$

with solutions

$$\mu_\pm = p \pm \sqrt{p^2 - 1} \, .$$

From (5.13) we find

$$\lambda_\pm = \frac{1}{T} \ln \mu_\pm = \pm \frac{1}{T} \cosh^{-1} p \, .$$

The behaviour of the solution for $t \to \infty$ depends on the $\lambda_\pm$. These exponents are known if p, i.e. $\mathrm{Tr}\,\mathbf{C}$, is known. So, it suffices to integrate the system over one period T. After that the solutions are known for all t.

7. Classification of planar, autonomous systems

The planar, homogeneous, autonomous systems

(7.1) $\dot{\mathbf{x}} = \mathbf{A}\,\mathbf{x}$

with $\mathbf{A}$ a constant 2×2 matrix can be classified according to the character of its stationary point, the origin. A similar classification is also possible for systems of higher order, but it does not add much extra information and the clarity gets lost. It is convenient to first transform $\mathbf{A}$ into Jordan form (see Appendix C and [49]). So let $\mathbf{S}$ be a nonsingular matrix, such that

(7.2) $\mathbf{J} := \mathbf{S}\,\mathbf{A}\,\mathbf{S}^{-1}$

has the Jordan block structure. The *similarity transformation* (7.2) preserves the eigenvalues. The matrix $\mathbf{J}$ can have only a few typical forms. Our classification will be done accordingly.

(i) J is a diagonal matrix:

(7.3) $\mathbf{J} = \begin{pmatrix} \lambda_1 & 0 \\ 0 & \lambda_2 \end{pmatrix} \, .$

In this case $\mathbf{A}$ has two linearly independent eigenvectors $\mathbf{v}_1$ and $\mathbf{v}_2$. Following (4.4) the solutions can be written as

(7.4) $\mathbf{x}(t) = c_1\,e^{\lambda_1(t-t_0)}\,\mathbf{v}_1 + c_2\,e^{\lambda_2(t-t_0)}\,\mathbf{v}_2 \, .$

We distinguish between three cases:

(ia) λ_1, λ_2 are real and different, $\lambda_1 < \lambda_2$ say.

$\lambda_1, \lambda_2 < 0$: the origin is a stable node. See Fig. 4a.
$\lambda_1, \lambda_2 \geq 0$: the origin is an unstable node. See Fig. 4b.
$\lambda_2 > 0,\ \lambda_1 < 0$: the origin is a saddle point. See Fig. 4c.

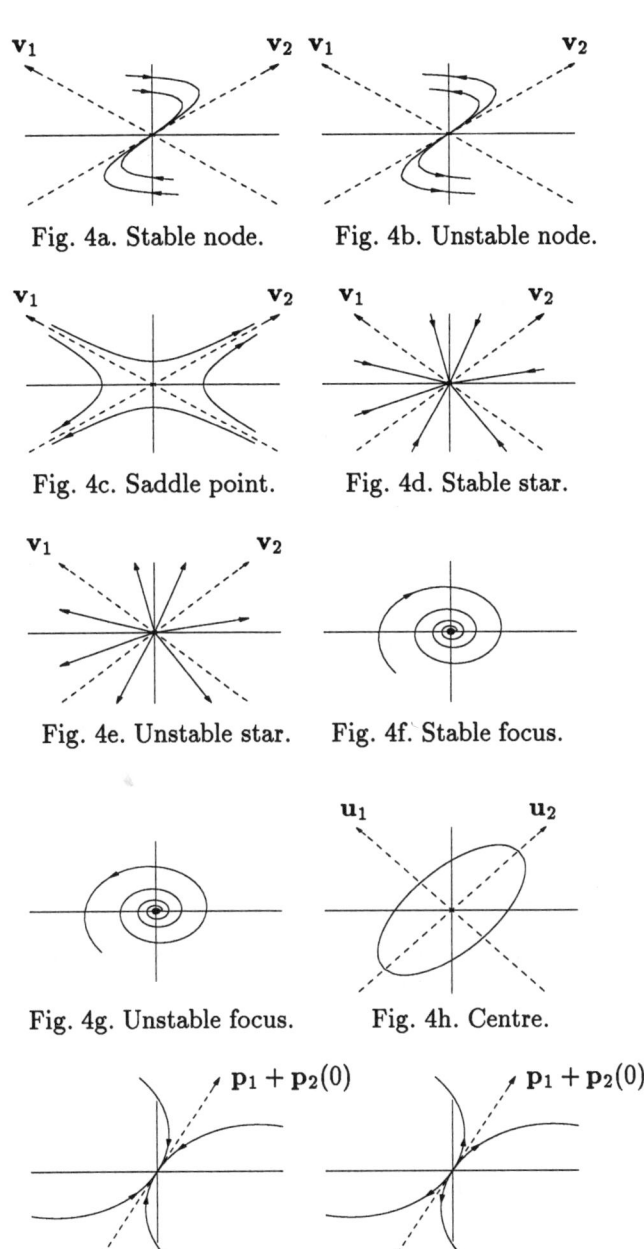

Fig. 4a. Stable node.   Fig. 4b. Unstable node.

Fig. 4c. Saddle point.   Fig. 4d. Stable star.

Fig. 4e. Unstable star.   Fig. 4f. Stable focus.

Fig. 4g. Unstable focus.   Fig. 4h. Centre.

Fig. 4i. Stable improper node.   Fig. 4j. Unstable improper node.

(ib) λ_1, λ_2 are real and equal, $\lambda_1 = \lambda_2 = \lambda$ say.

$\lambda < 0$: the origin is a stable star. See Fig. 4d.
$\lambda \geq 0$: the origin is an unstable star. See Fig. 4e.

(ic) λ_1, λ_2 are complex conjugates, say $\lambda_1 = \lambda$ and $\lambda_2 = \bar{\lambda}$. Then also the eigenvectors are complex conjugates, say $v_1 = v$ and $v_2 = \bar{v}$. Because the solutions $x(t)$ are real, we find that the constants c_1, c_2 in (7.4) are complex conjugates, say $c_1 = c$ and $c_2 = \bar{c}$. Equation (7.4) then reads

$$(7.5)\qquad x(t) = c\, e^{\lambda(t-t_0)}\, v + \bar{c}\, e^{\bar{\lambda}(t-t_0)}\, \bar{v} =$$

$$= 2\, \mathrm{Re}\left(c\, e^{\lambda(t-t_0)}\, v\right).$$

Introducing

$$(7.6)\qquad \lambda := \alpha + i\beta\,,\qquad v := u_1 + iu_2\,,\qquad c := d_1 + id_2$$

with α, β, u_1, u_2, d_1, d_2 real, we may express $x(t)$ in terms of real quantities only:

$$(7.7)\qquad x(t) = 2e^{\alpha(t-t_0)}\left\{\left(d_1\,\cos\,\beta(t-t_0) - d_2\,\sin\,\beta(t-t_0)\right)u_1 - \right.$$

$$\left. -\left(d_2\,\cos\,\beta(t-t_0) + d_1\,\sin\,\beta(t-t_0)\right)u_2\right\}.$$

If the imaginary part β of λ is non-vanishing, this causes the solutions to spiral around the origin. The real part α determines whether they approach it $(\alpha > 0)$, leave it $(\alpha < 0)$, or keep circling around it $(\alpha = 0)$ for $t \to \infty$. The sign of α is the classification parameter here :

$\alpha < 0$: the origin is a stable focus. See Fig. 4f.
$\alpha > 0$: the origin is an unstable focus. See Fig. 4g.
$\alpha = 0$: the origin is a centre. See Fig. 4h.

The matrices A and J are connected through the transformation (7.2). If the eigenvalues are complex, one often applies an alternative similarity transformation, viz. $S := (u_1 \,|\, u_2)$. In that case A is similar to the real, anti-symmetric matrix

$$(7.8)\qquad S\,A\,S^{-1} = \begin{pmatrix} \alpha & \beta \\ -\beta & \alpha \end{pmatrix}.$$

(ii) J is a non-diagonal matrix:

$$(7.9)\qquad J = \begin{bmatrix} \lambda & 1 \\ 0 & \lambda \end{bmatrix},$$

with λ real. The matrix A, and also J, has only one eigenvector, v say. The solutions of the *canonical* form

(7.10) $\dot{\mathbf{y}} = \mathbf{J}\,\mathbf{y}$

of (7.1) are given by

$$(7.11) \quad \mathbf{y}(t) = e^{\mathbf{J}(t-t_0)}\,\mathbf{y}_0 = e^{\lambda(t-t_0)}\{y_1 \begin{bmatrix} 1 \\ 0 \end{bmatrix} + y_2 \begin{bmatrix} t-t_0 \\ 1 \end{bmatrix}\}$$

for arbitrary $\mathbf{y}_0 = (y_1, y_2)^T$ (cf. Appendix C). The solutions of (7.1) then are given by

$$(7.12) \quad \mathbf{x}(t) = \mathbf{S}\,\mathbf{y}(t) = e^{\lambda(t-t_0)}\left(\mathbf{p}_1 + \mathbf{p}_2(t)\right)$$

with $\mathbf{p}_1$ a constant vector and $\mathbf{p}_2(t)$ a linearly independent vector, whose elements are polynomials in t of degree ≤ 1, i.e. $\mathbf{p}_2(t)$ is a linear function of t. The sign of λ determines the character of the system:

$\lambda < 0$: the origin is a stable improper node. See Fig. 4i.
$\lambda > 0$: the origin is an unstable improper node. See Fig. 4j.

8. Difference equations

The theory of linear Δ-equations

$$(8.1) \quad \mathbf{x}_{i+1} = \mathbf{A}_i\,\mathbf{x}_i + \mathbf{b}_i$$

shows many similarities with the theory of linear ODE. We shall therefore only show some basic principles. From these the reader should be able to derive the analogues of the results in the preceding sections. For convenience we assume the recursion to start at $i = 0$ with initial value $\mathbf{x}_0$. If the matrices $\mathbf{A}_i$ are all non-singular, equation (8.1) can be inverted and solutions can be obtained for both negative and positive i values. If (some of) the $\mathbf{A}_i$ are singular, the recursion can be performed in the forward direction only.

The solutions of the *homogeneous equation*

$$(8.2) \quad \mathbf{x}_{i+1} = \mathbf{A}_i\,\mathbf{x}_i$$

form a linear vector space V. Introducing the evolution matrix function

$$(8.3) \quad \mathbf{Y}_{i,j} = \begin{cases} \mathbf{A}_{i-1}...\mathbf{A}_j \ , & i > j \geq 0 \\ \mathbf{I} & , \quad i = j \geq 0 \ , \end{cases}$$

the solutions of (8.2) can be indicated by an expression similar to (2.11) for ODE, viz.

$$(8.4) \quad \mathbf{x}_i = \mathbf{Y}_{i,0}\,\mathbf{x}_0 \ .$$

The solutions of the *inhomogeneous equation* are the sum of the general solution of the homogeneous equation (8.2) and a particular solution, $\{\mathbf{p}_i\}$ say. The latter is the solution of (8.1) starting at the origin, thus with $\{\mathbf{x}_0\} = \mathbf{0}$. Iteration of (8.1) with $\mathbf{x}_0 = \mathbf{0}$ yields

$$(8.5) \qquad \mathbf{p}_i = \sum_{j=0}^{i-1} \mathbf{Y}_{i,j}\, \mathbf{b}_j \ .$$

The solution of (8.1) with arbitrary initial value $\mathbf{x}_0$ is given by the sum

$$(8.6) \qquad \mathbf{x}_i = \mathbf{Y}_{i,0}\, \mathbf{x}_0 + \mathbf{p}_i \ ,$$

which is similar to (3.5) with the integral replaced by a summation.

If $\mathbf{A}$ is *constant*, the evolution matrix function is given by

$$(8.7) \qquad \mathbf{Y}_{i,j} = \mathbf{A}^{i-j} \ , \qquad i \ge j \ .$$

If $\mathbf{A}$ has n different eigenvectors $\mathbf{v}_j$, expression (8.4) can be written as

$$(8.8) \qquad \mathbf{x}_i = \sum_{j=1}^{n} c_j\, \lambda_j^i\, \mathbf{v}_j$$

with the coefficients c_j determined by the unique expansion of the initial value $\mathbf{x}_0$ in terms of the basis functions $\mathbf{v}_j$:

$$(8.9) \qquad \mathbf{x}_0 = \sum_{j=1}^{n} c_j\, \mathbf{v}_j \ .$$

If $\mathbf{A}$ has p eigenvectors with $p < n$, we proceed as in (4.5)–(4.8). If $\mathbf{J}$ is the Jordan form of $\mathbf{A}$, i.e.

$$(8.10) \qquad \mathbf{A} = \mathbf{S}\,\mathbf{J}\,\mathbf{S}^{-1} \ ,$$

then the evolution matrix function is given by

$$(8.11) \qquad \mathbf{Y}_{i,j} = \mathbf{A}^{i-j} = \mathbf{S}\,\mathbf{J}^{i-j}\,\mathbf{S}^{-1} \ .$$

The matrix product $\mathbf{J}^i$ has the form (see Appendix C)

$$(8.12) \qquad \mathbf{J}^i = \begin{pmatrix} \mathbf{J}_1^i & & \\ & \ddots & \\ & & \mathbf{J}_p^i \end{pmatrix}$$

with the $\mathbf{J}_k$ the Jordan blocks of size $r_k \times r_k$. We may write $\mathbf{J}_k$ in the form

$$(8.13) \qquad \mathbf{J}_k = \lambda_k \begin{pmatrix} 1 & \lambda_k^{-1} & & \emptyset \\ & \ddots & \ddots & \\ & & \ddots & \lambda_k^{-1} \\ \emptyset & & & 1 \end{pmatrix} =: \lambda_k\, \mathbf{K}_k \ .$$

The i-th power of $\mathbf{K}_k$ reads

$$(8.14) \quad (\mathbf{K}_k)^i =$$

$$\begin{pmatrix}
1 & i\lambda_k^{-1} & \binom{i}{2}\lambda_k^{-2} & \cdots & \binom{i}{r_j-1}\lambda_k^{-(r_j-1)} \\
& \ddots & \ddots & \ddots & \vdots \\
& & \ddots & \ddots & \ddots & \binom{i}{2}\lambda_k^{-2} \\
& & & \ddots & \ddots & i\lambda_k^{-1} \\
\emptyset & & & & & 1
\end{pmatrix} .$$

Substituting the $(\mathbf{J}_k)^{i-j}$ into (8.12) we find an expression for $\mathbf{J}^{i-j}$, and thus for $\mathbf{A}^{i-j}$. The solution of the homogeneous equation (8.2) reads in the general case thus as:

$$(8.15) \quad \mathbf{x}_i = \mathbf{S}\,\mathbf{J}^i\,\mathbf{S}^{-1}\,\mathbf{x}_0 .$$

Note the similarity between (8.15) and (4.8). The inhomogeneous case (cf. equations (4.9)–(4.18)) and the periodic systems (cf. §5) can be treated in a completely similar way.

Exercises Chapter IV

1. Let $\mathbf{A}$ be symmetric.

    a) Give the Jordan form of $\mathbf{A}$.

    b) Show that the eigenvectors of $\mathbf{A}$ are mutually orthogonal.

    c) Let $\mathbf{v}$ be an eigenvector of $\mathbf{A}$. Show that, if $\mathbf{x}_0 \perp \mathbf{v}$, then $\mathbf{x}(t) = e^{\mathbf{A}t}\mathbf{x}_0 \perp \mathbf{v}$ for all $t \in \mathbb{R}$.

2. Consider the planar linear system $\dot{\mathbf{x}} = \mathbf{A}\,\mathbf{x}$, where $\mathbf{x} := (x_1, x_2)^T$ and

$$\mathbf{A} = \begin{bmatrix} 2 & 1 \\ 0 & -3 \end{bmatrix} .$$

Classify the system according to the classification given in §7. Find the solution for arbitrary initial condition $\mathbf{x}_0$.

3. Consider the planar system $\dot{\mathbf{x}} = \mathbf{A}\mathbf{x}$ with

$$\mathbf{A} = \begin{bmatrix} \alpha & \beta \\ -\beta & \alpha \end{bmatrix}.$$

Introduce polar coordinates and show that the solutions are spiralling around the origin. What effect has a change of sign of β?

4. Determine the solutions of the three-dimensional system $\dot{\mathbf{x}} = \mathbf{A}\mathbf{x}$ with

$$\mathbf{A} = \begin{bmatrix} 3 & 0 & 0 \\ 0 & -1 & -2 \\ 0 & 2 & -1 \end{bmatrix}.$$

Sketch the flow in phase space near the origin.

5. Determine the Wronskian $W(t)$ of two linearly independent solutions of the IVP

$$\dot{\mathbf{x}} = \begin{bmatrix} 0 & f(t) \\ g(t) & 0 \end{bmatrix} \mathbf{x}$$

in terms of the (continuous) functions $f(t)$, $g(t)$, and $W(t_0)$.

6. Consider the planar, periodic system $\dot{\mathbf{x}} = \mathbf{A}(t)\mathbf{x}$ with

$$\mathbf{A}(t) = \begin{bmatrix} -\sin 2t & -1 + \cos 2t \\ 1 + \cos 2t & \sin 2t \end{bmatrix}.$$

a) Show that

$$\mathbf{y}(t) = \begin{bmatrix} e^t(\cos t - \sin t) & e^{-t}(\cos t + \sin t) \\ e^t(\cos t + \sin t) & e^{-t}(-\cos t + \sin t) \end{bmatrix}$$

is a fundamental matrix of this system.

b) Determine the matrices $\mathbf{C}$ and $\mathbf{B}$ in (5.8) and the characteristic multipliers and exponents.

7. Determine the solutions for arbitrary initial values $x(t=0) = x_0$, $\dot{x}(t=0) = v_0$, of

a) $\ddot{x} + x = \sin t + \cos 3t$.

b) $\ddot{x} + x = t\,e^t - t^4$.

c) $\ddot{x} - x = e^{-t}$.

Hint: Use the results from §4. It is convenient to bring the operator e^{At} into Jordan form (see Appendix C) using the fact that the matrix S needed to apply the similarity transformation consists of eigenvectors of A.

8. Derive expression (6.12) for volume preserving planar systems from (6.11) using the general expression (1.36).

9. Show the correctness of Property 6.13.

10. Find the solution of the scalar IVP

$$
\begin{cases}
x_{i+1} & = & 2x_i - 3x_{i-1} + 2 , \qquad i \geq 1 , \\
x_0 & = & 1 , \\
x_1 & = & 1 .
\end{cases}
$$

11. Find the solution of the planar IVP

$$
\begin{cases}
\mathbf{x}_{i+1} & = & \begin{pmatrix} 3 & 1 \\ 1 & 3 \end{pmatrix} \mathbf{x}_i + \begin{pmatrix} 1 \\ 1 \end{pmatrix} , \\
\mathbf{x}_0 & = & \begin{pmatrix} 1 \\ 1 \end{pmatrix} .
\end{cases}
$$

V

Stability

In this chapter we study the behaviour of the solutions of IVP for $t \to \infty$. In §1 the various possibilities for the long term behaviour of solutions are mentioned. One possibility is chaotic behaviour. To this phenomenon a separate chapter (VI) is devoted. In the present chapter we focus on convergence to stationary points and periodic solutions. In §2 we point out that there are many notions of stability and give the most common definitions. Linear systems play a special rôle in stability analysis, because for those systems explicit results can often be found. This is worked out for systems with constant coefficients in §3. For nonlinear systems one can rely on two different approaches: linearisation and Lyapunov functions. The technique of linearisation is presented in §4. It uses the fact that most nonlinear vector fields can be approximated by a linear one in the vicinity of a stationary or periodic solution. Lyapunov functions are introduced in §5. Both linearisation and Lyapunov functions yield local information. A global analysis of the stability properties of nonlinear systems is difficult in general. For planar systems one can investigate the problems in more detail, as is illustrated in §6. The important case of periodic systems is considered in §7. Finally, we deal with the stability of Δ-equations in §8.

1. Introduction

In this section we study the effect of perturbations of the IVP

$$(1.1) \qquad \begin{cases} \dot{\mathbf{x}} = \mathbf{f}(t, \mathbf{x}) \,, \\ \mathbf{x}(t_0) = \mathbf{x}_0 \end{cases}$$

on its solution $\mathbf{x}$. Perturbations may concern both the initial value, the initial time, and the vector field. The *perturbed problem* then reads as $(t_1 \geq t_0)$

$$(1.2) \qquad \begin{cases} \dot{\mathbf{y}} = \mathbf{f}(t, \mathbf{y}) + \mathbf{r}(t, \mathbf{y}) \,, \\ \mathbf{y}(t_1) = \mathbf{x}(t_1) + \mathbf{z}_1 \,, \end{cases}$$

with solution $\mathbf{y}$. Stability considerations now deal with the behaviour of the difference $\mathbf{y}(t) - \mathbf{x}(t)$ as a function of time.

In §II.4 we studied only the effect of the perturbations $\mathbf{r}$ and $\mathbf{z}_1$, taking $t_1 = t_0$. Theorem II.4.7 then provides us with an upper bound on the distance $\|\mathbf{y}(t) - \mathbf{x}(t)\|$ for t in some bounded interval $I := [t_0, T]$:

$$(1.3) \qquad \|\mathbf{y}(t) - \mathbf{x}(t)\| \leq e^{L(t-t_0)} \|\mathbf{y}_0 - \mathbf{x}_0\| + \frac{1}{L} \left(e^{L(t-t_0)} - 1 \right) \|\mathbf{r}\|_{I \times \Omega} \, ,$$

where Ω is a domain large enough to contain at least $\mathbf{x}(t)$ and $\mathbf{y}(t)$, $t \in I$. This result tells us what possible effects perturbations of the data may have on the (possibly numerically obtained) solution. This relates to a physically induced notion of stability: if the effects are fairly limited in a bounded time interval, we call the system *stable*. Theorems II.5.2 and II.5.3 show that, under mild conditions, the error depends continuously (or even differentiably) on the magnitude of the perturbations. This implies that this error can be made arbitrarily small just by decreasing the magnitude of the perturbation terms.

Here we are interested in the effect of perturbations as $t \to \infty$. We therefore assume the solution of the unperturbed problem (1.1) to exist for $t \geq t_0$. The theorems of Chapter II do not provide the required information. In practice a wide variety of stability concepts are used. The most general one is:

Definition 1.4 (Total stability). *The solution $\mathbf{x}$ of (1.1) is called totally stable if, for each $\varepsilon > 0$ and for each $t_1 \geq t_0$, a $\delta(\varepsilon, t_1)$ exists such that*

$$\{\|\mathbf{z}_1\| < \delta(\varepsilon, t_1) \text{ and } R_1 < \delta(\varepsilon, t_1)\} \Rightarrow \{\|\mathbf{y}(t) - \mathbf{x}(t)\| < \varepsilon \text{ for all } t \geq t_1\} \, ,$$

where $R_1 := \sup_{t \geq t_1} \|\mathbf{r}(t, \mathbf{x}(t))\|$.

If δ does not depend on t_1 the solution $\mathbf{x}$ of (1.1) is called *uniformly totally stable*. Though useful in its own right, and in particular in view of numerical computations, one often needs a stability concept that does not relate to perturbations of the vector field. In the rest of this chapter we shall mainly dwell on perturbations of the initial value, and only come back to the concept of total stability in §8. To structure our stability analysis it is useful to list the various possibilities for solutions $\mathbf{x}$ of (1.1) for $t \to \infty$. It may happen that

(i) $\mathbf{x}$ approaches a stationary point (or is such a point itself).

(ii) $\mathbf{x}$ approaches a (quasi-)periodic solution (or is such a solution itself).

(iii) $\mathbf{x}$ becomes unbounded.

(iv) $\mathbf{x}$ behaves otherwise.

Case (iii) clearly relates to instability, a situation of little practical interest. In the following we shall concentrate on cases (i) and (ii). Case (iv), which includes so-called *chaotic* behaviour, will be dealt with in Chapter VI. Chaotic solutions are inherently unstable, whatever stability definition is used; they exhibit an extreme kind of sensitivity with respect to the initial condition. See also [31].

2. Stability definitions

In the following we assume the vector field to be unperturbed. The perturbed IVP (1.2) then reduces to $(t_1 \geq t_0)$

$$(2.1) \qquad \begin{cases} \dot{\mathbf{y}} = \mathbf{f}(t, \mathbf{y}) \,, \\ \mathbf{y}(t_1) = \mathbf{x}(t_1) + \mathbf{z}_1 \,. \end{cases}$$

The solution $\mathbf{x}$ of (1.1) is called (*Lyapunov*) *stable* if the solution $\mathbf{y}$ of (2.1) remains in the vicinity of $\mathbf{x}$ for $t \to \infty$, for all t_1 and all (small) perturbations $\mathbf{z}(t_1)$. More formally we have:

Definition 2.2. *The solution* $\mathbf{x}$ *of (1.1) is Lyapunov stable if, for every* $\varepsilon > 0$ *and for every* $t_1 \geq t_0$*, a* $\delta(\varepsilon, t_1)$ *exists such that*

$$(2.2a) \qquad \{\|\mathbf{z}_1\| < \delta(\varepsilon, t_1)\} \Rightarrow \{\|\mathbf{y}(t) - \mathbf{x}(t)\| < \varepsilon \text{ for all } t \geq t_1\} \,,$$

where $\mathbf{y}$ *is the solution of (2.1).*

For a stronger form of stability one requires the quantity δ in Definition 2.2 not to depend on t_1. This is the case if, for given ε,

$$(2.3) \qquad \delta(\varepsilon) := \inf_{t_1 \geq t_0} \delta(\varepsilon, t_1)$$

exists and is positive. Then, $\mathbf{x}$ is called *uniformly (Lyapunov) stable*.

An even stronger form of stability requires the solution $\mathbf{y}$ of (2.1) not only to remain in a neighbourhood of $\mathbf{x}$, but also to approach it for $t \to \infty$:

Definition 2.4. *The solution* $\mathbf{x}$ *of (1.1) is asymptotically stable if* $\mathbf{x}$ *is stable, and if, for every* $t_1 \geq t_0$*, a* $\delta(t_1)$ *exists such that*

$$(2.4a) \qquad \{\|\mathbf{z}_1\| < \delta(t_1)\} \Rightarrow \{\|\mathbf{y}(t) - \mathbf{x}(t)\| \to 0 \text{ if } t \to \infty\} \,.$$

For clarity we explicitly mention the meaning of the right-hand side of (2.4a): for every $\varepsilon > 0$ a $T(\varepsilon, t_1)$ exists such that $\|\mathbf{y}(t) - \mathbf{x}(t)\| < \varepsilon$ for all $t \geq t_1 + T(\varepsilon, t_1)$.

Finally, the strongest form of stability also requires $T(\varepsilon, t_1)$ not to depend on t_1. This is the case if

$$(2.5) \qquad T(\varepsilon) := \sup_{t_1 \geq t_0} T(\varepsilon, t_1)$$

exists, i.e. $T(\varepsilon) < \infty$. If this is so, and $\mathbf{x}$ is also uniformly stable, then $\mathbf{x}$ is called *uniformly asymptotically stable*. For autonomous systems stability and asymptotic stability are always uniform.

Definition 2.6. *The solution $\mathbf{x}$ of (1.1) is unstable if it is not (Lyapunov) stable.*

If $\mathbf{x}$ is unstable, small perturbations may cause the system to follow a completely different orbit in the long run. In practical systems small perturbations are always present, while in numerical calculations the round-off errors act as a source of perturbations. Therefore, unstable solutions are hard to handle and often undesired. For example, mechanical systems are nearly always tuned such that they operate in asymptotically stable modes, so that uncontrollable changes of the conditions have little effect. Though the stability definitions above apply to general $\mathbf{x}$, they are mostly used in situations where $\mathbf{x}$ is a stationary point, i.e. if $\mathbf{x}(t) \equiv \mathbf{x}_0$ normally. In Fig. V.1 the definitions have been illustrated for this case.

For autonomous systems it makes sense to introduce the following notion:

Definition 2.7. *A domain of attraction of an asymptotically stable stationary point $\mathbf{x}_0$ of the system $\dot{\mathbf{x}} = \mathbf{f}(\mathbf{x})$ consists of points for which solutions starting at these points approach $\mathbf{x}_0$ for $t \to \infty$.*

Usually one is interested in the maximum domain of attraction, which is called the *basin of attraction*. For nonlinear problems it is often difficult to determine these domains. They have to be estimated either numerically or through methods such as those dealt with in §5 and §XI.4.

In the following example various stability properties are illustrated.

Example V.1.
Consider the scalar, non-autonomous, nonlinear equation

$$\dot{x} = a(t)\, x^3 \ .$$

We may ask which stability properties the stationary point $\mathbf{x} = 0$ has. Because of the simple structure of this equation, we can solve this IVP by separation of variables. A solution x starting at $t_1 \geq t_0$ with $x(t_1) = x_1$ satisfies

$$\int_{x_1}^{x} \frac{1}{x^3}\, dx = \int_{t_1}^{t} a(s)\, ds \ .$$

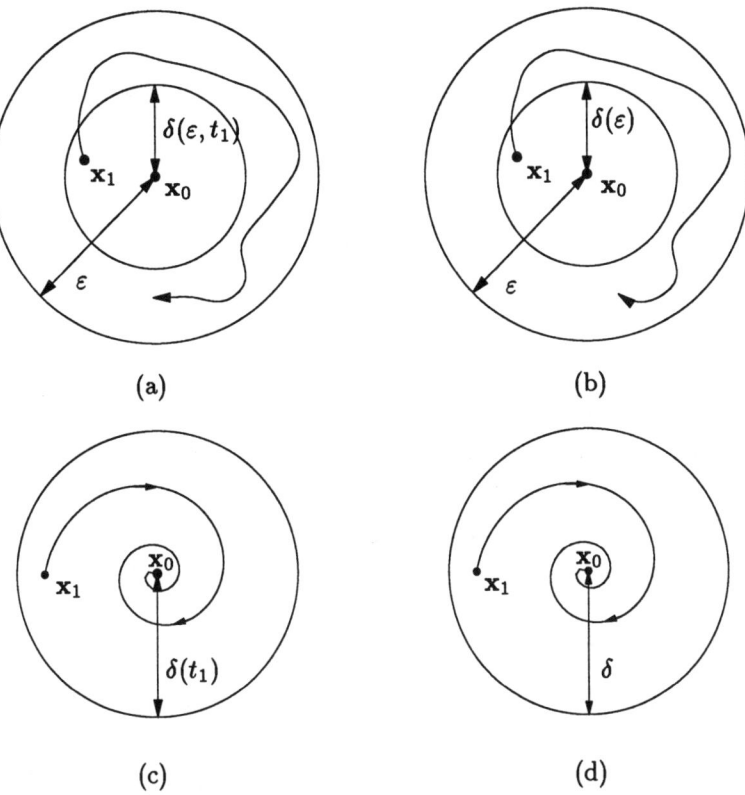

Figure V.1 (a) Lyapunov stability, (b) uniform Lyapunov stability, (c) asymptotic Lyapunov stability, (d) uniform asymptotic Lyapunov stability.

From this we find

$$x(t) = x_1 \left(1 - 2x_1^2 A(t, t_1)\right)^{-\frac{1}{2}}$$

with

$$A(t, t_1) = \int_{t_1}^{t} a(s) \, ds .$$

We consider the following cases:

(i) $a(t) = a < 0$. Then $A(t, t_1) = a(t - t_1) \leq 0$, $t \geq t_1$, and $x(t) \to 0$
      if $t \to \infty$. Since x depends on the difference $t - t_1$ only, the origin
      is uniformly asymptotically stable. The basin of attraction is the
      whole real axis.

(ii) $a(t) = 0$. In this case $x(t) \equiv x_1$, and the origin is uniformly stable.

(iii) $a(t) = a > 0$. Then the origin is unstable because $x(t) \to \infty$ for all
      $x_1 \neq 0$.

(iv) $a(t) = -(t + 1)^{-1}$. Then $A(t, t_1) = \ln(t_1 + 1) - \ln(t + 1) \leq 0$,
      $t \geq t_1$. We see that $x(t) \to 0$ as $t \to \infty$ for all x_1. Furthermore,
      $x(t)$ is monotonically decreasing as t increases. The origin is thus
      asymptotically stable. The stability is not uniform, because the
      rate of convergence to the zero solution depends on t_1 and $T(\varepsilon)$ in
      (2.5) does not exist in this case.

(v) $a(t) = -(t+1)^{-2}$. Then $A(t, t_1) = (t+1)^{-1} - (t_1+1)^{-1} \leq 0, t \geq t_1$.
      In this case $x(t)$ approaches a constant value as $t \to \infty$. This value
      depends on x_1 and t_1. Because $x(t) \leq x_1$ for $t \geq t_1$, the origin is
      stable. We can get the limiting value of $x(t)$ arbitrarily small by
      taking x_1 smaller and smaller. The value of t_1 is not relevant in
      that procedure, so the stability is uniform.

(vi) $a(t) = \sin t$. Then $A(t, t_0) = \cos t_1 - \cos t$. Then A is periodic and
      changes sign unless $\cos t_1 = 1$. Because A is bounded, x exists
      and is periodic if x_1 is sufficiently small. The origin is uniformly
      stable. One could question whether the periodic solution starting
      at $x_1 \neq 0$ is itself stable. We leave it to the reader to show that
      the solutions starting at x_1 and $x_1 + \delta$ approach each other for all
      $t \geq t_1$ as $\delta \to 0$. The periodic solution is stable.

3. Stability of linear systems

We shall work out the stability concepts, introduced in §2, for the special case
of linear systems. Let us perturb the linear IVP

(3.1)
$$\begin{cases} \dot{\mathbf{x}} = \mathbf{A}(t)\,\mathbf{x} + \mathbf{b}(t)\ , \\ \mathbf{x}(t_0) = \mathbf{x}_0 \end{cases}$$

with respect to the initial value $(t_1 \geq t_0)$

(3.2)
$$\begin{cases} \dot{\mathbf{y}} = \mathbf{A}(t)\,\mathbf{y} + \mathbf{b}(t)\ , \\ \mathbf{y}(t_1) = \mathbf{x}(t_1) + \mathbf{z}_1\ . \end{cases}$$

Stability of the solution $\mathbf{x}$ of (3.1) is related to the dependence of the difference
$\mathbf{z}(t) := \mathbf{y}(t) - \mathbf{x}(t)$ on the perturbation $\mathbf{z}_1$. This error satisfies the homogeneous
IVP

$$(3.3) \qquad \begin{cases} \dot{\mathbf{z}} = \mathbf{A}(t)\,\mathbf{z} \ , \\ \mathbf{z}(t_1) = \mathbf{z}_1 \ . \end{cases}$$

In view of the stability Definitions 2.2 and 2.4, we find that the stability properties of the solution of any linear IVP are given by the stability properties of the zero solution in the corresponding homogeneous problem. This zero solution is always a stationary point of the homogeneous problem (3.3). Because of this, we may speak of *the stability of a linear system*, rather than of the stability of a specific solution. In Chapter IV explicit expressions have been derived for the solution $\mathbf{z}$ of (3.3) if $\mathbf{A}(t)$ is constant or periodic.

If $\mathbf{A}$ is constant, $\mathbf{z}$ can be written as

$$(3.4) \qquad \mathbf{z}(t) = \mathbf{S}\, e^{\mathbf{J}(t-t_0)}\, \mathbf{S}^{-1}\, \mathbf{z}_0 \ .$$

The matrix exponent in (3.4) is given by (IV.4.6) with (IV.4.7) From the latter expressions we may deduce directly the following theorem:

Theorem 3.5. *The stability of the linear IVP (3.1) with $\mathbf{A}$ constant is determined by the eigenvalues λ_i of $\mathbf{A}$. The following cases may occur:*

(i)   *uniform asymptotic stability, if $\mathrm{Re}(\lambda_i) < 0$ for all λ_i.*

(ii) *uniform (Lyapunov) stability, if $\mathrm{Re}(\lambda_i) \leq 0$ for all λ_i provided that the dimension of the Jordan block for λ_i with $\mathrm{Re}(\lambda_i) = 0$ equals one.*

(iii) *instability, if the conditions (i) and (ii) are not fulfilled.*

If $\mathbf{A}$ is not constant, there exists no direct, general relation between the eigenvalues of $\mathbf{A}(t)$ at successive times and the stability of (3.1), as is illustrated by the following example.

Example V.2.
Consider the equation

$$\dot{\mathbf{x}} = \mathbf{A}(t)\,\mathbf{x} \ ,$$

with

$$\mathbf{A}(t) = \begin{bmatrix} -1 + \gamma \cos^2 t & 1 - \gamma \sin t \cos t \\ -1 - \gamma \sin t \cos t & -1 + \gamma \sin^2 t \end{bmatrix} .$$

The eigenvalues of $\mathbf{A}(t)$ are time independent and given by

$$\lambda_{\pm} = \tfrac{1}{2}\left(\gamma - 2 \pm \sqrt{\gamma^2 - 4}\right) .$$

So, we have $\mathrm{Re}(\lambda_{\pm}) < 0$ if $\gamma < 2$.
Next, we apply a coordinate transformation such that the new coordinate frame rotates as time t evolves:

$$\mathbf{y}(t) = \mathbf{P}(t)\,\mathbf{x}(t) \ ,$$

with

$$\mathbf{P}(t) := \begin{bmatrix} \cos t & -\sin t \\ \sin t & \cos t \end{bmatrix} .$$

Then, $\mathbf{y}(t)$ satisfies the equation

$$\dot{\mathbf{y}} = \mathbf{P}(t)\,\mathbf{A}(t)\,\mathbf{P}^{-1}(t)\,\mathbf{y} = \begin{bmatrix} \gamma - 1 & 0 \\ 0 & -1 \end{bmatrix} \mathbf{y} \ .$$

So, $\mathbf{y}$, and thus $\mathbf{x}$, is Lyapunov stable for $\gamma = 1$. For $\gamma < 1$ we even find asymptotic stability, while for $\gamma > 1$ the system is unstable. The real part of $\lambda_{\pm}$, given above, did not suggest this!

If $\mathbf{A}(t)$ is **periodic**, we find from (IV.5.12) that $\mathbf{z}$ can be written as

$$(3.6) \qquad \mathbf{z}(t) = \mathbf{F}(t)\,\mathbf{S}\,e^{\mathbf{J}(t-t_0)}\,\mathbf{S}^{-1}\,\mathbf{z}_0$$

In the matrix exponent in (3.6) the characteristic exponents of $\mathbf{A}(t)$ occur. Because of the similarity between expressions (3.4) and (3.6) we may conclude:

Theorem 3.7. *If $\mathbf{A}(t)$ in (3.1) is periodic, the same stability properties as in Theorem 3.5 apply, provided that the λ_i are interpreted as the characteristic exponents of $\mathbf{A}(t)$.*

If $\mathbf{A}(t)$ in (3.1) is not constant or periodic, the situation may be quite complex. In the literature a lot of special cases have been investigated. For cases where $\mathbf{A}(t)$ is of the form $\mathbf{A}(t) = \mathbf{A} + \mathbf{B}(t)$, with $\mathbf{A}$ constant and $\mathbf{B}$ satisfying special conditions, results are known. The reader interested in these details is referred to e.g. [38]. Here, we restrict ourselves to a result concerning asymptotic stability. The idea is to generalise some obvious conclusions for the scalar case

$$(3.8) \qquad \dot{x} = ax \ ,$$

which has the evolution matrix function (see (IV.2.12))

$$(3.9) \qquad Y(t,s) = e^{a(t-s)} \ .$$

From this expression we may conclude that system (3.8) is uniformly asymptotically stable if and only if $a < 0$. The higher dimensional analogue is contained in Theorem 3.12. It will be used in §4 which deals with the stability properties of nonlinear systems. The theorem is based on the insight, derived in §IV.2, that the solution $\mathbf{x}$ of a homogeneous linear IVP can be written as

(3.10) $\mathbf{x}(t) = \mathbf{Y}(t, t_0)\, \mathbf{x}_0$,

where the evolution matrix $\mathbf{Y}$ satisfies the multiplication property

(3.11) $\mathbf{Y}(t, s)\, \mathbf{Y}(s, u) = \mathbf{Y}(t, u)$.

Theorem 3.12. *The linear IVP (3.1) is uniformly asymptotically stable if and only if positive numbers α, β exist such that*

(3.12a) $\|\mathbf{Y}(t, t_0)\| \le \beta\, e^{\alpha(t - t_0)}$,   *for all* $t \ge t_0$.

Proof: The 'if' statement directly follows if (3.10) and (3.12a) are combined. The 'only if' part of the proof is more delicate. Assume that the system (3.1) is uniformly asymptotically stable. Because of the linearity, this also holds for the corresponding homogeneous system

$$\dot{\mathbf{x}} = \mathbf{A}(t)\, \mathbf{x} .$$

Because any solution $\mathbf{x}$ of the homogeneous problem can be written in the form (3.10), we may conclude that

$$\|\mathbf{Y}(t, t_0)\| \to 0 \qquad \text{if } t \to \infty, \text{ uniform in } t_0 .$$

Thus, for every $\varepsilon > 0$ a $T(\varepsilon)$ exists such that

(∗) $\|\mathbf{Y}(t, t_0)\| \le \varepsilon$ if $t \ge t_0 + T(\varepsilon)$.

The uniformity of the limit above implies that $T(\varepsilon)$ may be taken independent of t_0. Let us make a special choice for ε, namely $\varepsilon = e^{-1}$. The corresponding $T(\varepsilon)$ is simply denoted by T in what follows. For convenience, we introduce the grid points

$$t_n := t_0 + nT , \qquad n = 0, 1, \dots .$$

Using (3.11), we may write

$$\mathbf{Y}(t_n, t_0) = \mathbf{Y}(t_n, t_{n-1})\, \mathbf{Y}(t_{n-1}, t_{n-2}) \dots \mathbf{Y}(t_1, t_0) .$$

Application of (∗) then yields

(∗∗) $\|\mathbf{Y}(t_n, t_0)\| \le e^{-n}$.

For t in the interval $[t_n, t_{n+1}]$ we may write

$$\mathbf{Y}(t, t_0) = \mathbf{Y}(t, t_n)\, \mathbf{Y}(t_n, t_0) .$$

Using (∗∗) we thus find

$$\|\mathbf{Y}(t, t_0)\| \le \gamma\, e^{-n} \; ,$$

with γ given by

$$\gamma := \max_{s \in [t_n, t_{n+1}]} \|\mathbf{Y}(s, t_n)\| \; .$$

The upper bound γ exists and is independent of n, because, by assumption, the system is uniformly stable. Since $t_n \le t \le t_{n+1}$, we have

$$n \ge \frac{t - t_0}{T} - 1 \; .$$

Hence for all $t \ge t_0$

$$\|\mathbf{Y}(t, t_0)\| \le \beta\, e^{-\alpha(t - t_0)} \; ,$$

where

$$\beta := \gamma\, e \text{ and } \alpha := \frac{1}{T} \; . \qquad\qquad \square$$

4. Nonlinear systems; linearisation

In the preceding section we saw that the stability properties of autonomous linear systems are fairly easy to investigate. The conclusions in Theorems 3.5 and 3.7 apply not only for specific solutions, but even for the system as a whole. For nonlinear systems such general results are not available. Then stability analyses are valid only locally, i.e. in the neighbourhood of a specific solution. Nonlinear problems are often studied by investigating the properties of linearised problems. We first point out the concept of *linearisation* around an arbitrary solution $\mathbf{x}$, and then apply it for $\mathbf{x}$ being constant or periodic. Let $\mathbf{x}$ be the solution of

$$(4.1) \qquad \begin{cases} \dot{\mathbf{x}} = \mathbf{f}(t, \mathbf{x}) \; , \\ \mathbf{x}(t_0) = \mathbf{x}_0 \end{cases}$$

and $\mathbf{y}$ a solution starting at $t_1 \ge t_0$ in the vicinity of $\mathbf{x}$:

$$(4.2) \qquad \begin{cases} \dot{\mathbf{y}} = \mathbf{f}(t, \mathbf{y}) \; , \\ \mathbf{y}(t_1) = \mathbf{x}(t_1) + \mathbf{z}_1 \; . \end{cases}$$

The difference $\mathbf{z}(t) := \mathbf{y}(t) - \mathbf{x}(t)$ satisfies, for $t \ge t_1$,

$$(4.3) \qquad \begin{cases} \dot{\mathbf{z}} = \mathbf{f}(t, \mathbf{y}) - \mathbf{f}(t, \mathbf{x}) \; , \\ \mathbf{z}(t_1) = \mathbf{z}_1 \; . \end{cases}$$

From the stability Definitions 2.2 and 2.4 we find that the stability properties of the solution $\mathbf{x}$ of (4.1) are given by the asymptotic behaviour of the solution $\mathbf{z}$ of (4.3) as a function of $\mathbf{z}_1$ in the neighbourhood of $\mathbf{z}_1 = \mathbf{0}$. Throughout this section we assume that $\mathbf{f}$ is differentiable with respect to its second argument around $\mathbf{x}(t)$. Then we may write

$$(4.4) \qquad \mathbf{f}\Big(t, \mathbf{y}(t)\Big) = \mathbf{f}\Big(t, \mathbf{x}(t)\Big) + \mathbf{J}\Big(t, \mathbf{x}(t)\Big)\mathbf{z}(t) + \mathbf{g}\Big(t, \mathbf{z}(t)\Big) ,$$

with $\mathbf{J}$ the Jacobian matrix of $\mathbf{f}$. The nonlinear term $\mathbf{g}(t, \mathbf{z})$ satisfies

$$(4.5) \qquad \lim_{\|\mathbf{z}\| \to 0} \frac{\|\mathbf{g}(t, \mathbf{z})\|}{\|\mathbf{z}\|} = 0 .$$

The *linearisation* of (4.1) around the solution $\mathbf{x}$ is obtained by truncating (4.4) after the linear term. Substitution of this truncated expansion into (4.3) yields

$$(4.6) \qquad \dot{\mathbf{z}} = \mathbf{J}\Big(t, \mathbf{x}(t)\Big)\mathbf{z} .$$

This homogeneous, linear equation is of the form (3.3) with $\mathbf{A}(t) \equiv \mathbf{J}(t, \mathbf{x}(t))$. Theorems 3.5 and 3.7 provide us with information about its stability if $\mathbf{A}(t)$ is constant or periodic. The solution of (4.6) will be a reliable approximation of the solution of (4.3) if $\|\mathbf{z}_1\|$ and $|t - t_1|$ are small. The question arises whether conclusions concerning the stability of (4.6) are also valid for the solution $\mathbf{x}$ of the original, nonlinear problem (4.1). The following famous theorem gives a partial answer.

Theorem 4.7 (Poincaré-Lyapunov). *If the system (4.6) is uniformly asymptotically stable, and if the vector field $\mathbf{g}(t, \mathbf{z})$ in (4.4) satisfies (4.5) uniformly in t and is moreover Lipschitz continuous in a neighbourhood of $\mathbf{z} = \mathbf{0}$, then the solution $\mathbf{x}$ of (4.1) is also uniformly asymptotically stable.*

Proof: We have to prove that the zero solution of (4.3) is uniformly asymptotically stable, because then this also holds for the solution of (4.1). We consider therefore (4.3) with $\mathbf{z}_1$ a small perturbation of the zero solution. Since $\mathbf{g}(t, \mathbf{z})$ is Lipschitz continuous, Theorem II.2.3 implies that (4.3) has a unique solution for some time interval around t_1. Following (IV.3.5), we may write solution $\mathbf{z}$ of (4.3) in the form

$$\mathbf{z}(t) = \mathbf{Y}(t, t_1)\mathbf{z}_1 + \int_{t_1}^{t} \mathbf{Y}(t, s)\mathbf{g}\Big(s, \mathbf{z}(s)\Big) ds ,$$

where $\mathbf{Y}$ is the evolution matrix function of ODE (4.6). This expression only makes sense within the time interval in which $\mathbf{z}(t)$ exists. This interval is not known in advance, but we can estimate its length from this expression itself. Using Theorem 3.12 we find

$(*)$ $$\|\mathbf{z}(t)\| \leq \|\mathbf{Y}(t,t_1)\| \, \|\mathbf{z}_1\| + \int_{t_1}^{t} \|\mathbf{Y}(t,s)\| \, \left\| \mathbf{g}\big(s,\mathbf{z}(s)\big) \right\| \, ds \leq$$

$$\leq \beta \, e^{-\alpha(t-t_1)} \, \|\mathbf{z}_1\| + \beta \int_{t_1}^{t} e^{-\alpha(t-s)} \, \left\| \mathbf{g}\big(s,\mathbf{z}(s)\big) \right\| \, ds$$

for certain $\alpha, \beta > 0$. From (4.5) it follows that for every $\varepsilon > 0$ a $\delta > 0$ exists such that

$$\|\mathbf{g}(t,\mathbf{z})\| < \varepsilon \, \|\mathbf{z}\| \qquad \text{if } \|\mathbf{z}\| < \delta \ .$$

We choose $\|\mathbf{z}_1\| < \delta$, and denote the first time point that $\|\mathbf{z}(t)\|$ becomes equal to δ by t_2, so $\|\mathbf{z}(t_2)\| = \delta$ for some $t_2 > t_1$. It will turn out that $t_2 = \infty$ if $\mathbf{z}_1$ is chosen sufficiently small. We may write inequality $(*)$ as

$(**)$ $$e^{\alpha t} \, \|\mathbf{z}(t)\| \leq h(t) \ ,$$

with the scalar function $h(t)$ defined by

$$h(t) := \beta \, e^{\alpha t_1} \, \|\mathbf{z}_1\| + \beta \varepsilon \int_{t_1}^{t} e^{\alpha s} \, \|\mathbf{z}(s)\| \, ds \ .$$

Differentiating h with respect to t, we find that $h(t)$ satisfies the differential inequality

$$\dot{h}(t) = \beta \, \varepsilon \, e^{\alpha t} \, \|\mathbf{z}(t)\| \leq \beta \, \varepsilon \, h(t) \ .$$

In the last step we again used $(**)$. Lemma II.1.8 then yields

$$h(t) \leq e^{\beta \varepsilon (t-t_1)} \, h(t_1) \ .$$

Combining this with the inequality $(**)$ we obtain for $\|\mathbf{z}(t)\|$ the bound

$(***)$ $$\|\mathbf{z}(t)\| \leq \beta \, e^{(\beta \varepsilon - \alpha)(t-t_1)} \, \|\mathbf{z}_1\| \ , \qquad t_1 \leq t \leq t_2 \ .$$

By choosing ε so small that $\beta \varepsilon - \alpha < 0$, we can guarantee that the right-hand side has a maximum $\beta \, \|\mathbf{z}_1\|$ for $t = t_1$ and decays to zero for increasing t values. After having fixed the value of ε, we may fix a value for δ such that $\|\mathbf{g}(t,\mathbf{z})\| < \varepsilon \, \|\mathbf{z}\|$ if $\|\mathbf{z}\| < \delta$. For this we choose the initial value $\mathbf{z}_1$ so small that $\|\mathbf{z}_1\| < \delta/\beta$. Inequality $(***)$ then states that $\|\mathbf{z}(t)\| < \delta$ for all $t \leq t_2$, while, from the definition, we have $\|\mathbf{z}(t_2)\| = \delta$. So we conclude that $t_2 = \infty$ for small ε and $\mathbf{z}_1$. Inequality $(***)$ also guarantees that $\|\mathbf{z}(t)\| \to 0$ if $t \to \infty$. This limit is independent of t_1, because α and β are independent of t_1. The zero solution of (4.3) is thus uniformly asymptotically stable. $\square$

Let us apply the linearisation ideas outlined above to the investigation of the stability of a stationary point $\hat{\mathbf{x}}$ of an autonomous system. For convenience a translation $\mathbf{x} \rightarrow \mathbf{x} - \hat{\mathbf{x}}$ is performed, so that the origin coincides with the stationary point under consideration. The linearised equation (4.6) is in this case

$$(4.8) \qquad \dot{\mathbf{z}} = \mathbf{J}(0)\,\mathbf{z}\ ,$$

with the Jacobian matrix being constant. Then, Theorem 4.7 applies. If we combine it with case (i) of Theorem 3.5 the following corollary results:

Corollary 4.9. *A stationary point of an autonomous system is uniformly asymptotically stable if all eigenvalues of the Jacobian matrix in that point have negative real parts.*

This corollary of Theorem 3.5 has a counterpart when one of the eigenvalues has a positive real part. Though the result is intuitively very reasonable, the proof is nontrivial. We only mention the theorem here. For a proof see e.g. [19] (Chapter 13).

Theorem 4.10. *A stationary point of an autonomous system is unstable if the Jacobian matrix in that point has at least one eigenvalue with a positive real part.*

If none of the eigenvalues of $\mathbf{J}(0)$ has a vanishing real part, the stationary point is called *hyperbolic*. For hyperbolic points the eigenvalues alone serve to determine its stability, as Corollary 4.9 and Theorem 4.10 show.

If some eigenvalues of $\mathbf{J}(0)$ have vanishing real parts and the others negative real parts, linearisation does not lead to a decisive insight about the stability of the stationary point. The following example illustrates this.

Example V.3.
Consider the system (cf. Example II.4)

$$\dot{x} = -x^p\ , \qquad p \text{ integer and } > 1\ .$$

It has the Jacobian 'matrix'

$$J(x) = -p\,x^{p-1}\ ,$$

and thus $J(0) = 0$. The solution of the ODE with $x(t_0) = x_0$ is

$$x(t) = x_0\left(1 - (1-p)\,x_0^{p-1}(t-t_0)\right)^{\frac{1}{1-p}}\ .$$

If p is odd, thus $p \geq 3$, we find that $x(t) \to 0$ if $t \to \infty$ for all $x_0 \neq 0$. In this case the origin is uniformly asymptotically stable. If p is even we find that for $x_0 < 0$ the solution has a finite existence interval, because

$$x(t) \to \infty \qquad \text{if } t \to t_0 + \frac{1}{1-p} x_0^{1-p} .$$

Hence the origin is unstable for p even.

For hyperbolic points an important result is known concerning stable and unstable manifolds. The *stable manifold* W^s of a stationary point $\hat{\mathbf{x}}$ of the autonomous system

$$(4.11) \qquad \begin{cases} \dot{\mathbf{x}} = \mathbf{f}(\mathbf{x}) , \\ \mathbf{x}(t_0) = \mathbf{x}_0 \end{cases}$$

is defined by

$$(4.12) \qquad W^s(\hat{\mathbf{x}}) := \{\mathbf{x}_0 \,|\, \text{solution } \mathbf{x} \text{ of } (4.11) \text{ approaches } \hat{\mathbf{x}} \text{ for } t \to \infty\} .$$

The *unstable manifold* $W^u(\hat{\mathbf{x}})$ is similarly defined but then for $t \to -\infty$.
The linearised version of (4.11) around $\hat{\mathbf{x}}$ is

$$(4.13) \qquad \dot{\mathbf{z}} = \mathbf{J}(\hat{\mathbf{x}})\,\mathbf{z} .$$

Also for (4.13) we can determine stable and unstable manifolds and we denote them by $E^s(\hat{\mathbf{x}})$ and $E^u(\hat{\mathbf{x}})$ respectively. All solutions of (4.13) together form the *tangent space* $\mathcal{T}(\hat{\mathbf{x}})$ at $\hat{\mathbf{x}}$. Because we can choose n linearly independent initial values for (4.13), the tangent space is n-dimensional. $E^s(\hat{\mathbf{x}})$ and $E^u(\hat{\mathbf{x}})$ are linear subspaces of $\mathcal{T}(\hat{\mathbf{x}})$. $E^s(\hat{\mathbf{x}})$ is the linear vector space spanned by the solutions of (4.13) that start in the directions of those eigenvectors of $\mathbf{J}(\hat{\mathbf{x}})$ that have eigenvalues with negative real parts; $E^u(\hat{\mathbf{x}})$ is spanned by the solutions of (4.13) that start in the directions of those eigenvectors of $\mathbf{J}(\hat{\mathbf{x}})$ that have eigenvalues with positive real parts. In the plane a hyperbolic point is a saddle point if neither $E^s(\hat{\mathbf{x}})$ nor $E^u(\hat{\mathbf{x}})$ is empty. One can show that the nonlinearity of (4.11) does not completely destroy the structure of the manifolds E^s and E^u of the linearised version (4.13). For a proof of the following theorem see e.g. [45] (Chapter 9).

Theorem 4.14 (Stable and unstable manifolds). *Let $\hat{\mathbf{x}}$ be a hyperbolic point of the autonomous system (4.11) with linearisation (4.13) around $\hat{\mathbf{x}}$. The stable manifold $E^s(\hat{\mathbf{x}})$ of (4.13) has the same dimension as the stable manifold $W^s(\hat{\mathbf{x}})$ of (4.11) and $E^s(\hat{\mathbf{x}})$ and $W^s(\hat{\mathbf{x}})$ are tangent at $\hat{\mathbf{x}}$. The same holds for the unstable manifolds $E^u(\hat{\mathbf{x}})$ and $W^u(\hat{\mathbf{x}})$.*

In practice one is often interested in the positions of W^s and W^u of (4.11). In nearly all cases they have to be determined numerically. Theorem 4.14 may then be of help. E^s and E^u are given by the eigenvectors of $\mathbf{J}(\hat{\mathbf{x}})$. To obtain W^s one calculates E^s and starts the integration of (4.11) very close to $\hat{\mathbf{x}}$ in the direction of E^s and with $t \to -\infty$. To obtain W^u the integration of (4.11) is performed in the direction of E^u and with $t \to +\infty$.

If $\hat{\mathbf{x}}$ is a stationary point of a non-autonomous equation

$$(4.15) \quad \dot{\mathbf{x}} = \mathbf{f}(t, \mathbf{x}) \; ,$$

with linearisation

$$(4.16) \quad \dot{\mathbf{z}} = \mathbf{J}(t, \hat{\mathbf{x}}) \, \mathbf{z} \; ,$$

then the stability of $\hat{\mathbf{x}}$ with respect to (4.15) and the stability of the origin with respect to (4.16) are generally not directly related. However, if $\mathbf{J}(t, \hat{\mathbf{x}})$ has the form

$$(4.17) \quad \mathbf{J}(t, \hat{\mathbf{x}}) = \mathbf{A} + \mathbf{B}(t) \; ,$$

with $\mathbf{A}$ constant we have the following useful theorem. For a proof see [38].

Theorem 4.18. *If the Jacobian matrix around the stationary point $\hat{\mathbf{x}}$ of a non-autonomous system can be decomposed as in (4.17) and $\mathbf{A}$ has only eigenvalues with negative real parts, while $\|\mathbf{B}(t)\| \to 0$ as $t \to \infty$, then $\hat{\mathbf{x}}$ is uniformly asymptotically stable.*

It is rather the exception than the rule that a non-autonomous system has a stationary point and we therefore do not pay further attention to it. Example V.1 is an illustration of such a system. In those cases linearisation does not help much, because, even in the linear case, no general theorems are available.

5. Nonlinear systems; Lyapunov functions

The linearisation procedure dealt with in §4 has the attractive property that its application is straightforward. However, the method does not allow us to conclude (in)stability in all cases. Here, we treat an alternative stability analysis developed by and named after Lyapunov. Sometimes it is also called the *direct method*. It may be applicable in cases in which linearisation does not yield results. Moreover, it may provide an estimate of the basin of attraction of a stable stationary point. The direct method has much in common with the theory of Hamiltonian systems, which will be considered in §XI.3. In Hamiltonian systems the concept of energy function plays a central rôle. In the direct method this rôle is played by the so-called *Lyapunov function*. A serious disadvantage of the method is that no systematic, general procedure is known to construct Lyapunov functions.

We only aim at giving a brief introduction here, and restrict ourselves to autonomous systems to start with. Extension to non-autonomous systems is possible, but not straightforward, see e.g. [77] (§10.5).

We first introduce some definitions. If $\mathbf{x}$ is a solution of the autonomous system

$$(5.1) \qquad \dot{\mathbf{x}} = \mathbf{f}(\mathbf{x})$$

and $V(\mathbf{x})$ a scalar, differentiable function defined on phase space, then the *total* or *orbital derivative* of $V(\mathbf{x})$ is defined by

$$(5.2) \qquad \mathcal{D}_{\mathbf{f}}\, V(\mathbf{x}) := \frac{d}{dt}\, V(\mathbf{x}) = \frac{dV}{d\mathbf{x}} \cdot \dot{\mathbf{x}} = \frac{dV}{d\mathbf{x}} \cdot \mathbf{f}\left(\mathbf{x}\right)$$

with $dV/d\mathbf{x}$ the gradient of V and $\cdot$ denoting the inner product. Equation (5.2) indicates that the time derivative of V along a solution of (5.1) is given by the projection of the gradient of V on the vector field.

Definition 5.3. *A function $V(\mathbf{x})$ is positive (semi-)definite around the origin if a neighbourhood $\mathcal{D}$ of the origin exists such that*

(i) $V(\mathbf{x})$ *is differentiable on $\mathcal{D}$.*

(ii) $V(0) = 0.$

(iii) $V(\mathbf{x}) > 0$ *(positive definite) or $V(\mathbf{x}) \geq 0$ (positive semi-definite) for all* $\mathbf{x} \in \mathcal{D},\ \mathbf{x} \neq 0$.

The definition of *negative (semi-)definiteness* is similar.

Definition 5.4. $V(\mathbf{x})$ *is called a Lyapunov function for a vector field $\mathbf{f}$, which has the origin as stationary point, if a bounded neighbourhood $\mathcal{D}$ of the origin exists such that*

(i) $V(\mathbf{x})$ *is positive definite on $\mathcal{D}$.*

(ii) $\mathcal{D}_{\mathbf{f}}\, V(\mathbf{x})$ *is negative semi-definite on $\mathcal{D}$* .

Definition 5.5. $V(\mathbf{x})$ *is a called a strong Lyapunov function if in Definition 5.4 (i) holds and (ii) is replaced by*

(ii) $\mathcal{D}_{\mathbf{f}}\, V(\mathbf{x})$ *is negative definite on $\mathcal{D}$* .

Note that the latter requirement implies that a strong Lyapunov function has no local maxima or minima in $\mathcal{D}$. The existence of a Lyapunov function makes it possible to show stability:

Theorem 5.6. *The origin is a (Lyapunov) stable, stationary point of (5.1), if a Lyapunov function for* **f** *exists.*

Proof: For a ball around the origin we shall use the notation

$$B(\varepsilon) := \{\mathbf{x} \,|\, \|\mathbf{x}\| < \varepsilon\} \,,$$

and for a sphere

$$S(\varepsilon) := \{\mathbf{x} \,|\, \|\mathbf{x}\| = \varepsilon\} \,.$$

In view of Definition 2.2 we have to show that for any $\varepsilon > 0$ a $\delta(\varepsilon)$ exists such that all solutions starting within $B(\delta)$ remain in $B(\varepsilon)$ for $t \geq t_0$. First we take the largest $\varepsilon_0 > 0$ such that $B(\varepsilon_0)$ and $S(\varepsilon_0)$ are fully contained in $\mathcal{D}$, and next we show that $\delta(\varepsilon)$ exists for all $\varepsilon \leq \varepsilon_0$.

The minimum of the Lyapunov function V on $S(\varepsilon_0)$ is denoted by

$$V_{\min} := \min_{\mathbf{x} \in S(\varepsilon_0)} V(\mathbf{x}) \,.$$

Because V is positive definite, we have $V_{\min} > 0$. Next, we consider the subset of $B(\varepsilon_0)$ at which V is smaller than $V_{\min}$:

$$\mathcal{D}(\varepsilon_0) := \{\mathbf{x} \,|\, \mathbf{x} \in B(\varepsilon_0), \; V(\mathbf{x}) < V_{\min}\} \,.$$

Solutions starting in $\mathcal{D}(\varepsilon_0)$ cannot leave it, because the value of V cannot increase along solutions, due to the fact that the orbital derivative of V is semi-negative:

$$V\Big(\mathbf{x}(t)\Big) = V\Big(\mathbf{x}(t_0)\Big) + \int_{t_0}^{t} \mathcal{D}_{\mathbf{f}} V\Big(\mathbf{x}(s)\Big) \, ds \leq V\Big(\mathbf{x}(t_0)\Big) \,.$$

Because $V(0) = 0$, the origin is in $\mathcal{D}(\varepsilon_0)$, and because of the continuity of $V(\mathbf{x})$, we can find a $\delta(\varepsilon_0) > 0$ such that

$$B\Big(\delta(\varepsilon_0)\Big) \subset \mathcal{D}(\varepsilon_0) \,.$$

The same arguments apply for all $\varepsilon < \varepsilon_0$.
For $\varepsilon > \varepsilon_0$ we can take $\delta(\varepsilon) = \delta(\varepsilon_0)$. $\qquad\qquad\square$

For asymptotic stability a similar result holds:

Theorem 5.7. *The origin is an asymptotically stable, stationary point of (5.1), if a strong Lyapunov function for* **f** *exists.*

Proof: Theorem 5.6 guarantees that the origin is stable. We have to show that a $\delta > 0$ exists such that all solutions starting in $B(\delta)$ converge to the origin for $t \to \infty$. We show that $\delta(\varepsilon_0)$, introduced in the proof of Theorem 5.6, satisfies. Let us assume that a solution exists which starts in $B(\delta(\varepsilon_0))$, but does not converge to the origin as $t \to \infty$. Then $0 < \delta < \delta(\varepsilon_0)$ exists such that this solution does not enter the ball $B(\delta)$. Let V be the strong Lyapunov function. Consider the maximum of the orbital derivative on the 'shell' between $S(\delta)$ and $S(\delta(\varepsilon_0))$:

$$M := \max_{\delta < \|\mathbf{x}\| < \delta(\varepsilon_0)} \mathcal{D}_{\mathbf{f}} V(\mathbf{x}) \ .$$

Because $\delta > 0$, we have $M < 0$. For solutions starting in this shell it holds that

$$V\Big(\mathbf{x}(t)\Big) \leq V\Big(\mathbf{x}(t_0)\Big) + M(t - t_0) \ .$$

So, for $t \to \infty$ the value of $V(\mathbf{x}(t))$ along those orbits continues to decrease and at some moment $\mathbf{x}(t)$ must enter $B(\delta)$. This contradicts the assumption. We thus conclude that all solutions starting in $B(\delta(\varepsilon_0))$ converge to the origin as $t \to \infty$.   □

The proofs of Theorems 5.6 and 5.7 show that a (strong) Lyapunov function contains information about a domain of attraction of the stationary point under consideration. This is expressed by the following theorem:

Theorem 5.8. *Let $V(\mathbf{x})$ be a strong Lyapunov function for (5.1) on a neighbourhood $\mathcal{D}$ of the origin. Let V_{min} be defined as in Theorem 5.6, and the subset $\mathcal{D}_0$ of $\mathcal{D}$ by*

(5.8a) $\mathcal{D}_0 := \{\mathbf{x} \mid \mathbf{x} \in \mathcal{D}, \ V(\mathbf{x}) < V_{min}\} \ .$

Then, $\mathcal{D}_0$ is a domain of attraction of the origin.

Proof: We can follow similar arguments to those in the proofs of Theorems 5.6 and 5.7. The difference is that in the present theorem we do not have to find spherical neighbourhoods of the origin. In fact, any solution starting in $\mathcal{D}_0$ cannot leave $\mathcal{D}_0$. The arguments in the proof of Theorem 5.7 guarantee that all solutions starting in $\mathcal{D}_0$ converge to the origin as $t \to \infty$.   □

We conclude this section with two remarks.

Remark 5.9. The shape of $\mathcal{D}_0$ depends, of course, on the Lyapunov function under consideration. Different Lyapunov functions may yield different domains of attraction and their union is an improved estimate of the basin of attraction.

Remark 5.10. In trying to construct a Lyapunov function, one usually starts with a positive definite test function containing some parameters, and tries to adjust the parameter values such that the orbital derivative $\mathcal{D}_f V(\mathbf{x})$ becomes negative (semi-)definite. A first choice might be

$$V(\mathbf{x}) = \mathbf{x}^T \mathbf{A} \mathbf{x}$$

with $\mathbf{A}$ a positive definite matrix.

Both remarks are illustrated in the following example. See also Example V.5.

Example V.4.
Consider the scalar equation

$$\dot{x} = -x^3 \ .$$

Linearisation yields a vanishing Jacobian matrix for $x = 0$. A strong Lyapunov function is e.g. $V_1(x) = \sin^2 x$. This function is positive definite if $|x| < \pi$. The corresponding orbital derivative

$$\mathcal{D}_f V_1(x) = -x^3 \sin 2x \ ,$$

is negative definite if $|x| < \frac{1}{2}\pi$. So, V_1 yields the interval $(-\pi/2, \pi/2)$ as a domain of attraction. Following Remark 5.10 we find that the function $V_2(x) = x^2$ is also a strong Lyapunov function with orbital derivative

$$\mathcal{D}_f V_2(x) = -2x^4 < 0 \ , \qquad x \neq 0 \ .$$

The origin is asymptotically stable and its basin of attraction appears to be the whole real axis.

6. Global analysis of the phase plane

In §3 we found that for linear systems all solutions have the same stability character and the analysis could therefore be restricted to the (in)stability of the origin. If the origin of a linear system is asymptotically stable, the domain of attraction is the whole phase space. For nonlinear systems the (in)stability of a solution has a local character. The basin of attraction of a solution will generally be a neighbourhood of this solution. In most cases the exact form of such a basin can be determined only numerically. The direct method may yield estimates for it. Analysis of the linearised problem does not provide any information on this point.

For autonomous systems of dimension ≤ 2 much insight can be gained by observing the direction of the vector field. For these so-called *planar systems* we mention the method of isoclines. Let the system be given by

(6.1)
$$\begin{cases} \dot{x}_1 = f_1(x_1, x_2) \\ \dot{x}_2 = f_2(x_1, x_2) \ . \end{cases}$$

The curves defined by $f_1 = c\, f_2$ with c some constant are called *isoclines*. The vector field has constant direction along isoclines. The *vertical isocline* $f_1 = 0$ is intersected vertically by solutions of (6.1), while the *horizontal isocline* $f_2 = 0$ is intersected horizontally. The intersections of the vertical and horizontal isoclines are the stationary points. The isoclines divide the phase plane into parts in which the signs of $\dot{x}_1$ and $\dot{x}_2$ are fixed. Consequently, also the directions of the solutions are more or less fixed: the vector field in such a part of the plane points in the north-east, north-west, south-west, or south-east direction.

In the following example we illustrate several elements of the stability analysis in this and the preceding sections. The system under consideration is a Hamiltonian system. The general theory of such systems is dealt with in §XI.3.

Example V.5.
Consider the planar, autonomous system

$$\dot{x}_1 = x_2^2 - 1$$
$$\dot{x}_2 = x_1 \ .$$

The Jacobian matrix depends only on x_2:

$$\mathbf{J}(x_2) = \begin{bmatrix} 0 & 2x_2 \\ 1 & 0 \end{bmatrix} \ .$$

The stationary points are $(0, \pm 1)^T$. In $(0, 1)^T$ the eigenvalues of $\mathbf{J}$ are $\lambda_\pm = \pm\sqrt{2}$ with eigenvectors

$$\mathbf{v}_\pm = \begin{bmatrix} \pm 1 \\ \frac{1}{2}\sqrt{2} \end{bmatrix} \ .$$

The situation is sketched in Fig. V.2. In $(0, -1)^T$ the eigenvalues of $\mathbf{J}$ are $\pm i\,\sqrt{2}$ with complex-valued eigenvectors. So, we conclude that after linearisation $(0, 1)^T$ is a saddle point, and $(0, -1)^T$ a centre. Theorem 4.10 states that $(0, 1)^T$ is also an unstable stationary point of the full problem. For $(0, -1)^T$ we cannot draw conclusions yet, because a centre of the linearised system could correspond to a stable spiral, a centre, or an unstable spiral of the nonlinear system.

The stable and unstable manifolds E^s and E^u of the saddle point in the linearised problem are the straight lines through $\mathbf{v}_-$ and $\mathbf{v}_+$ respectively. Theorem 4.14 states that the stable and unstable manifolds W^s and W^u

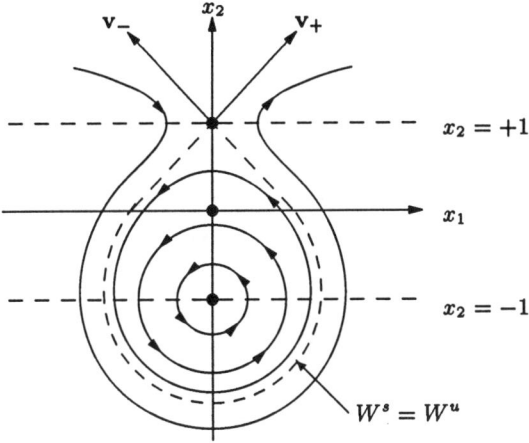

Figure V.2

are curves, which are tangent to v_- and v_+ respectively. The position of W^s and W^u can be roughly determined using the method of isoclines. The horizontal isocline is the x_2-axis, and the vertical isocline consists of the lines $x_2 = \pm 1$. The isoclines divide the plane into 6 parts, in which the direction of the velocity of the solutions is roughly known. Let us follow a solution starting at

$$ x_0 = \begin{bmatrix} 0 \\ 1 \end{bmatrix} - \varepsilon \, v_+ \; , $$

with $0 < \varepsilon \ll 1$. This solution will move in the south-west direction, cross the $x_2 = -1$ isocline vertically, turn around $(0,-1)^T$, and cross the $x_1 = 0$ isocline horizontally. The rest of the orbit follows from reflection with respect to the x_2-axis, because this is a symmetry axis of the system. The smaller ε, the more the solution observed will be close to the unstable manifold of the saddle point. We conclude that the stable and unstable manifolds W^s and W^u coincide as for the parts below the saddle point. Above the saddle point they are different and extend to infinity.

Because the (un)stable manifolds enclose the stationary point $(0,-1)^T$, we find that the parts of W^s and W^u below $(0,1)$ form the boundary of a neighbourhood of $(0,-1)^T$, from which solutions cannot escape. However, this insight on its own does not imply that $(0,-1)^T$ is Lyapunov stable: orbits starting close to $(0, -1)^T$ might leave it. Still, we can prove stability, because the function

$$V(x_1, x_2) = -\tfrac{1}{3} x_2^3 + x_2 + \tfrac{1}{2} x_1^2 + \tfrac{2}{3}$$

is a Lyapunov function around $(0, -1)^T$. For all points $(x_1, x_2)^T$ we have

$$\mathcal{D}_f V(x_1, x_2) = 0 .$$

This implies that solutions follow level lines of $V(x_1, x_2)$. Furthermore, $V(0, -1) = 0$ and V is positive in a neighbourhood of $(0, -1)^T$. Application of Theorem 5.6 yields that $(0, -1)^T$ is (Lyapunov) stable. In fact, the parts of W^s and W^u below $(0, 1)^T$ are a level line of V, and given by the equation

$$V(x_1, x_2) = V(0, 1) = \tfrac{4}{3} .$$

The following theorem enables us to draw conclusions about asymptotic stability only by observing the direction of the vector field in the vicinity of a stationary point in the plane. The extension of this theorem to higher dimensions is straightforward, but less useful in practice.

A closed curve in the plane is called a *Jordan curve*, if it does not intersect itself. The interior of such a curve is a bounded, open set.

Theorem 6.2 (Contracting neighbourhoods). *The stationary point $\mathbf{x}_0$ of (6.1) is uniformly asymptotically stable, if a family $C(s)$ of Jordan curves with s in some interval I exists such that*

(i)   *$C(s)$ depends continuously on s.*
(ii)   *$\mathbf{x}_0$ is in the interior of all $C(s)$, $s \in I$.*
(iii) *$C(s)$ and $C(\hat{s})$ do not intersect if $s \neq \hat{s}$.*
(iv)   *$\bigcap_{s \in I} O(s) = \mathbf{x}_0$, where $O(s)$ is defined as the interior of $C(s)$.*
(v)   *the vector field $\mathbf{f}$ points on $C(s)$ into $O(s)$.*

Proof: The idea of the theorem is easily understood from Fig. V.3. Without loss of generality we may assume $I = (0, 1]$ with $C(1)$ enclosing all other curves. Let us take an arbitrary initial value $\mathbf{x}_1 \in O(s = 1)$. Because of (i) we can find an $s_1 < 1$ such that $\mathbf{x}_1 \in C(s_1)$. If time evolves, the solution $\mathbf{x}$ of (6.1), starting at $\mathbf{x}_1$, enters the interior $O(s_1)$ because of (v). So, a time $t_2 > t_1$ exists such that $\mathbf{x}_2 := \mathbf{x}(t_2) \in O(s_1)$. Next, we can find an $s_2 < s_1$ such that $\mathbf{x}_2 \in C(s_2)$. Repeating this argument we obtain a series $\mathbf{x}_n := \mathbf{x}(t_n)$ with $\mathbf{x}_n \in O(s_{n-1})$ and $s_n < s_{n-1}$, $n = 1, 2, \ldots$. Because of (iv), we thus have $\mathbf{x}_n \to \mathbf{x}_0$, $n \to \infty$ and so $\mathbf{x}(t) \to \mathbf{x}_0$, $t \to \infty$. We thus find that $C(1)$ is a domain of attraction of $\mathbf{x}_0$. □

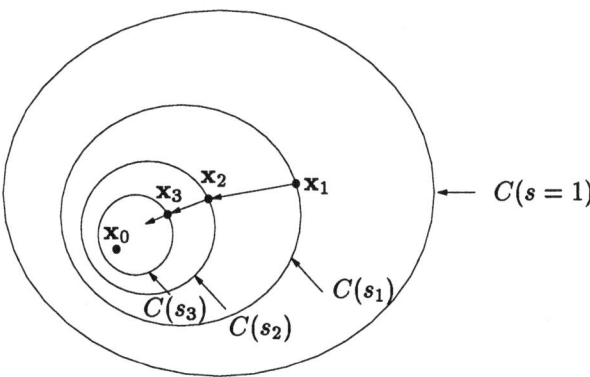

Figure V.3

7. Periodic ODE

For periodic solutions the notion of stability is somewhat different from the definitions given above. We recall that periodic solutions are solutions of autonomous or periodic systems with $\mathbf{x}(t + T) = \mathbf{x}(t)$ for some period T. A periodic solution follows a closed curve in phase space. For autonomous systems this curve does not intersect itself, whereas for periodic systems self-intersection may happen. Let us denote by C the set of points visited by the periodic solution $\mathbf{x}$ during one period:

$$C := \{\mathbf{x}(s) \mid t \leq s < t + T\}$$

for arbitrary $t \geq t_0$. It may happen that a solution starting in the vicinity of C remains close to C as $t \to \infty$, but that it is still not (Lyapunov) stable, because a *phase shift* occurs between the periodic solution and the perturbed one. For this reason we introduce a more appropriate notion of stability, which is often referred to as *Poincaré* or *orbital stability*. The distance from a point $\mathbf{x}_0$ to C is defined as

$$d(x_0, C) := \min_{\mathbf{x} \in C} \|\mathbf{x}_0 - \mathbf{x}\| .$$

Definition 7.1. *A periodic solution with closed orbit C in phase space is orbitally stable, if for each $\varepsilon > 0$ and each $t_1 \geq t_0$ a $\delta(\varepsilon, t_1)$ exists such that a solution $\mathbf{y}$ of (2.1), starting at t_1 in the vicinity of C, remains in the vicinity of C:*

$$\{d(\mathbf{y}(t_1), C) < \delta(\varepsilon, t_1)\} \Rightarrow \{d(\mathbf{y}(t), C) < \varepsilon \text{ for all } t \geq t_1\} .$$

The orbital stability is *uniform*, if δ in Definition 7.1 does not depend on t_1. A stronger form of orbital stability is:

Definition 7.2. *A periodic solution with a closed orbit C in phase space is asymptotically orbitally stable, if it is orbitally stable and if for every $t_1 \geq t_0$ a $\delta(t_1)$ exists such that a solution* $\mathbf{y}$, *starting at t_1 in the vicinity of C, approaches C for $t \to \infty$:*

$$\{d(\mathbf{y}(t_1), C) < \delta(t_1)\} \Rightarrow \{\lim_{t \to \infty} d(\mathbf{y}(t), C) = 0\} \ .$$

The asymptotic orbital stability is *uniform*, if δ in Definition 7.2 does not depend on t_1.

Let $\hat{\mathbf{x}}(t)$ be a periodic solution of

(7.3) $\dot{\mathbf{x}} = \mathbf{f}(t, \mathbf{x})$

with period $T > 0$:

(7.4) $\hat{\mathbf{x}}(t) = \hat{\mathbf{x}}(t + T)$.

The existence of a periodic solution implies that the vector field $\mathbf{f}$ is either autonomous or periodic with period T. Following the same steps as in (4.1)–(4.6) we obtain the linearised equation with respect to $\hat{\mathbf{x}}(t)$:

(7.5) $\dot{\mathbf{z}} = \mathbf{J}\left(t, \hat{\mathbf{x}}(t)\right) \mathbf{z}$,

with the Jacobian matrix $\mathbf{J}(t, \hat{\mathbf{x}}(t))$ being T-periodic. The general solution of the periodic linear system (7.5) has been considered in §IV.5 and §V.3. In the latter section it is shown that periodic linear systems are very similar to autonomous linear systems as far as instability is concerned: instead of eigenvalues of a constant matrix one obtains the information from the characteristic exponents of a periodic matrix. In view of this we can immediately formulate the counterpart of Corollary 4.9 for periodic solutions:

Corollary 7.6. *A periodic solution of a periodic system is uniformly asymptotically stable, if all characteristic exponents of the linearised system around that solution have negative real parts.*

It should be realised that Corollary 7.6 is not relevant for autonomous systems. Asymptotic stability cannot occur in these systems. This can be understood as follows. Let C be the closed curve in phase space of the periodic solution. If we choose two initial values on C, the resulting solutions follow the same curve C, but with a constant time delay. This is a direct consequence of the shift property (I.3.4) for autonomous systems. Another way of looking at this property is the following. If $\hat{\mathbf{x}}$ is a solution of the autonomous equation

(7.7) $\dot{\mathbf{x}} = \mathbf{f}(\mathbf{x})$,

then its time derivative is a solution of the linearised equation

(7.8) $\dot{\mathbf{z}} = \mathbf{J}\big(\hat{\mathbf{x}}(t)\big)\,\mathbf{z}$,

as is easily checked by differentiation of (7.7). If $\hat{\mathbf{x}}$ is periodic, then this also holds for its time derivative, while the periods are the same. The solutions of the homogeneous linear periodic system (7.8) are explicitly given in (3.6). From that expression we may directly conclude when (7.8) has a periodic solution. This is the case if all characteristic exponents have negative real parts except for one or more exponents that have vanishing real parts and Jordan blocks of size one. The essential point is that a periodic solution of an autonomous system always corresponds to at least one of the characteristic exponents having a vanishing real part. So, asymptotic stability is excluded. Instead of asymptotic stability, orbital stability may occur as expressed in the following theorem:

Theorem 7.9. *A periodic solution $\hat{\mathbf{x}}$ of an autonomous system is asymptotically orbitally stable, if all characteristic exponents of the linearised system around that solution except one have negative real parts.*

Solutions starting in some neighbourhood of $\hat{\mathbf{x}}(t_0)$, with t_0 arbitrary, get for $t \to \infty$ a constant phase difference with the periodic solution starting in $\hat{\mathbf{x}}(t_0)$ itself. We illustrate this in the following example.

Example V.6.
We consider a system in the plane. In terms of polar coordinates (r, φ) the equations read as

$$\begin{cases} \dot{r} = 1 - r \\ \dot{\varphi} = r \ . \end{cases}$$

This system has the periodic solution $r = 1$, $\varphi = \varphi_0 + (t - t_0)$, corresponding to the unit circle. For convenience we choose $\varphi_0 = 0$ and $t_0 = 0$ in the following. The first equation can be solved by separation of variables. If we write the initial value r_0 as $r_0 = 1 + \delta$, with $\delta > -1$, the solution $r(t)$ is given by

$$r(t) = 1 + \delta\,e^{-t} \ .$$

From this it is clear that every solution starting off the unit circle will converge to it as $t \to \infty$. We can find $\varphi(t)$ from integrating the second equation:

$$\varphi(t) = t + \delta(1 - e^{-t}) \ .$$

From this we observe that the two solutions starting at $(r_0, \varphi_0) := (1, 0)$ and $(r_0, \varphi_0) := (1 + \delta, 0)$ both follow the unit circle as $t \to \infty$, but with a constant phase shift δ. Theorem 7.9 predicts that this should happen, provided that one characteristic exponent of the periodic solution would have a vanishing real part and the other a negative real part. The present system is autonomous, so the former condition is fulfilled. Both conditions can be checked explicitly. The system under consideration is linear, inhomogeneous, and has constant coefficients. In the standard notation $\dot{z} = A z + b$ with $z := (r, \varphi)^T$ and the matrix A given by

$$A := \begin{bmatrix} -1 & 0 \\ 1 & 0 \end{bmatrix} \ .$$

From §3 we know that the stability of this system is determined by the stability of the zero solution of the homogeneous problem $\dot{z} = A z$. Because the system is linear the characteristic exponents are given by the eigenvalues of A. These are $\lambda_1 = 0$ and $\lambda_2 = -1$. The conditions of Theorem 7.9 are thus satisfied.

8. Stability of Δ-equations

The preceding theory for continuous systems has many analogues for discrete problems (cf. [2, 52]). In particular for Δ-equations arising from discretised ODE the two main perturbation sources, viz. rounding errors and discretisation errors, make the concept of total stability meaningful. Consider the IVP for a k-step (vector) Δ-equation

(8.1) $\qquad \begin{cases} x_{i+1} = F_i(x_i, x_{i-1}, \dots, x_{i-k+1}) \ , & i \geq k \ , \\ x_0, \dots, x_{k-1} \quad \text{given initial values} \ , \end{cases}$

and its perturbed version

(8.2) $\qquad \begin{cases} y_{i+1} = F_i(y_i, y_{i-1}, \dots, y_{i-k+1}) + e_i \ , & i \geq K + k \ , \\ y_j = x_j + e_j \ , & j = K, \dots, K + k - 1 \ , \end{cases}$

where $\{e_i\}$ are some (small) vectors and K is some positive integer. Assume both (8.1) and (8.2) have solutions, the latter for all choices of $\{e_i\}$ with $\|e_i\|$ sufficiently small. Then $\{x_i\}$ is called *uniformly totally stable*, if for all $\varepsilon > 0$ and all $K > 0$ a $\delta > 0$ exists such that

(8.3) $\qquad \max_{i \geq K} \|e_i\| < \delta \ \Rightarrow \ \max_{i \geq K} \|y_i - x_i\| \leq \varepsilon \ .$

We did not need this concept in Chapter III, as we were only interested in finite time intervals there. For relatively long time intervals, as will be encountered in Chapter VIII, total stability is an important numerical concept too. It is also interesting to note here that this concept is *essential* for understanding growth behaviour of basis solutions of boundary value problems; these problems are dealt with in Chapter IX.

As in §2 we can give definitions of (asymptotic) stability of solutions of (8.1); since we can reduce multistep Δ-equations to one-step ones (cf. (I.5.11)) we restrict ourselves to the latter case.

Definition 8.4. *Let $\hat{x}$ be a stationary solution of the Δ-equation*

(8.4a) $x_{i+1} = f_i(x_i)$.

If for every $\varepsilon > 0$ and $i_0 > 0$ a $\delta(\varepsilon, i_0)$ exists such that

$$\{\|x_{i_0}\| < \delta(\varepsilon, i_0)\} \Rightarrow \{\|x_i - \hat{x}\| < \varepsilon \text{ for all } i \geq i_0\} ,$$

then $\hat{x}$ is Lyapunov stable.
If $\hat{x}$ is Lyapunov stable, and for every $i_0 > 0$, a $\delta(i_0)$ exists such that

$$\{\|x_{i_0}\| < \delta(i_0)\} \Rightarrow \{\|x_i - \hat{x}\| \to 0 \text{ if } i \to \infty\} ,$$

then $\hat{x}$ is called asymptotically stable.

If the system is autonomous we have in (8.4a) $f_i \equiv f$ for all i. The iteration process is then also referred to as *successive substitution*. For such systems a necessary condition for asymptotic stability is

(8.5) $\rho(J(\hat{x})) < 1$,

where J is the Jacobian matrix of f and ρ is the spectral radius (see Appendix B).

If f_i in (8.4a) is linear and homogeneous, we have

(8.6) $x_{i+1} = A_i x_i$.

Obviously stability then implies boundedness of $\{\|\prod_{j=0}^{i} A_j\|\}$ for all $i > 0$, whereas asymptotic stability would require $\lim_{i \to \infty} \|\prod_{j=0}^{i} A_j\| = 0$. If $A_i \equiv A$ for some A, i.e. the autonomous case, all stability properties reduce to eigenvalue properties. If $\rho(A) < 1$ the zero solution is asymptotically stable. Analogously to Theorem 3.5, we see that stability holds if $\rho(A) \leq 1$, with the eigenvalues of modulus 1 having Jordan blocks of size one.

We conclude this section with a remark: in this book we almost exclusively encounter Δ-equations resulting from *discretised* ODE. Only in the next chapter do we encounter a few exceptions to this. Hence, as is customary in numerical analysis, one tries to find properties of the (numerical) *method*, rather than of the specific Δ-equation which is the result of applying it to an ODE. One such property is 'stability'. In order to let this be a meaningful concept one assumes that the method is applied to a suitable class of problems, like constant coefficient ODE or ones with a polynomial scalar field. In both Chapter VII and Chapter VIII we will therefore encounter stability notions of a method.

Exercises Chapter V

1. Determine the stability properties of the planar, linear systems as classified in §IV.7.

2. Show that for the linear systems uniformly asymptotic stability implies uniform total stability.
   Hint: Use the fact that the solution of $\dot{x} = A(t)\,x + b(t)$ can be written as in (IV.3.5) in combination with the result of Theorem 3.12.

3. Show that the solution of the following IVP is totally stable:

$$\dot{x} = \begin{bmatrix} -3 & 1 \\ 1 & -3 \end{bmatrix} x + \begin{bmatrix} 1 \\ 5 \end{bmatrix}$$

$$x(0) = \begin{bmatrix} 1 \\ 2 \end{bmatrix} .$$

4. Determine the stability properties of the linear systems

   a) $\dot{x} = \begin{bmatrix} 0 & \sin t \\ 0 & 0 \end{bmatrix} x .$

   b) $\dot{x} = \begin{bmatrix} 0 & 0 \\ 1 & 1 \end{bmatrix} x .$

   c) $\dot{x} = \begin{bmatrix} -2 & 1 \\ 1 & -2 \end{bmatrix} x + \begin{bmatrix} 1 \\ -2 \end{bmatrix} e^t .$

   d) $\ddot{x} + e^{-t} \dot{x} + x = 0 .$

5. Determine the stability properties of the stationary points of

$$\ddot{x} = x^2 + a , \qquad a \in \mathbb{R} ,$$

   and sketch the flow in phase space.

6. Show that the origin is an unstable stationary point of

$$\ddot{x} - \dot{x}^2 \operatorname{sign}(\dot{x}) + x = 0 .$$

7. Find Lyapunov functions around the origin for the vector fields

a) $f(x) = -x$.

b) $f(x) = -x(1-x)$.

c) $f(x) = A x$ with $A := \begin{bmatrix} 0 & 1 \\ -1 & -a \end{bmatrix}$, $a \geq 0$.

8. Find a Lyapunov function for the zero solution of the damped pendulum equation

$$\ddot{x} + c \dot{x} + \sin x = 0 , \qquad c \geq 0 .$$

9. Consider the system $(a, b > 0)$

$$\dot{x}_1 = x_1(x_2 - b) ,$$
$$\dot{x}_2 = x_2(x_1 - a) .$$

a) Find a strong Lyapunov function around the origin.

b) Show that all solutions starting within the ellipse

$$\left(\frac{x_1}{a}\right)^2 + \left(\frac{x_2}{b}\right)^2 = 1$$

approach the origin for $t \to \infty$.

10. Consider the ODE

$$\dot{x} = (t-2)(x^2 - 4) , \qquad t \geq 0 .$$

Find the initial values x_0 for which the solution exists for $t \to \infty$ and determine in those cases the value of $x(t)$ for $t \to \infty$.
Hint: Determine first the isoclines $\dot{x} = c$ with c a constant.

11. Consider the difference equation

$$x_{i+1} = x_i \sqrt{1 - x_i} .$$

a) Show that this equation has two stationary points.

b) Show that the origin is asymptotically stable.

12. Consider the linear difference equation

$$x_{i+1} = A x_i + b ,$$

with $|\lambda_j| < 1$ for all eigenvalues λ_j of A. Show that a stationary point exists, and that it is totally stable.

VI

Chaotic Systems

In this chapter we deal with systems which have bounded solutions for $t \to \infty$, but do not converge to stationary or (quasi-)periodic solutions. In §1 we introduce the concept of sensitive dependence of solutions on the initial conditions. Due to this phenomenon chaotic systems are not predictable in the long run, although they are deterministic. A convenient measure for the sensitivity is provided by the concept of Lyapunov exponents given in §3. In §4 strange and chaotic attractors are considered. There we also discuss why numerically obtained solutions, which unavoidably contain errors, still provide useful information about the properties of these attractors. In §5 the concepts of generalised and fractal dimension are introduced. It is shown that attractors can be characterised by a variety of dimension definitions. Each of these definitions is concerned with a different aspect of the object. The reconstruction of chaotic attractors from experimental data is the subject of §6. The reconstruction technique makes it possible to predict chaotic time series, although within a certain horizon only. The prediction algorithms based on insights from chaos theory are presented in §7.

1. Introduction

In Chapter V we analysed solutions remaining bounded for $t \to \infty$ and converging to point or periodic attractors. Until fairly recently other types of globally bounded solutions were assumed to be untreatable because of lack of structure. That is why the corresponding systems are usually referred to as being *chaotic*, but this term is somewhat misleading. Chaos theory is based on the insight that the time evolution of those systems in phase space still possesses a definite structure. In this chapter we shall introduce the main concepts of this theory and discuss their consequences for both the analytical and the numerical treatment of ODE. The history of chaos theory is described in [30]. In literature on chaos theory systems described by ODE are commonly referred to as *dynamical systems*, and the time evolution of their solutions is usually called the *dynamics* of the system. Where appropriate, we shall follow this terminology. In this introduction we restrict ourselves to some general aspects of chaotic systems, and introduce the concepts mainly through examples. For a more practical application we may refer to §XII.10, where a dripping faucet is analysed.

In Chapter IV we have shown that the solutions of linear, autonomous ODE either become unbounded or converge to point or periodic solutions. Chaotic behaviour can therefore be found in the solutions of nonlinear ODE only. As mentioned in §I.3, every non-autonomous system can be transformed into an autonomous one by increasing the order; hence we shall consider autonomous systems only. One should be aware, however, that by this augmentation linear systems usually become nonlinear, and may thus show chaotic behaviour. A typical property of chaotic systems is their unpredictability in the long run. At first sight this may seem contradictory to the achievements of Chapter II. There we found that, under rather mild conditions, every ODE has a unique solution which is completely determined by the initial value. So, whenever information is given at some time, the time evolution of the system is determined, in principle. Despite this property, chaotic systems are intrinsically unpredictable. We shall return to this paradox in §7.

Throughout this chapter the following maps will be used over and again.

Example VI.1.

(i)   The *tent map* is the one-dimensional Δ-equation

$$x_{i+1} = f(x_i) := 2\mu \begin{cases} x_i & , \text{ if } 0 \le x_i < \frac{1}{2} \\ 1 - x_i & , \text{ if } \frac{1}{2} \le x_i \le 1 . \end{cases}$$

The *iteration function* $f(x)$ is plotted in Fig. VI.1 (solid line) for $\mu = 1$. For $0 \le \mu \le \frac{1}{2}$ the tent map has one attracting stationary point, the origin. For $\mu = \frac{1}{2}$ the interval $[0,1]$ is mapped on $[0,\frac{1}{2}]$, and all points in the latter interval are stationary. For $\mu > \frac{1}{2}$ a second stationary point x^* is found, which is given by

$$x^* := \frac{2\mu}{1 + 2\mu} .$$

Both stationary points are unstable, since $|df/dx| > 1$ at these points. Iterates of the tent map are given in Fig. VI.2 as a function of μ. The initial value of the iterations is chosen arbitrarily and the first 200 iterates have been discarded. The orbits converge to subsets of $[0,1]$ which contain infinitely many points.

(ii)   The *doubling map* is the one-dimensional Δ-equation

$$x_{i+1} = f(x_i) = 2x_i \bmod(1) .$$

The iteration function $f(x)$ is plotted in Fig. VI.1 (dotted line). An appropriate interpretation of this map can be obtained from identifying x with an angle φ through $\varphi = 2\pi x$. Then the modulo term comes in naturally.

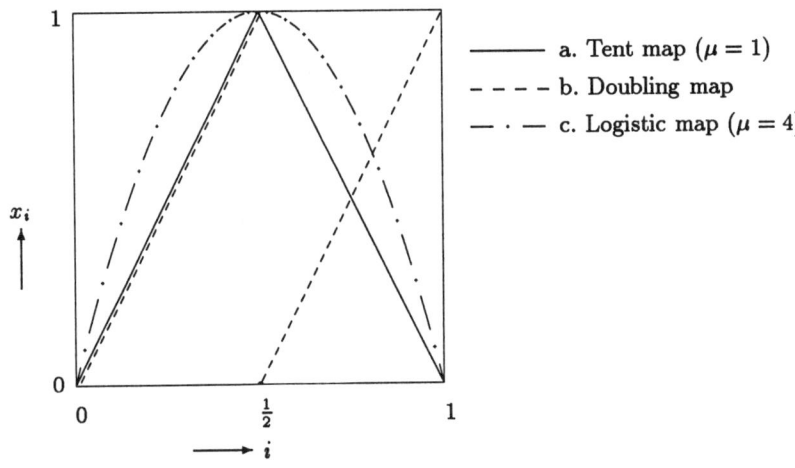

Figure VI.1

(iii) The *logistic map* is the one-dimensional Δ-equation

$$x_{i+1} = f(x_i) = \mu x_i(1 - x_i) , \qquad 0 \leq \mu \leq 4 .$$

The iteration function $f(x)$ is plotted in Fig. VI.1 (dotted broken line) for $\mu = 4$. Iterates of the logistic map are given in Fig. VI.3 as a function of μ. The initial value is chosen arbitrarily and the first 150 iterates have been discarded. For $\mu \leq 3$ a period-1 orbit is found consisting of a stationary point of the map. Increasing μ leads to a period-2 orbit. At $\mu_2 = 3.44949...$ a transition to a period-4 mode is observed, which is followed by a transition to a period-8 mode at $\mu_3 = 3.54409...$. The μ_i values accumulate at $\mu_\infty = 3.56994...$. This convergence is geometric. The constant

$$\delta := \frac{\mu_k - \mu_{k-1}}{\mu_{k+1} - \mu_k} = 4.669202...$$

is called the *Feigenbaum* constant after its discoverer [27]. This constant is called *universal*, because it does not depend on the details of the logistic map, but only on the fact that its iteration function has a quadratic maximum. Many other universal properties of this kind of maps are found. The *period doubling* phenomenon is one of the best known *routes to chaos*. For $\mu > \mu_\infty$ the aperiodic orbits converge to subsets of $[0, 1]$ containing infinitely many points. For some windows, however, again periodic orbits and period doubling are observed.

(iv) The *Henon map* is the two-dimensional Δ-equation

$$x_{i+1} = 1 - a\, x_i^2 + y_i$$
$$y_{i+1} = b\, x_i .$$

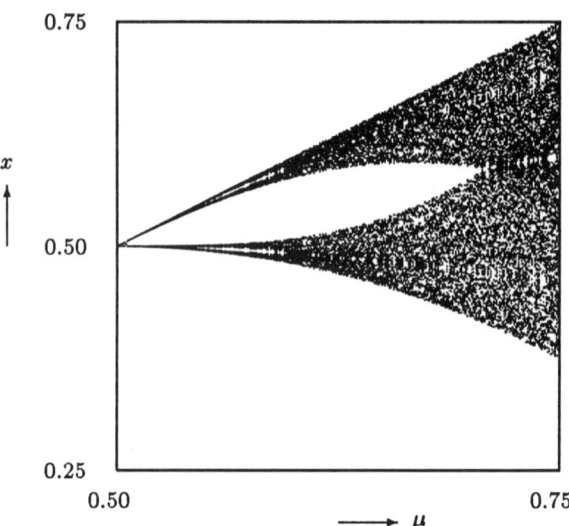

Figure VI.2   Iterates of the tent map as a function of μ. The first 200 iterates
have been discarded.

If $b = 0.3$, and a is varied, a period doubling cascade is observed
starting at $a = 0.3675$ with a transition to chaos at $a = 1.0580...$
(see [23]). A plot of the iterates for parameter values $a = 1.4$,
$b = 0.3$ is given in Fig. VI.4 and Fig. VI.5, where the initial 100
iterates have been discarded.

(v)   The *Lorenz equations*

$$\begin{aligned}
\dot{x} &= \sigma(y - x) \\
\dot{y} &= -xz + rx - y \\
\dot{z} &= xy - bz
\end{aligned}$$

describe a few characteristic features of the climate (see [71]). In this
model the spatial dependence is replaced by a description in terms
of 'modes', i.e. general types of weather. For the parameter values
$\sigma = 10$, $b = 8/3$, and $r > 24.74$ chaotic solutions are obtained. The
numerical integration of the Lorenz equations is performed with a
Runge–Kutta procedure (see §III.1). A projection of the limit set
onto the (x, z) plane is given in Fig. VI.6. For obvious reasons the
attractor is called the *butterfly of Lorenz*.

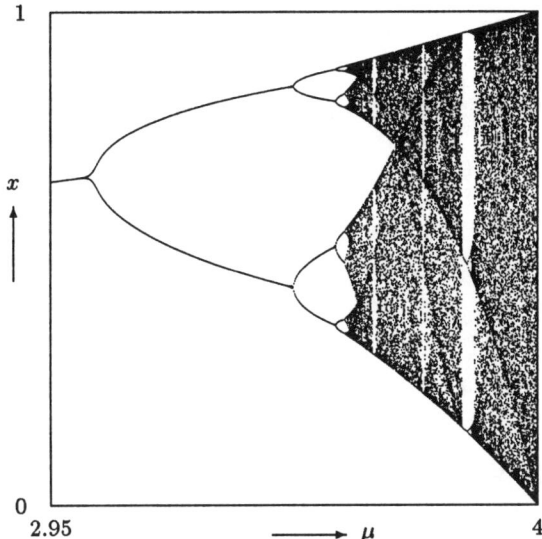

Figure VI.3   Iterates of the logistic map as a function of μ. The first 150 iterates
have been discarded.

2. Local divergence

The examples in §1 teach us that small changes in the initial conditions of ODE
may give rise to dramatic changes in the solutions. We shall formalise this in
the next section. In this section we introduce the notion of *local divergence*;
this is of great use in understanding the phenomenon of *sensitive dependence
on the initial conditions*, which is characteristic of chaotic systems and leads
to the unpredictability of these systems.

Definition 2.1. *The system*

$$(2.1a) \qquad \dot{\mathbf{x}} = \mathbf{f}(\mathbf{x})$$

is said to be locally divergent at $\mathbf{x}_0$*, if the solution starting at* $\mathbf{x}_0$ *and the solution starting at an arbitrary point in the vicinity of* $\mathbf{x}_0$ *diverge exponentially.*

To complete this definition we have to specify the notions 'arbitrary point'
and 'exponential divergence'. To that end we linearise around $\mathbf{x}_0$, obtaining
the equation

$$(2.2) \qquad \dot{\mathbf{x}} = \mathbf{J}(\mathbf{x}_0)\,\mathbf{x}$$

with $\mathbf{J}$ the Jacobian matrix of $\mathbf{f}$. Let $\mathbf{J}(\mathbf{x}_0)$ have eigenvalues λ_j and
corresponding eigenvectors $\mathbf{v}_j$ for $j = 1, ..., n$. For convenience we assume

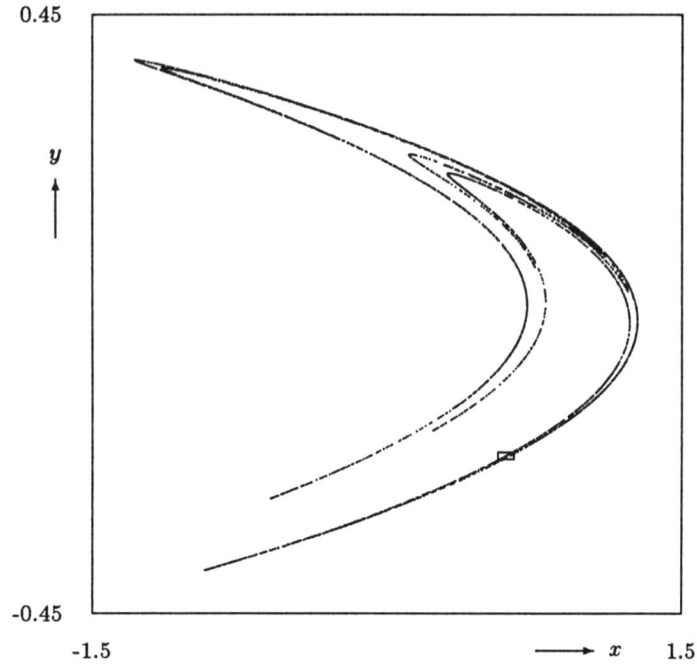

0.45

y

-0.45

-1.5 x 1.5

Figure VI.4   Iterates of the Henon map with $a = 1.4$ and $b = 0.3$. The first 100 iterates have been discarded.

that all λ_j are different, so that the $\mathbf{v}_j$ are linearly independent. This is not restrictive for the argument below. The solutions of (2.2), with initial value $\mathbf{x}_0$, lie in the so-called *tangent space* $T_{\mathbf{x}_0}$. This linear vector space has x_0 as origin. Since (2.2) is linear, we can describe the growth behaviour of its solutions quite well (cf. Chapter IV). Since any initial condition $\mathbf{x}_1$ in the vicinity of $\mathbf{x}_0$ can be written as

$$(2.3) \qquad \mathbf{x}_1 = \mathbf{x}_0 + \sum_{j=1}^{n} c_j \mathbf{v}_j \ ,$$

the solution $\mathbf{x}_1(t)$ of (2.2) with initial condition $\mathbf{x}_1$ is given by

$$(2.4) \qquad \mathbf{x}_1(t) = \mathbf{x}_0(t) + \sum_{j} c_j e^{\lambda_j t} \mathbf{v}_j \ ,$$

where $\mathbf{x}_0(t)$ is the solution of (2.2) with initial value $\mathbf{x}_0$. The solutions $\mathbf{x}_0(t)$ and $\mathbf{x}_1(t)$ thus diverge exponentially if *at least one of the λ_j has a positive*

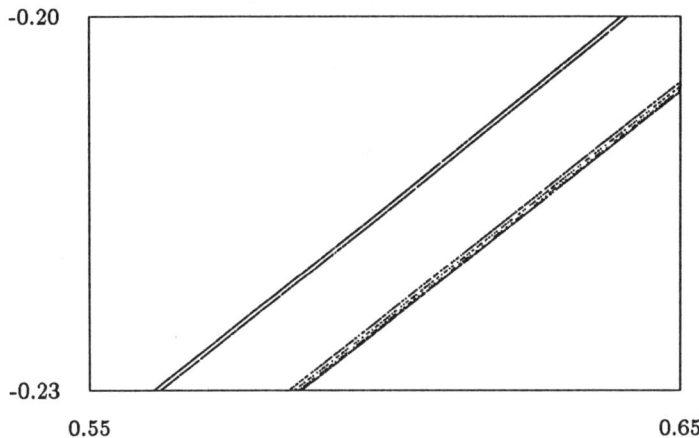

Figure VI.5 Magnification of the box indicated in **Figure VI.4**.

real part and if the vector $\mathbf{x}_1$ is *arbitrary* in the sense that none of the c_j is zero. From a numerical point of view the latter condition is quite natural as $\mathbf{x}$ is generated from $\mathbf{x}_0$ including fairly random rounding errors. The *solutions* $\mathbf{x}_0(t)$ and $\mathbf{x}_1(t)$ of (2.2) are tangent to the solutions of the original nonlinear equation (2.1a) with initial values $\mathbf{x}_0$ and $\mathbf{x}_1$. The solutions of the nonlinear equation are then said to *diverge exponentially*, too. However, we should real-ise that the latter divergence occurs in a restricted time interval after t_0 only. The divergence cannot be maintained for $t \to \infty$, because the solutions are assumed to be bounded for $t \to \infty$.

Definition 2.1 and the subsequent considerations are given for continuous time systems. For discrete time systems, i.e. maps described by Δ-equations, a completely similar derivation can be given. The analogue of (2.4) is in that case

$$(2.5) \qquad \mathbf{x}_i = \mathbf{x}_0 + \sum_{j=1}^{n} c_j \lambda_j^i \mathbf{v}_j \ ,$$

where the λ_j are the eigenvalues of the Jacobian matrix evaluated at $\mathbf{x}_0$. We conclude that in the discrete time case divergence occurs if *at least one of the λ_j has absolute value larger than one*.

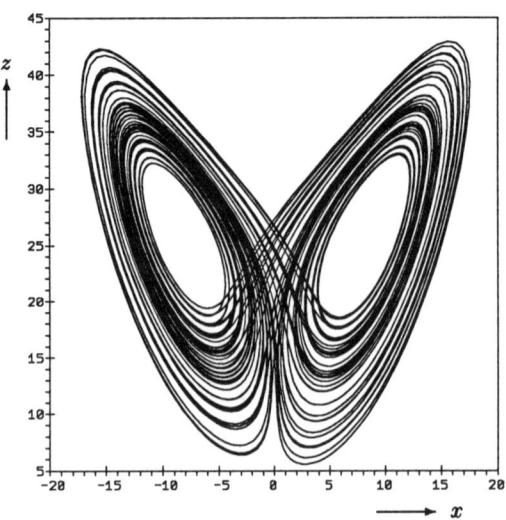

Figure VI.6   Orbit of the Lorenz equations projected on the $y = 0$ plane.

In fact, we have introduced little new in this section. The technique of linearisation has already been dealt with in §V.4, where we used it to determine the stability properties of a stationary point. The present context is a little more general in that $\mathbf{x}_0$ in Definition 2.1 is not necessarily a stationary point, but just an arbitrary point in phase space. If $\mathbf{x}_0$ in Definition 2.1 is a stationary point, local divergence is equivalent to instability in the sense of Theorem V.4.10.

Example VI.2.
We illustrate local divergence using the systems presented in Example VI.1.

(i)   For the *tent map* we have

$$|f'(x)| = 2\mu , \qquad x \neq \frac{1}{2} .$$

This implies that for $\mu > \frac{1}{2}$ this map is locally divergent at all x_0 in $[0, 1]$. This also holds for $x_0 = \frac{1}{2}$, although f' is not defined there, because any orbit starting there leaves this point.

(ii)   For the *doubling map* we have

$$|f'(x)| = 2 , \qquad x \neq \frac{1}{2} .$$

This implies that the map is locally divergent at all x_0 in $[0, 1]$. The dynamics of the doubling map is most appropriately investigated by *symbolic dynamics*. In this approach the variable x is represented as binary. We leave this approach as an exercise. See Exercise 11.

(iii) The *logistic map* has the derivative

$$f'(x) = \mu(1 - 2x) .$$

So, it depends on both μ and x whether $|f'(x)| > 1$. From Fig. VI.3 we know that convergence to periodic solutions may occur, e.g., if $\mu < \mu_\infty \approx 3.57$. For $1 < \mu < \mu_\infty$ this map is locally divergent near $x_0 = 0$ and $x_0 = 1$, but not around $x_0 = \frac{1}{2}$. The stable periodic orbits are mainly located in regions without local divergence.

(iv) The Jacobian matrix of the *Henon map* is given by

$$\mathbf{J}(x) := \begin{bmatrix} -2ax & 1 \\ b & 0 \end{bmatrix}$$

with eigenvalues

$$\lambda_\pm = -ax \pm \sqrt{a^2 x^2 + b} .$$

Because λ_+ is always positive, this map is locally divergent in the whole plane. The Henon map has two stationary points. We leave it to the reader (see Exercise 3) to show that both points are unstable for $b = 0.3$ and $a > 1.0580....$.

(v) The graph in Fig. VI.6 for the *Lorenz equations* suggests that the divergence is small in the wings of the 'butterfly', but large in the central region. The Jacobian matrix of the Lorenz equations is

$$\mathbf{J}(x, y, z) := \begin{bmatrix} -\sigma & +\sigma & 0 \\ r - z & -1 & -x \\ y & x & -b \end{bmatrix} .$$

A point at the outer side of the wings is, e.g., $(17.1, 20.5, 35.7)^T$. The eigenvalues at this point are $0.8 + 20.8i$, $0.8 - 20.8i$, and -15.3. The real eigenvalue, which is strongly negative, indicates that orbits starting outside the wings are attracted to it very rapidly. This takes place more or less along the corresponding eigenvector. The relatively small real parts of the complex eigenvalues indicate that the divergence in the wings is small indeed.

A point in the central region of the attractor is $(1.7, 1.0, 20.8)^T$ with real eigenvalues 3.6, -2.2, and -15.1. The positive eigenvalue confirms that the divergence is considerable.

3. Lyapunov exponents

In §2 we already remarked that chaotic behaviour is characterised by *sensitive dependence on the initial conditions*: two solutions which start very close rapidly diverge from each other. It is intuitively clear that this may happen if the system is locally divergent along an orbit. However, it is not clear in advance whether this local divergence has to apply everywhere along the orbit or not. In this section we make these things precise.

In Chapter V we studied the stability properties of stationary points and periodic solutions. We showed that these properties could be expressed in terms of eigenvalues of the Jacobian matrix and of characteristic exponents respectively. For chaotic systems these notions have to be extended. The generalisations of eigenvalues and characteristic exponents are the Lyapunov exponents. They are indicative of the contraction and/or stretching of small intervals, surfaces, volumes, and higher dimensional subsets around an orbit while time proceeds. Instead of giving the definition of Lyapunov exponent in the most general form right away, we consider special cases first, so that the reader becomes acquainted with the ideas.

Let us consider the scalar Δ-equation

$$(3.1) \qquad x_{i+1} = f(x_i) \ .$$

A small interval $[x_0, x_0 + \varepsilon_0]$ having length $\varepsilon_0 > 0$ will be contracted and/or stretched upon iteration. After one iteration its width, ε_1 say, will approximately be given by

$$(3.2) \qquad \varepsilon_1 \doteq |f'(x_0)| \varepsilon_0 \ .$$

The symbol $\doteq$ denotes that equality holds in the limit $\varepsilon_0 \downarrow 0$ only.

After i iterations we have

$$(3.3) \qquad \varepsilon_i \doteq |f'(x_{i-1})| \ldots |f'(x_0)| \varepsilon_0 \ .$$

It is convenient to write

$$(3.4) \qquad |f'(x_i)| =: e^{\nu_i} \ ,$$

i.e. $\nu_i := \ln |f'(x_i)|$. If $\nu_i > 0$ (< 0) the interval is stretched (contracted) at the i-th iteration. From (3.3) we find

$$(3.5) \qquad \lim_{\varepsilon_0 \downarrow 0} \frac{\varepsilon_i}{\varepsilon_0} = \exp \left(\sum_{j=0}^{i-1} \nu_j \right) \ .$$

Definition 3.6. *The Lyapunov exponent (LE) $\bar{\lambda}(x_0)$ of a one-dimensional discrete time system is the average of the ν_j in the limit $i \to \infty$:*

$$(3.6a) \qquad \bar{\lambda}(x_0) := \lim_{i \to \infty} \frac{1}{i} \sum_{j=0}^{i-1} \nu_j \ .$$

If $\bar{\lambda}(x_0) > 0$, any interval around x_0 will be stretched in the long run. If $\bar{\lambda}(x_0) = 0$, the interval length remains the same on average. If $\bar{\lambda}(x_0) < 0$, the interval will shrink and tend to coalesce with the orbit starting at x_0.

Next we consider n-dimensional Δ-equations. After one iteration an interval $[\mathbf{x}_0, \mathbf{x}_0 + \boldsymbol{\varepsilon}_0]$, with $\|\boldsymbol{\varepsilon}_0\|$ small, will be transformed into the interval $[\mathbf{f}(\mathbf{x}_0), \mathbf{f}(\mathbf{x}_0) + \boldsymbol{\varepsilon}_1]$ with $\boldsymbol{\varepsilon}_1$ given by

$$(3.7) \qquad \boldsymbol{\varepsilon}_1 \doteq \mathbf{J}(\mathbf{x}_0)\,\boldsymbol{\varepsilon}_0 \ .$$

After i iterations we have

$$(3.8) \qquad \boldsymbol{\varepsilon}_i \doteq \mathbf{J}(\mathbf{x}_{i-1}) \dots \mathbf{J}(\mathbf{x}_0)\,\boldsymbol{\varepsilon}_0 \ .$$

The interpretation of the corresponding LE is straightforward for systems with constant Jacobian matrix $\mathbf{J}$, such as the tent map and the doubling map. Then we have

$$(3.9) \qquad \boldsymbol{\varepsilon}_i \doteq \mathbf{J}^i\,\boldsymbol{\varepsilon}_0 \ ,$$

and the LE can be identified with the eigenvalues of $\mathbf{J}$, ordered according to their real parts, viz. $\mathrm{Re}(\lambda_1) \geq \mathrm{Re}(\lambda_2) \geq \dots \geq \mathrm{Re}(\lambda_n)$. Assuming for convenience these λ_j to be different, so that the eigenvectors $\mathbf{v}_j$ are linearly independent, we may expand $\boldsymbol{\varepsilon}_0$ in terms of the $\mathbf{v}_j$:

$$(3.10) \qquad \boldsymbol{\varepsilon}_0 = \sum_{j=1}^{n} c_j\,\mathbf{v}_j \ .$$

Then we find that

$$(3.11) \qquad \boldsymbol{\varepsilon}_i = \sum_{j=1}^{n} c_j\,\lambda_j{}^{n}\,\mathbf{v}_j \ .$$

If $\boldsymbol{\varepsilon}_0$ is arbitrarily chosen, then all $c_j \neq 0$, and the term related to the eigenvalue with the largest real part will dominate. If λ_1 is real, we thus have

$$(3.12) \qquad \boldsymbol{\varepsilon}_i \rightarrow c_1\,\lambda_1{}^{i}\,\mathbf{v}_1 \ , \qquad n \rightarrow \infty \ .$$

If $\boldsymbol{\varepsilon}_0$ is chosen such that $c_1 = 0$ and $\boldsymbol{\varepsilon}_0$ thus has no component along $\mathbf{v}_1$, the λ_2 term in (3.11) will dominate. In general, if the eigenvalues are real and $\boldsymbol{\varepsilon}_0$ is chosen such that it has no components along $\mathbf{v}_1, \dots, \mathbf{v}_j$, the λ_{j+1} term in (3.11) will dominate. This is not a realistic situation in numerical practice, because numerical errors will inevitably generate rounding errors, thus causing c_1 to be nonzero effectively. However, from a theoretical point of view we may define the LE in the case of an n-dimensional Δ-equation as follows.

Definition 3.13. *The Lyapunov exponents of higher dimensional discrete time systems are defined as*

$$(3.13a) \quad \bar{\lambda}(\mathbf{x}_0, \boldsymbol{\varepsilon}_0) := \lim_{\|\boldsymbol{\varepsilon}_0\| \downarrow 0} \lim_{i \to \infty} \frac{1}{i} \ln \frac{\|\boldsymbol{\varepsilon}_i\|}{\|\boldsymbol{\varepsilon}_0\|} = \lim_{i \to \infty} \frac{1}{i} \ln \|\boldsymbol{\varepsilon}_i\| ,$$

where $\boldsymbol{\varepsilon}_i$ is given by the right-hand side of (3.8).

For constant $\mathbf{J}$ we have seen that $\bar{\lambda}(\mathbf{x}_0, \boldsymbol{\varepsilon}_0)$ can attain at most n different values if $\boldsymbol{\varepsilon}_0$ is varied over the tangent space $T_{\mathbf{x}_0}$. Definition 3.13 also applies if the system has no constant Jacobian matrix. Also in that case at most n distinct values for $\bar{\lambda}(\mathbf{x}_0, \boldsymbol{\varepsilon}_0)$ are found. The value obtained depends on the subspace in which $\boldsymbol{\varepsilon}_0$ is chosen. In most cases the value of the LE does not depend on $\mathbf{x}_0$, but on the governing equation only. Then we may omit the dependence on $\mathbf{x}_0$.

The numerical computation of the largest LE, $\bar{\lambda}_1$ say, is quite easy. For arbitrary initial values $\mathbf{x}_0$ and $\boldsymbol{\varepsilon}_0$ one iterates the system

$$(3.14) \quad \begin{cases} \mathbf{x}_{i+1} = \mathbf{f}(\mathbf{x}_i) , \\ \boldsymbol{\varepsilon}_{i+1} = \mathbf{J}(\mathbf{x}_i) \, \boldsymbol{\varepsilon}_i . \end{cases}$$

In principle $\bar{\lambda}_1$ then follows from (3.13a). However, this naive approach is not practical, if only because of computer overflow. So we had better use normalised iterates, normalising the interval length after each iteration step. Denoting the interval length after step i before normalising by ε_i, we have

$$(3.15) \quad \varepsilon_i = \varepsilon_{i-1} \ldots \varepsilon_0 .$$

The largest LE then follows from the analogue of (3.6a):

$$(3.16) \quad \bar{\lambda}_1 := \lim_{i \to \infty} \bar{\lambda}_1^{(i)} , \qquad \bar{\lambda}_1^{(i)} := \frac{1}{i} \sum_{j=0}^{i-1} \ln \varepsilon_j .$$

Example VI.3.

(i) For the *tent map* the derivative $f'(x)$ is constant for all x in $[0, 1]$. From (3.4) and (3.6) we directly find that the (only) Lyapunov exponent is given by

$$\bar{\lambda} = \ln 2\mu$$

with μ the tent map parameter.

(ii) For the *doubling map* we similarly conclude that

$$\bar{\lambda} = \ln 2 .$$

(iii) The *logistic map* has no constant derivative. The Lyapunov exponent can only be obtained numerically via direct application of (3.6a). It turns out that $\bar{\lambda}$ is quite wildly varying as a function of the parameter μ. We leave its calculation to the reader. See Exercise 12.

(iv) The largest Lyapunov exponent $\bar{\lambda}_1$ of the *Henon map* can be calculated from the iterative process (3.14) with the Jacobian matrix given in Example VI.2(iv). The convergence of the limiting process in (3.16) appears to be quite slow. In Fig. VI.7 the successive estimates of $\bar{\lambda}_1$ are shown for $a = 1.4$ and $b = 0.3$. The true values of the Lyapunov exponents are $\bar{\lambda}_1 = 0.41$ and $\bar{\lambda}_2 = -1.58$, cf. [66].

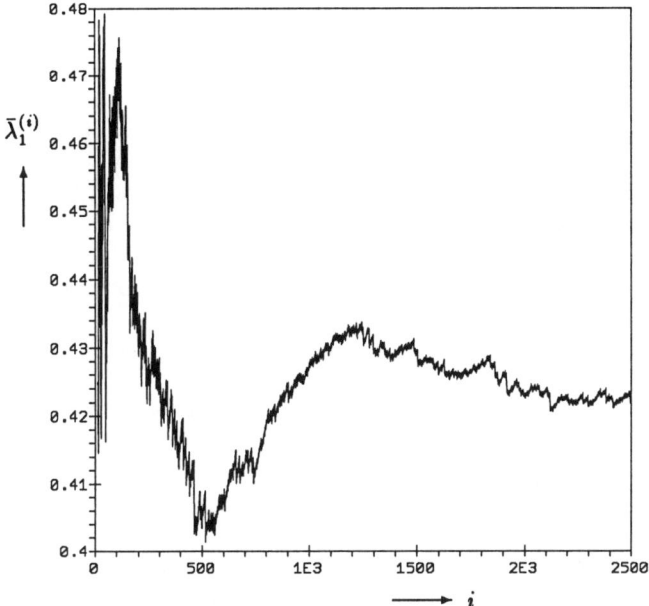

Figure VI.7   Estimates of the largest Lyapunov exponent of the Henon map calculated from (3.16).

(v) The Lyapunov exponents of the Lorenz equations with parameters $\sigma = 16$, $b = 4$, and $r = 40$ are $\bar{\lambda}_1 = 1.37$, $\bar{\lambda}_2 = 0$, and $\bar{\lambda}_3 = -22.37$, cf. [66].

Instead of a one-dimensional interval one can study the behaviour of higher dimensional neighbourhoods of $\mathbf{x}_0$ under the dynamics of the system. It turns out, cf. [50], that the stretching and contracting phenomena of these neighbourhoods are completely characterised by the LE defined in (3.13). The volume V_k of an arbitrarily chosen k-dimensional ($k \leq n$) neighbourhood of $\mathbf{x}_0$ behaves like

$$(3.17) \qquad V_k(i) \doteq V(0) \exp\left(i \sum_{j=1}^{k} \mathrm{Re}(\bar{\lambda}_j)\right), \quad i \to \infty.$$

From this we conclude that a map is *volume preserving* if

$$(3.18) \qquad \sum_{j=1}^{n} \mathrm{Re}(\bar{\lambda}_j) = 0,$$

where we assume the LE $\bar{\lambda}_j$ to be independent of the choice of $\mathbf{x}_0$. Volume preserving systems are called *conservative*. Hamiltonian systems, e.g., have this property, as shown in Chapter XI. Systems with

$$(3.19) \qquad \sum_{j=1}^{n} \mathrm{Re}(\bar{\lambda}_j) < 0$$

are called *dissipative*. For these systems the volume of phase space is contracting on average. Dissipative systems converge to an attracting set, i.e. a proper subspace of the phase space. Dissipative systems with $\mathrm{Re}(\bar{\lambda}_j) > 0$ for at least one j converge to so-called *chaotic attractors*, which are dealt with in §4. Dissipative systems that converge to periodic attractors have $\mathrm{Re}(\bar{\lambda}_j) = 0$ for some j and $\mathrm{Re}(\bar{\lambda}_j) < 0$ for the other ones. Dissipative systems with $\mathrm{Re}(\bar{\lambda}_j) < 0$ for all j converge to point attractors.

Now we turn to the definition of Lyapunov exponents for differential equations. They can be introduced in a similar fashion to that for Δ-equations. We take a perturbation $\mathbf{x}_0 + \boldsymbol{\varepsilon}_0$ of $\mathbf{x}_0$ and solve the linearised system

$$(3.20) \qquad \begin{cases} \dot{\mathbf{y}} = \mathbf{J}\left(\mathbf{x}(t)\right)\mathbf{y}, \\ \mathbf{y}(t_0) = \boldsymbol{\varepsilon}_0, \end{cases}$$

where $\mathbf{x}$ is the solution of the original system $\dot{\mathbf{x}} = \mathbf{f}(\mathbf{x})$, starting at $\mathbf{x}_0$.

Definition 3.21. *The Lyapunov exponent of a continuous time system is defined as*

$$(3.21a) \qquad \bar{\lambda}(\mathbf{x}_0, \boldsymbol{\varepsilon}_0) := \lim_{t \to \infty} \frac{1}{t} \ln \|\mathbf{y}(t)\|,$$

where $\mathbf{y}$ *is the solution of (3.20).*

If $\boldsymbol{\varepsilon}_0$ is arbitrarily chosen, (3.21a) yields the largest Lyapunov exponent $\bar{\lambda}_1$. Special choices of $\boldsymbol{\varepsilon}_0$ yield at most n different values $\bar{\lambda}_j$. Instead of an interval $\boldsymbol{\varepsilon}_0$ we could follow the evolution of a k-dimensional neighbourhood V_k around $\mathbf{x}_0$. In analogy to (3.17) this evolution is given by

$$(3.22) \qquad V_k(t) \doteq V_k(0) \exp\left(t \sum_{j=1}^{k} \mathrm{Re}(\bar{\lambda}_j)\right), \quad t \to \infty.$$

The calculation of $\bar{\lambda}_1$ from (3.14) or (3.20) is straightforward. However, this procedure can only reproduce approximations of the largest Lyapunov exponent (cf. the power method in [33]). A more general approach is based on subspace iteration; here one computes a second (and third etc.) iterate which is orthonormalised each step (cf. Gram-Schmidt procedure, Appendix G). So (3.14) or (3.20) is solved for two linearly independent perturbations ε_0 and ε_1 of the initial value $\mathbf{x}_0$; the time evolution of the area spanned by $\varepsilon_0(t)$ and $\varepsilon_1(t)$ is preserved if the solution vectors are orthogonalised. If the solutions are not only orthogonalised, but also normalised, the evolution of the area $V_2(t)$ in time can be deduced from (3.22):

$$(3.23) \qquad \bar{\lambda}_1 + \bar{\lambda}_2 = \lim_{t \to \infty} \frac{1}{t} \ln V_2(t) \ .$$

Because $\bar{\lambda}_1$ is already known, $\bar{\lambda}_2$ follows from (3.23). The same approach can be used to calculate $\bar{\lambda}_j$, $j > 2$, by calculating sums of $\bar{\lambda}_j$ and peeling off the individual $\bar{\lambda}_j$ one by one. See also [64, 40].

4. Strange and chaotic attractors

In the preceding section we introduced dissipative systems as systems for which condition (3.19) holds. Those systems are 'space contracting' as time evolves. This implies that an orbit starting from an arbitrary point will not 'visit' certain parts of the phase space after some transient period. This phenomenon is clearly present in the vicinity of asymptotically stable stationary points and stable periodic solutions. All orbits starting in the *basin of attraction* of such solutions converge to it. Here we are interested in dissipative systems with solutions which do not converge to point solutions or periodic solutions. For the present, introductory, context it suffices to characterise an attractor in the following, general way:

Definition 4.1. *A subset X of the phase space is an attractor for some ODE if*

(i)   *X is invariant, i.e. solutions of the ODE with initial value in X remain in X.*

(ii)   *there exists a neighbourhood of X, the basin of attraction, such that orbits starting in this basin converge to X.*

(iii) *X is connected.*

This definition states that orbits starting in the neighbourhood of an attractor are attracted to it, while orbits starting at the attractor remain in it.

Furthermore, X cannot be decomposed into smaller attractors. One and the same system can have different types of attractors, and the number and types of attractors may strongly depend on the value of one or more parameters. If the attractor type suddenly changes when the parameter passes a critical value, this phenomenon is called a *bifurcation*. From Example VI.2 it is clear that the logistic map has lots of bifurcation points, where either the periodicity of the attractor changes or the type of the attractor switches from periodic to chaotic or vice versa.

An attractor is called *strange* if its dimension is *fractal*, i.e. non-integer. The notion of dimension is deferred to the next section. Here, the 'capacity' or the 'Hausdorff' dimension is meant. An attractor is called *chaotic* if the largest Lyapunov exponent of the solutions of the related ODE is positive at nearly all points of X. This implies that on a chaotic attractor two randomly chosen orbits which are initially close, will diverge exponentially as time proceeds. It is as though the orbits repel each other. This repulsive character provides an argument why continuous time systems cannot have a chaotic attractor in dimension 1 or 2. Because of uniqueness, solutions cannot cross each other. If X has dimension ≤ 2, the tendency of the orbits to diverge will cause them to push each other to the boundary of X. But if all orbits are crowding at the boundary, they cannot be diverging, and this leads to a contradiction. If X has dimension > 2, these geometrical arguments no longer hold. Many chaotic attractors have a complicated structure due to the mutual repulsion of orbits. For example, they may consist of an infinite number of layers. In the layers stretching takes place. Orbits meeting the boundary of the attractor continue their way on a neighbouring layer.

Discrete time systems may have chaotic attractors of any dimension. For these systems the orbits consist of points instead of continuous curves, and the geometrical argument of passing orbits does not apply.

Example VI.4.

(i)   The *tent map* has an attractor for $0 \leq \mu < 1$. From Fig. VI.2 we conclude that for $\mu > \frac{1}{2}$ this attractor is chaotic, because the Lyapunov exponent $\ln 2\mu$ is positive. For $\frac{1}{2} < \mu < 1$ it can be shown that the attractor has a fractal dimension and thus is strange. For $\mu = 1$ the orbits visit the whole interval $[0,1]$, and an attractor is no longer present. In this case the dynamics on $[0,1]$ is chaotic, because the Lyapunov exponent is positive.

(ii) The *doubling map* has the interval $[0,1]$ both as domain and as range. So, no attractor is present. The dynamics on $[0,1]$ is chaotic, for the Lyapunov exponent is positive.

(iii) The *logistic map* has a behaviour similar to that of the tent map. From Fig. VI.3 we see that this map has an attractor for $0 \leq \mu < 4$. For $\mu = 4$ the orbits fill the interval $[0,1]$ and no attractor is present. In that case the dynamics is chaotic, because the Lyapunov exponent is positive. For $\mu < \mu_\infty = 3.56994..$, the attractor is periodic and the corresponding Lyapunov exponent is negative. At the accumulation point μ_∞ the attractor is not periodic, but also not chaotic, because the Lyapunov exponent is vanishing there. For $\mu = \mu_\infty$ the attractor is strange, because its dimension is fractal.

(iv) The attractor of the *Henon map*, depicted in Fig. VI.4, has a fractal dimension of about 1.26, and largest Lyapunov exponent $\bar{\lambda}_1 = 0.41$. The attractor is thus both chaotic and strange. The same holds for the attractor of the *Lorenz equations* shown in Fig. VI.6. Its dimension is about 2.1 and $\bar{\lambda}_1 = 1.37$ (see [24]).

Most attractors can be provided with a measure in a natural way. Such a *measure* μ assigns to each subset A of X a number $\mu(A) \in [0,1]$. If A_1 and A_2 are non-overlapping subsets of X, then $\mu(A_1 + A_2) = \mu(A_1) + \mu(A_2)$; furthermore $\mu(X) = 1$. Such a measure μ is generated in a natural way by the time evolution of the solutions on X, if we interpret $\mu(A)$ as the probability that the system is in A when observed at a randomly chosen time. Let us assume that the system has an orbit $\mathbf{x}$ on X which visits all regions of X, i.e. the orbit gets arbitrarily close to all points on X if $t \to \infty$. Then we can define the measure μ by

$$(4.2) \qquad \mu(A) := \lim_{t \to \infty} \frac{1}{t} \{\text{total time that } \mathbf{x}(t) \in A\} \ .$$

This definition is only useful if such an orbit exists and if there is no dependence on the specific orbit used. Attractors for which this is the case are called *ergodic*. Most chaotic systems appear to be ergodic. For ergodic attractors nearly every point is the starting point of an orbit visiting the whole attractor. Ergodic attractors have Lyapunov exponents which are independent of the position on the attractor and thus really characterise the dynamics on X globally.

Let us indicate the time evolution of the solutions on X by the flow φ_t, i.e. $\mathbf{x}(t) := \varphi_t \, \mathbf{x}_0$ (see §I.3). In terms of this flow the measure defined by (4.2) has the important property that

$$(4.3) \qquad \mu(A) = \mu\left(\varphi_t^{-1}(A)\right) ,$$

for all subsets $A \subset X$ and all $t > 0$. In view of this the measure induced by (4.2) is said to be *invariant*. For invertible maps (4.3) is equivalent to $\mu(A) = \mu(\varphi_t(A))$. Continuous time systems are invertible and so there we can use both formulations. Discrete time systems, however, may be non-invertible, and then this equivalence does not hold.

In most cases the measure μ can only be calculated numerically. Some exceptions are dealt with in Example VI.5. In practice the measure μ is estimated using the so-called *box-counting* method. This method starts with the generation of an arbitrary, long orbit on X. If we assume that the system is ergodic, the results do not depend on this choice. Because of the discrete nature of numerical algorithms, the orbit always consists of discrete points at successive times, even in the continuous time case. Next, the relevant part of the phase space is covered with a grid of non-overlapping boxes. The relative number of orbital points in a box provides an estimate of the measure of that box. By refining the grid and lengthening the orbit, the estimate can gradually be improved.

Example VI.5.
In most cases the measure μ on an attractor has to be determined numerically. Exceptions are the *tent map with $\mu = 1$*, the *doubling map*, and the *logistic map with $\mu = 4$*. The dynamics of these three systems are very similar, because they can be transformed into each other in a smooth way. Setting $x = \sin^2 \varphi$ in the logistic equation

$$x_{i+1} = 4x_i(1 - x_i) \, ,$$

we obtain after some algebra

$$\sin^2 \varphi_{i+1} = \sin^2 2\varphi_i,$$

which is equivalent to the doubling map. On the other hand, on applying the transformation $x = \sin^2(\frac{1}{2}\pi y)$ we obtain the Δ-equation

$$\sin^2\left(\frac{\pi y_{i+1}}{2}\right) = \sin^2(\pi y_i) \, .$$

This leads to $\frac{1}{2}\pi y_{i+1} = \pm\pi y_i + m\pi$, with m integer. Because y lies in $[0, 1]$, the choice of m and the sign of πy_i can be determined. We obtain the tent map

$$y_{i+1} = \begin{cases} 2y_i & , \quad 0 \leq y_i < \frac{1}{2} \\ 2(1 - y_i) & , \quad \frac{1}{2} \leq y_i \leq 1 \, . \end{cases}$$

It is possible to derive the natural measure for the tent map analytically. The natural measures for the doubling and logistic maps then follow from the transformations given above. The natural measure μ has to be invariant under the dynamics, i.e.

$$\mu(A) = \mu\left(\varphi^{-1}(A)\right),$$

with A any subset of the domain of the map. For the maps under consideration the domain and range coincide. Let A be the interval

$A := [a, b]$, $0 \leq a \leq b \leq 1$. Then it can easily be checked that for the tent map with $\mu = 1$ the measure

$$\mu(A) := b - a$$

satisfies. So, the measure of an interval is its length. For example, if $A = [\frac{1}{4}, \frac{1}{2}]$, then $\varphi^{-1}(A) = [\frac{1}{8}, \frac{1}{4}] \cup [\frac{3}{4}, \frac{7}{8}]$. The same measure also satisfies for the doubling map. To obtain the measure of the logistic map we have to invert the transformation mentioned above:

$$y = \frac{2}{\pi} \arcsin \sqrt{x},$$

and thus

$$dy = \frac{1}{\pi} \frac{1}{\sqrt{x(1-x)}} dx \ .$$

From this we conclude that for the logistic equation with $\mu = 4$ the measure is given by

$$\mu(A) = \frac{1}{\pi} \int_a^b \frac{1}{\sqrt{x(1-x)}} dx \ .$$

It is an important question whether a numerically computed orbit on X yields any information about a real orbit on X. In view of the sensitivity to the choice of the initial condition, we know in advance that the computed orbit is a bad approximation for the real orbit with the same starting point. Because numerical errors are unavoidable, these two orbits will certainly diverge exponentially. However, the numerical orbit still contains useful information. This insight relies on the theory of *pseudo-orbits*, which we shall briefly explain now.

In what follows we denote a true orbit with initial value x_i by $x_i(t)$, so $x_i(t_0) = x_i$. After a step Δt, the true orbit with initial value x_0 will be at $x_0(t_1 = t_0 + \Delta t)$. The corresponding numerical approximation at t_1 is denoted by x_1. In Chapter III we have seen that the numerical procedure effectively 'steps' onto a neighbouring true solution curve after a time step. The ('local') discretisation error at t_1 is denoted by $d_1 := x_0(t_1) - x_1$. For the next time step Δt, the numerical procedure starts from x_1. Again a numerical error is introduced: $d_2 := x_1(t_2) - x_2$ with x_2 the numerical outcome at $t_2 := t_0 + 2\Delta t$ and $x_1(t_2)$ the true orbit with $x_1(t_1) = x_1$. This procedure is repeated and the following definition applies.

Definition 4.4. *The series* x_i, $i = 1, ..., N$, *of numerically generated values is*

called an α-pseudo-orbit, if the errors $e_i := \|d_i\|$ satisfy $e_i < \alpha$, $i = 1, ..., N$,
for a given tolerance $\alpha > 0$.

In Fig. VI.8 we have illustrated the situation. It is clear that numerical integration yields pseudo-orbits with some tolerance α, which is determined by the step length Δt and the accuracy of the integration procedure used.

(a)
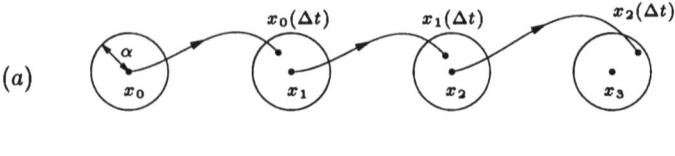

(b)
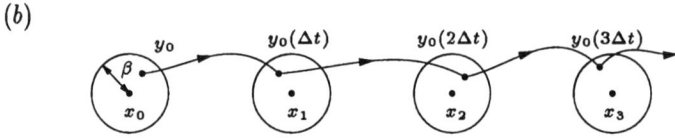

Figure VI.8   (a) Pseudo-orbit x_i, $i = 0,1,2 ...$. (b) β-shadowing of the pseudo-orbit in (a) by the true orbit starting at y_0.

Definition 4.5. *A series of points x_i, $i = 1, ..., N$, is called β-shadowed by a point y_0, if $\|x_i - y_0(t_0 + i\Delta t)\| < \beta$ for $i = 1, ..., N$, where $y_0(t)$ is the orbit with $y_0(t_0) := y_0$.*

The following theorem guarantees that every pseudo-orbit is close to a true orbit on X. The theorem applies to hyperbolic, ergodic attractors. An attractor is called *hyperbolic*, if the linearised vector field has no vanishing eigenvalues everywhere on the attractor.

Theorem 4.6. *Let X be a hyperbolic, ergodic attractor. Then for every $\beta > 0$ an $\alpha > 0$ exists, such that every α-pseudo-orbit in X is β-shadowed by some point on X.*

So, we can never calculate a true orbit exactly, due to discretisation and rounding errors, but nonetheless we always find an approximation to some orbit of the system. For the proof of this theorem and related topics, we refer to [14, 56, 44].

5. Fractal dimension

The notion of the dimension of a set X in $\mathbb{R}^n$ is not uniquely defined. In this section we give a short overview of current definitions. The dimension definition closest to intuition is the *capacity* d_{cap}. To introduce this definition we cover X with identical n-dimensional, non-overlapping boxes. Let l be some length scale characterising the box size. For the volume $V(l)$ of such a box we have

$$(5.1) \qquad V(l) \sim l^n \ .$$

This is called a *scaling relation*. The scaling symbol $\sim$ means that the relation holds in the limit $l \downarrow 0$, and that a possible constant of proportionality is ignored.

Next we give a scaling relation like (5.1) for X itself. For given l, let $N(l)$ be the minimal number of boxes needed to cover X. The *capacity* is then defined through

$$(5.2) \qquad N(l) \sim l^{-d_{\mathrm{cap}}} \ ,$$

which implies that

$$(5.3) \qquad d_{\mathrm{cap}} := -\lim_{l \downarrow 0} \frac{\ln N(l)}{\ln l} \ .$$

It is simple to check that d_{cap} has integer values for common objects like a point, a line, a square, or a cube. The numerical evaluation of d_{cap} for general X directly follows from its definition. Usually, X is characterised by a large number of points on X. This set of points is covered with boxes of characteristic length l as introduced above, and $N(l)$ is the number of boxes that contain at least one of the points. If $\ln N(l)$ is plotted as a function of $\ln l$, a curve results, whose slope provides an estimate for d_{cap}. The application of this *box-counting* procedure encounters several problems in practice:

– It does not make sense to take l smaller than the smallest distance between data points; so the limit $l \downarrow 0$ can only be approximated. It often appears that $\ln N(l)$, as a function of $\ln l$, is a straight line in a small interval of l-values only.

– The data are given with finite accuracy. This also implies a lower bound on l.

– The number of necessary boxes may increase very fast. In the box-counting procedure the object is covered with finer and finer coverings. If the mesh width l is halved at each refinement, the number of boxes increases by a factor of 2^n. After m refinements the covering contains at least 2^{nm} boxes.

However, it does not make sense to use a covering with more boxes than the number of data points. Since the number of data points is usually restricted, such a method is not quite useful in higher dimensions.

Definition 5.3 is not generally applicable, because the limit does not always exist. This is caused by the condition that all boxes must be equal. This problem is alleviated by the *Hausdorff dimension*. Here, for a given l, the boxes may be of different sizes $\leq l$; with such unequal boxes we try to cover X as efficiently as possible. Let $N(l)$ be the minimum number of boxes needed to cover X. The Hausdorff dimension then is the unique number d_{H} such that

$$(5.4) \qquad \lim_{l \downarrow 0} \sum_{i=1}^{N(l)} l_i^d = \begin{cases} 0 & \text{if } d > d_{\mathrm{H}} \text{ ,} \\ \infty & \text{if } d < d_{\mathrm{H}} \text{ ,} \end{cases}$$

where l_i is the characteristic size of the i-th box. Although this concept has obvious theoretical advantages, its practical use is very limited, because coverings with boxes of different sizes are hard to handle.

Other dimension concepts essentially use a measure μ on X. For attractors of dynamical systems we have already introduced such a measure in the preceding section. Let us cover X with identical boxes of size l. The measure

$$m_i := \mu(\text{box } i) \text{ ,}$$

is the probability of finding the system in box i when observed at an arbitrary time. In the following we shall use the *correlation function*

$$(5.5) \qquad P_q(l) := \sum_{i=1}^{N(l)} m_i^q \text{ .}$$

Note that the boxes with relatively large values of m_i contribute the most to $P_q(l)$ if q is increased. So, for large values of q the regions most visited by the system dominate the sum in (5.5). Following [46] and [35] we may introduce a family of dimensions D_q, with $q \geq 0$, as follows:

$$(5.6) \qquad D_q := \lim_{l \downarrow 0} \frac{1}{q-1} \frac{\ln P_q(l)}{\ln l} \text{ .}$$

It is clear that d_{cap} is also a member of this family, because $D_0 \equiv d_{\mathrm{cap}}$. We leave it to the reader (Exercise 5) to check that

$$(5.7) \qquad D_1 = \lim_{l \downarrow 0} \frac{I(l)}{\ln l}$$

with the *information* $I(l)$ defined as

$$(5.8) \qquad I(l) := \sum_{i=1}^{N(l)} m_i \ln m_i \text{ .}$$

D_1 is called the *information dimension*. It represents the amount of information needed to locate a point on X with accuracy l. In general $P_q(l)$ can be interpreted as the probability that the system is in one and the same box when observed at q arbitrary times. The so-called *two-point correlation function* $P_2(l)$ can be estimated as follows. Given a set of N points $\mathbf{x}_i$ on X, the distances $\|\mathbf{x}_i - \mathbf{x}_j\|$, $i, j = 1, .., N$, $i \neq j$, are calculated. Let $M(l)$ be the number of distances smaller than l. Then, $P_2(l)$ may be estimated from

$$(5.9) \qquad P_2(l) \doteq \frac{1}{N(N-1)} M(l) .$$

Substituting this estimate into (5.6) yields an estimate for the so-called *correlation dimension* D_2. By no means is it necessary to calculate all distances. In (5.6) we are only interested in small values of l, so that only distances between neighbouring points are relevant, cf. [74]. In practice, D_2 appears the most appropriate choice for the estimation of fractal dimension. Without proof we mention the following property [67]:

Property 5.10. *The family of dimensions D_q defined in (5.6) satisfies the inequality*

$$\frac{dD_q}{dq} \leq 0 ,$$

so that

$$(5.10a) \quad D_q \leq D_q' , \qquad if \ q > q' .$$

Equality holds only if the measures m_i in (5.5) are independent of i.

Example VI.6.
A famous example of a set with fractal value for its capacity D_0 is the middle-third *Cantor set*. This set is *self-similar*, i.e. it is constructed by means of repeated application of a refinement rule. The rule in the case of the Cantor set is the following: given an interval, remove the open middle third interval. We start with the unit interval $[0, 1]$. Application of the rule leaves the two intervals $[0, \frac{1}{3}]$ and $[\frac{2}{3}, 1]$. Application of the rule to these two intervals yields four intervals, etc. (see Fig. VI.9). This procedure is continued 'ad infinitum'. At each level of refinement the set resembles the set at other levels of refinement. Self-similarity is met in many fractal objects, although the refinement rule is not always as simple as is the case here. At refinement level i there are 2^i intervals, each of length $(\frac{1}{3})^i$. The total length of the intervals at level i becomes vanishing for $n \to \infty$, because

$$\lim_{i \to \infty} 2^i \left(\frac{1}{3}\right)^i = 0 .$$

The capacity of the Cantor set is easily found. The covering with one-dimensional boxes, i.e. intervals, is trivial, because the successive sets already consist of intervals. So, $N(l)$ in (5.2) is given by $N((\frac{1}{3})^i) = 2^i$. Application of (5.3) yields

$$d_{\text{cap}} = - \lim_{i \to \infty} \frac{\ln 2^i}{\ln (\frac{1}{3})^i} = \frac{\ln 2}{\ln 3} .$$

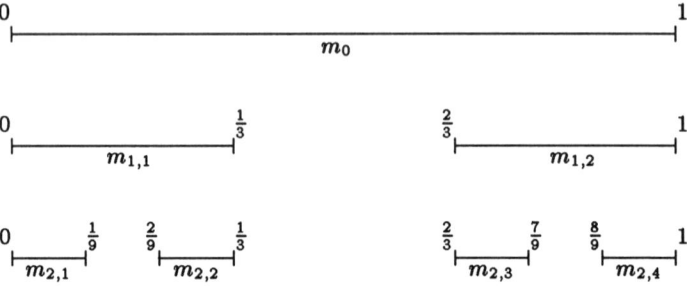

Figure VI.9 Construction of the middle-third Cantor set with measures $m_{i,j}$, where $m_{i,j}$ is the measure of the j-th interval at refinement level i.

Example VI.7.
In the preceding example we used $d_{\text{cap}} \equiv D_0$, which does not take into account any dynamical feature. To illustrate the calculation of $D_q, q > 1$, we provide the Cantor set in Example VI.6 with a measure, constructed via a simple redistribution rule. The initial interval $[0, 1]$ has measure $m_0 := 1$. After splitting up this interval, the left interval $[0, \frac{1}{3}]$ is given the measure $m_{1,1} := p$, with $0 \le p \le 1$, and the right interval $[\frac{2}{3}, 1]$ is given the measure $m_{1,2} := 1 - p$. This procedure is repeated at the next refinement level leading to measures, from left to right, $m_{2,1} := p^2$, $m_{2,2} := p(1 - p)$, $m_{2,3} := p(1 - p)$, $m_{2,4} := (1 - p)^2$, as indicated in Fig. VI.9.
Application of (5.5) at refinement level $l = (\frac{1}{3})^i$ yields

$$P_q((\tfrac{1}{3})^i) = \sum_{j=1}^{2^i} m_{i,j}^q = (p^q + (1 - p)^q)^i .$$

Substituting this into (5.6) we obtain

$$D_q(p) = - \frac{1}{q - 1} \frac{\ln(p^q + (1 - p)^q)}{\ln 3} .$$

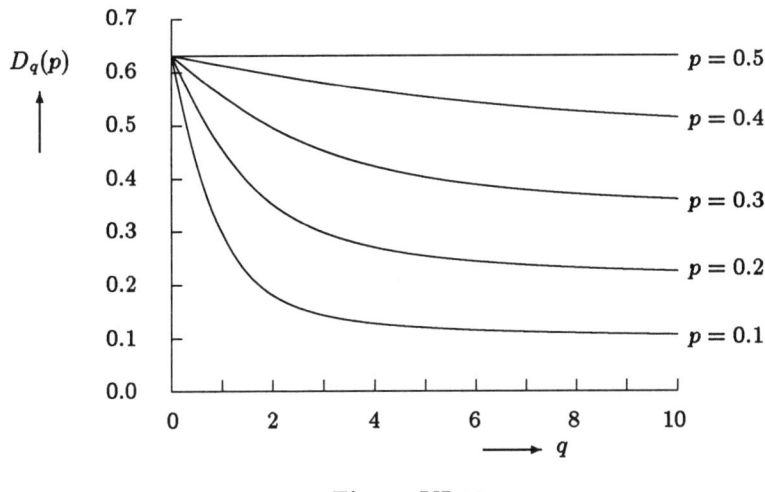

Figure VI.10

For $p = \frac{1}{2}$ the probability is uniformly distributed over the intervals of the Cantor set. In that case the D_q are identical for all q, which is in accord with Property 5.10. In Fig. VI.10 the behaviour of $D_q(p)$ as a function of q is given for several values of p.

Example VI.8.
Let us illustrate the numerical estimation of the correlation dimension D_2 by applying (5.6) with (5.9) to the 'butterfly' attractor of the Lorenz equations. We integrate the Lorenz equations in Example VI.1(v) using the Runge-Kutta method. The time step used is $\Delta t = 0.03$. Taking 1000 integration steps we obtain a series of 1000 three-dimensional points on the butterfly attractor. For this series the two-point correlation function $P_2(l)$ is calculated from (5.9). In Fig. VI.11 a log-log plot is given of $P_2(l)$ as a function of l. In the range $2 < l < 10$ the function shows linear behaviour with slope about 2.0. This value is a reasonable estimate for the true value $D_2 = 2.1$, obtained by using considerably more points.

6. Reconstruction

In many practical situations no appropriate model for the system under consideration is available. Even the number of degrees of freedom, and thus the dimension of the phase space, is not known in advance. All one can usually do is measure one or a few properties of the system. So, the data consist of one or

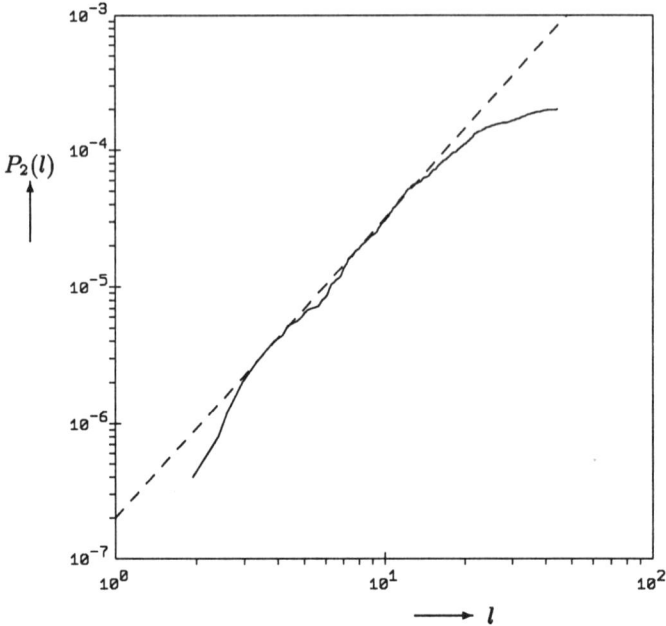

Figure VI.11   Correlation function $P_2(l)$ of the 'butterfly of Lorenz'.

more time series. From this information one should like to deduce whether the system is in a chaotic mode. If this were the case, one would be interested in the dimension of the possible attractor and the values of the Lyapunov exponents. In this section we point out how these questions can be answered. In §7 we shall show that the ideas presented here also form an essential ingredient of procedures for prediction based on measured series.

Let $\mathbf{x}(t)$ be the (unknown) state vector characterising the system at time t. Any measured property φ of the system is a function of $\mathbf{x}(t)$, and its dynamics is determined by the dynamics of $\mathbf{x}(t)$: $\varphi(t) = \varphi(\mathbf{x}(t))$. An important question is whether the state $\mathbf{x}(t)$ can be deduced from the measured $\varphi(t)$ data. Fortunately, a partial answer is given by the technique of *reconstruction*, cf. [73, 62].

Example VI.9.
To understand the idea behind reconstruction, one may consider a

free particle in space. Its state vector is composed of position, $\mathbf{z}(t)$ say, and velocity, $\mathbf{v}(t)$ say. If $\mathbf{z}(t)$ could be measured for all t, the other part $\mathbf{v}(t)$ of the state vector could be deduced from these data, because $\mathbf{v}(t) = \dot{\mathbf{z}}(t)$. In practice, only partial information about $\mathbf{z}(t)$ is available, because measurements are always performed at discrete times t_i, leading to data points $\mathbf{z}_i = \mathbf{z}(t_i)$. From these data estimates for the velocities $\mathbf{v}_i = \mathbf{v}(t_i)$ can be obtained, e.g., via the central differences $\mathbf{v}_i \doteq (\mathbf{z}_{i+1} - \mathbf{z}_{i-1})/(t_{i+1} - t_{i-1})$. So, successive data for some state space components may provide information about the other components.

The idea in this example has led to the so-called *embedding* procedure. Let us assume that the property φ is measured at all times. From these data a k-dimensional vector $\mathbf{y}(t)$ is constructed through

$$(6.1) \qquad \mathbf{y}(t) := (\varphi(t), \varphi(t+d), \varphi(t+2d), ..., \varphi(t+kd))^T .$$

The parameter d is called the *delay* and k the *embedding dimension*. The $\mathbf{y}$-series can be interpreted as an orbit in the *embedding space*. The key notion here is that the dynamics of the $\mathbf{y}$-series usually has the same characteristics as the dynamics of the state vector $\mathbf{x}$ in the phase space. This is formulated in the following theorem:

Theorem 6.2 (Embedding theorem). *Let the measured property φ be a differentiable function of the state of the system, and let the (integer) embedding dimension $k \geq 2D + 1$. If the state vector $\mathbf{x}$ follows an orbit on an attractor X in the phase space, then reconstruction vector $\mathbf{y}$ in (6.1) follows an orbit on an attractor Y in the embedding space. The attractors X and Y have the same dimensions D and the same Lyapunov exponents.*

For the proof we refer to [73]. In practice, $\varphi(t)$ is usually measured at equidistant times t_i leading to a series $\varphi_i = \varphi(t_i)$, $i = 1, 2...$. The vector $\mathbf{y}(t)$ is then available at discrete times t_i, i.e.

$$(6.3) \qquad \mathbf{y}_i := (\varphi_i, \varphi_{i+d}, \varphi_{i+2d}, ..., \varphi_{i+kd})^T .$$

This is no problem as long as the φ-series contains the essential information about the system, i.e. the time grid is sufficiently dense.

Reconstruction is a valuable tool to investigate whether a chaotic-looking time series stems from a chaotic or a stochastic system. In practice, the dimension is not known in advance. The embedding procedure is then applied for increasing values of k. For each value of k the dimension of the resulting attractor Y is calculated. For that purpose the correlation dimension is commonly used, but any other dimension may suffice as well. The method is said to *saturate* if the estimated dimension of Y becomes independent of k. If the underlying system is not deterministic but stochastic, no saturation will be found.

7. Prediction

Chaotic systems are not predictable in the long run because of the inherent sensitivity to the initial conditions. However, thanks to the deterministic character of those systems prediction is not completely impossible, but only within a certain time horizon T_{hor}. In general this horizon is determined by the largest Lyapunov exponent $\bar{\lambda}_1$. Let the system be numerically reproduced within an accuracy ε_0. The initial state of the system is then known with an error of order ε_0, and this initial error will on average be amplified as

$$(7.1) \qquad \varepsilon(t) \sim e^{\bar{\lambda}_1 t} \varepsilon_0 \;,$$

for a continuous time system. For discrete mappings a similar expression holds. If the errors are in percentages, we can find a rough estimate for the horizon T_{hor} from setting $\varepsilon(T_{\text{hor}}) \approx 100\%$:

$$(7.2) \qquad T_{\text{hor}} \approx -\frac{1}{\bar{\lambda}_1} \ln \frac{\varepsilon_0}{100} \;.$$

We emphasise that $\bar{\lambda}_1$ and thus T_{hor} contain information about the average divergence on the attractor only. Local divergences may differ considerably. For example, some types of weather are quite stable, so that forecasts are reliable for longer than a week, whereas other weather regimes do not allow a reliable prediction for even a few days. Similar (spatial) variation of local divergence is already met in the 'butterfly of Lorenz' in Example VI.2(v).

Let us now turn to the prediction of a chaotic series. Consider a time series φ_i, $i = 1, ..., M$, of measured values at equidistant times, where φ represents some property of the system. The insights presented above can be used to obtain a reliable estimate for φ_{M+j}, $j = 1, 2, ...$ The procedure to be presented is based on finding local portions of the time series in the past which closely resemble the final part of the series. The time evolution of those portions is known from the series itself, and this provides us with information about the evolution of the series in the future. The procedure for *one-step-ahead prediction* consists of the following steps:

(i) Apply reconstruction. The procedure dealt with in §6 will yield an estimate k for the appropriate embedding dimension. The φ_i-series is transformed into a series $\{\mathbf{y}_i, \; i = 1, ..., \hat{M}\}$ of k-dimensional vectors with $\hat{M} := M - (k - 1)$. Now the problem is to estimate $\mathbf{y}_{\hat{M}+1}$.

(ii) Search neighbours of $\mathbf{y}_{\hat{M}}$ in the $\mathbf{y}$-series.

(iii) Determine the evolution of these neighbours after one iteration.

(iv) Fit a mapping to the one-step-ahead evolution of the neighbours.

(v) Apply this map to $y_{\hat{M}}$. This yields an estimate for $y_{\hat{M}+1}$ and its last element is an estimate for x_{M+1}.

For practical reasons one usually fixes the number of neighbours searched for in step (ii). It would probably be better to select all neighbouring points within a sphere of fixed radius around $y_{\hat{M}}$. In step (iv) the use of a linear map is preferred. Higher order fits are not expected to yield better results when the prediction is only one-step-ahead.

We shall work out the procedure in the linear case. Let $p_1, p_2, ..., p_n$ be the set of neighbours of $y_{\hat{M}}$. After one iteration step they have evolved to $q_i := p_{i+1}$, $i = 1, ..., n$. In view of the construction (6.1), the q_i are obtained by shifting the p_i one position forward in the φ_i-series. The linear mapping sought for consists of a $k \times k$ matrix A and a k-dimensional vector b, which satisfy

(7.3) $A p_i + b = q_i$, $i = 1, ..., n$.

For the estimation of the A and b from these equations a number of numerical techniques are available, of which we mention the *singular value decomposition* and *QU-decomposition* techniques. For the technical details we refer to Appendix G and [33]. Having estimated A and b, and thus the linear mapping in (7.3), we then obtain the one-step-ahead prediction from

$$y_{\hat{M}+1} = A y_{\hat{M}} + b .$$

The predicted value for φ_{M+1} then is the k-th element of $y_{\hat{M}+1}$.

Multistep prediction may proceed in several ways. We mention three possibilities:

(i)   Apply the one-step-ahead procedure with the estimate φ_{M+j} as starting point to obtain a new estimate φ_{M+j+1}.

(ii) Modify step (iii) in the one-step-ahead procedure and determine the evolution of the neighbours after more than one iteration.

(iii) Calculate A and b with the one-step-ahead procedure and apply the linear mapping repeatedly.

Method (i) makes optimal use of the information in the series. However, this approach is computationally not attractive, because at each prediction step a near-neighbour search is needed and this is the most time-consuming part of the procedure. This disadvantage does not hold for methods (ii) and (iii). Method (iii) is clearly the cheaper of the two, but also yields the poorer results.

The prediction properties of a time series often provide a convenient tool to discriminate between chaotic and stochastic time series. Stochastic series are the more accurately predicted, the more auto-correlation they contain.

Chaotic time series can be well predicted, but only within a certain horizon. To discriminate between the two types, one may apply both the prediction procedure outlined above and an *autoregressive* (AR) predictor, which is only based on the assumption that the series may contain auto-correlation. The AR predictor of order l reads

$$(7.4) \qquad \varphi_i = \sum_{j=1}^{l} c_j \, \varphi_{i-j} \; .$$

The coefficients c_j are assumed to be independent of i, i.e. the position in the φ-series. They can easily be estimated by the same procedures as used for solving (7.3). The AR predictor thus contains information averaged over the whole series, whereas the procedure based on chaos theory tries to construct a predictor that is locally optimal. In the following example the latter approach is shown.

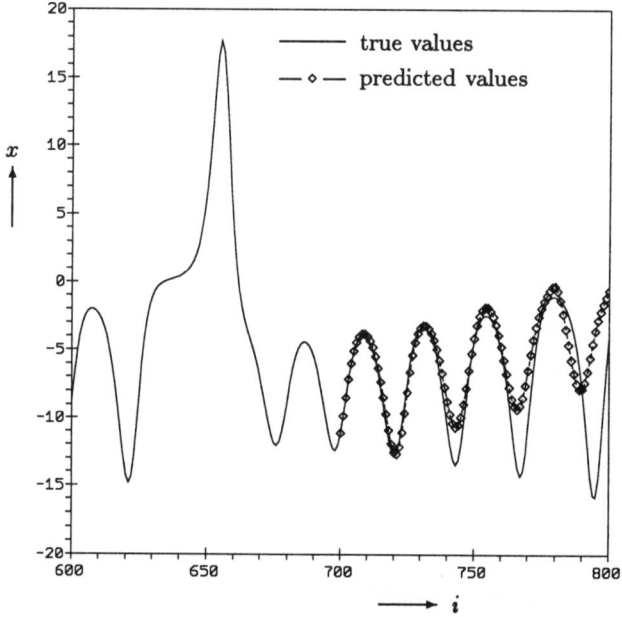

Figure VI.12   Prediction of the x-component of the Lorenz equations.

Example VI.10.
We illustrate the reconstruction and prediction techniques in §§6, 7 with a data series $\{x_i\}$ consisting of 800 calculated values of the x-component of the Lorenz equations. This time series is transformed into

a reconstructed series using embedding dimension 5 and delay 1. The
first 700 data points are used to estimate values for the rest of the
series. Because we know the true values, we may get an indication of the
reliability of the methods. Plotted in Fig. VI.12 are the predicted values
for $i = 701$–800 obtained using multistep prediction method (ii) with
starting point $i = 700$. The true values are shown, too. It is seen that
the reliability of the predictions is initially quite good, but inaccuracies
come in with increasing length of the prediction interval.

Exercises Chapter VI

1. Show that for a stationary point of a continuous-time system the Lyapunov
    exponents are precisely the eigenvalues of the Jacobian matrix evaluated
    at that point, whereas for a periodic orbit the Lyapunov exponents can
    be identified with the characteristic exponents introduced in §IV.5.

2. a) Show that the logistic map with $\mu > 3$ has a period-2 solution which
        jumps between x_+ and x_- given by

    $$x_\pm = \frac{1}{2\mu}\left(1 + \mu \pm \sqrt{\mu^2 - 2\mu - 3}\right).$$

    b) Calculate the Lyapunov exponent of this period-2 cycle (using the
        result of Exercise 1) and show that it is given by

    $$\bar{\lambda} = \tfrac{1}{2}\ln\left|-\mu^2 + 2\mu + 4\right|.$$

    c) Conclude from this result that the period-2 cycle is stable if

    $$3 < \mu < 1 + \sqrt{6}.$$

3. a) Show that the x-components $x_\pm$ of the fixed points of the Henon map
        are given by

    $$x_\pm = \frac{1}{2a}\left(b - 1 \pm \sqrt{b^2 - 2b + 4a + 1}\right),$$

    and the Lyapunov exponents of these points by

    $$\bar{\lambda}_\pm = -a\,x_\pm \pm \sqrt{a^2 x_\pm^2 + b}.$$

    b) Plot the $\bar{\lambda}_\pm$ for varying a, keeping the b-value fixed at $b = 0.3$,
        and investigate the stability properties of $x_\pm$ as a function of a.
        Calculate also the eigenvectors of the linearised Henon map at $x_\pm$ and
        investigate the positions of $x_\pm$ and the directions of these eigenvectors
        with respect to the attractor in Fig. VI.4.

4. Show that for the Lyapunov exponents of the Henon map it holds that

$$\bar{\lambda}_1 + \bar{\lambda}_2 = \ln |b| \ .$$

5. Derive equation (5.7) for the information dimension D_1 by applying a careful expansion of $P_q(l)$ in (5.5) around $q = 1$. One could also use the rule of l'Hôpital, which states that, under certain conditions, one may take the derivative with respect to q of both numerator and denominator in (5.6) before taking the limit $q \to 1$.

6. Take the limit $q \to 1$ for $D_q(p)$ of the generalised Cantor set in Example VI.7. See also the hint in Exercise 5.

7. Calculate the capacity D_0 of the Cantor set obtained by removing the middle interval of length $\frac{1}{2}$, instead of $\frac{1}{3}$ as in the usual Cantor set dealt with in Example VI.6.

8. Calculate the capacity D_0 of the so-called *Koch curve*, whose construction is depicted in Fig. VI.13.

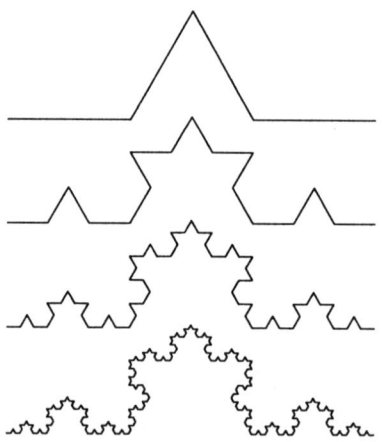

Figure VI.13   Construction of the Koch curve.

9. a) Show that the Hausdorff dimension of the rational numbers in the interval $[0, 1]$ is vanishing.

    Hint: Use the property that the rational numbers are countable and can be covered with intervals $\varepsilon_n = \varepsilon\, 2^{-n}$, where ε is given and n is the index of the numbers. The sum of these intervals vanishes in the limit $\varepsilon \downarrow 0$.

    b) Show also that the capacity of this set is equal to 1, due to the fact that in this case the covering must consist of equal intervals.

10. Consider the doubling map with additive noise, so that

$$x_{i+1} = 2x_i \bmod(1) + \text{noise} .$$

Assume that the noise is such that it changes randomly all the digits a_i in the binary representation of x for $i \geq 50$. The noise is therefore of the order $2^{-50} \sim 10^{-5}$. Let exact observations $\bar{x}_0, \bar{x}_1, ..., \bar{x}_T$ of the noisy orbit be given for a period $T \gg 50$. Show that an initial condition x_0 exists such that the exact true orbit $x_0, x_1, ..., x_T$, followed by the noiseless map, shadows the noisy orbit $\bar{x}_0, \bar{x}_1, ..., \bar{x}_T$. In particular, show that x_0 can be chosen such that $|\bar{x}_i - x_i| \leq 2^{-49}$ for all $i \leq T$. From [61].

11. Consider the doubling map given in Example VI.1(ii). Let x_0 be given in binary representation by

$$x_0 = 0\,.\,a_0 a_1 a_2 \ldots \qquad \text{with } a_i = 0 \text{ or } 1 .$$

The doubling map applied to x_0 has the effect that the a_i are shifted one position to the left and that the digit before the binary point is omitted. After i iterations we thus have

$$x_i = 0\,.\,a_i a_{i+1} a_{i+2} \cdots .$$

    a) After how many iterations is all information contained in x_0 lost, if the initial value is specified with an accuracy of n digits?

    b) Rational numbers have (binary) representations with periodic tails. What does this imply for the solutions with a rational number as initial value?

    c) Show that initial values of the form $x_0 = m/2^n (\bmod 1)$ with m and n integer give rise to solutions which converge to the origin.

12. Calculate the Lyapunov exponent of the logistic map as a function of its parameter μ in the range $[0,4]$. Relate the results to the information given in Fig. VI.3.

VII

Numerical Analysis of Multistep Methods

In §1 we introduce linear multistep methods (LMM) and show some important types. The useful concept of consistency is introduced in §2. Since LMM possess a richer (i.e. higher dimensional) solution space than the problem they discretise, it is important to ensure that the difference equations produce meaningful approximations; in particular one needs so-called root stability, cf. §3. This stability concept and also consistency are shown to be necessary and sufficient for convergence of the numerical solution to the exact one, as is worked out in §4. In §5 we consider the problem of how to obtain enough initial conditions to start the LMM and also give an asymptotically sharp estimate of the global error. For practical purposes it is important to implement implicit LMM jointly with explicit ones; this so-called predictor-corrector technique is dealt with in §6. Finally, in §7 we consider an important aspect of multistep implementation, viz. variable step, variable order algorithms.

1. Linear multistep methods

In §I.6 we have seen that discretising an ODE can also lead to a multistep method; for example, if the integral in the equation

$$(1.1) \qquad x(t_{i+1}) = x(t_i - \tau) + \int_{t_i - \tau}^{t_{i+1}} f\Big(s, x(s)\Big)\, ds$$

is approximated by interpolating quadrature, involving more than two grid points t_j, $j = i + 1, i, \dots$. If we use e.g. t_{i+1}, t_i and t_{i-1} and second order equispaced polynomial approximation (i.e. Simpson's rule), we obtain the formula

$$x_{i+1}^h - x_{i-1}^h = h \left[\tfrac{1}{3} f(t_{i+1}, x_{i+1}^h) + \tfrac{4}{3} f(t_i, x_i^h) + \tfrac{1}{3} f(t_{i-1}, x_{i-1}^h) \right].$$

We may also use a higher order interpolation polynomial on (t_{i-j}, t_{i+1}) or (t_{i-j}, t_i), $j \geq 1$, and integrate this over one time interval (t_i, t_{i+1}). A second order equispaced interpolation polynomial on t_{i-1}, t_i and t_{i+1} then results in

$$x_{i+1}^h - x_i^h = h \left[\tfrac{5}{12} f(t_{i+1}, x_{i+1}^h) + \tfrac{8}{12} f(t_i, x_i^h) + \tfrac{1}{12} f(t_{i-1}, x_{i-1}^h) \right] ,$$

which clearly is an implicit formula. If we use this polynomial on t_{i-2}, t_{i-1} and t_i instead, we obtain

$$x_{i+1}^h - x_i^h = h \left[\tfrac{23}{12} f(t_i, x_i^h) - \tfrac{16}{12} f(t_{i-1}, x_{i-1}^h) + \tfrac{5}{12} f(t_{i-2}, x_{i-2}^h) \right] .$$

A final source for constructing multisteps comes from using an interpolation formula for x and differentiating this at one of the grid points. Again for a second order interpolation polynomial on t_{i-1}, t_i and t_{i+1} we find the (implicit) formula

$$\frac{1}{h} \left[\tfrac{3}{2} x_{i+1}^h - 2x_i^h + \tfrac{1}{2} x_{i-1}^h \right] = f(t_{i+1}, x_{i+1}^h) ,$$

and using the same polynomial but now for approximating at t_i, we obtain the (explicit) formula

$$\frac{1}{h} \left[\tfrac{1}{2} x_{i+1}^h - \tfrac{1}{2} x_{i-1}^h \right] = f(t_i, x_i^h) .$$

Not all methods mentioned above are equally useful. In this chapter we shall analyse some general properties, such as convergence, and consider their application in variable step, variable order implementations.

A *linear multistep method* (LMM) is generally given by

$$(1.2) \qquad \sum_{j=0}^{k} \alpha_j \, x_{i-j+1}^h = h \sum_{j=0}^{k} \beta_j \, f(t_{i-j+1}, x_{i-j+1}^h) .$$

Note that all examples above can be written this way. This method is apparently a k-th order difference equation. If $\beta_0 = 0$ the method is explicit, otherwise it is implicit. Recalling the shift operator E (see §1.5) we can rewrite (1.2) as

$$\left(\sum_{j=0}^{k} \alpha_j \, E^{k-j} \right) x_{i-k+1}^h = h \left(\sum_{j=0}^{k} \beta_j \, E^{k-j} \right) f(t_{i-k+1}, x_{i-k+1}^h) ,$$

or

$$(1.3a) \qquad \rho(E) \, x_{i-k+1}^h = h \, \sigma(E) \, f(t_{i-k+1}, x_{i-k+1}^h) ,$$

where

$$(1.3b) \qquad \rho(\lambda) := \sum_{j=0}^{h} \alpha_j \, \lambda^{k-j} , \qquad \sigma(\lambda) := \sum_{j=0}^{k} \beta_j \, \lambda^{k-j} .$$

We review the most important classes of methods in the following example.

Example VII.1.

(i) Determine the coefficients of σ of the interpolation polynomial for f
at the points $t_{i-k+1}, ..., t_{i+1}$ and integrate the latter on the interval
(t_i, t_{i+1}). We then find the so-called *Adams-Moulton formula*

$$x_{i+1}^h - x_i^h = h \sum_{j=0}^{k} \beta_j \, f(t_{i-j+1}, x_{i-j+1}^h) \ .$$

Note that $\alpha_0 = 1$, $\alpha_1 = -1$ and $\alpha_j = 0$, $j = 2, ..., k$. For
$k = 0, 1, 2, 3$ the coefficients β_j are given in Table VII.1 (see
[42, 37]).

Table VII.1

k	β_0	β_1	β_2	β_3
0	1			
1	$\frac{1}{2}$	$\frac{1}{2}$		
2	$\frac{5}{12}$	$\frac{8}{12}$	$-\frac{1}{12}$	
3	$\frac{0}{14}$	$\frac{19}{24}$	$-\frac{5}{12}$	$\frac{1}{24}$

Note that for $k = 0$ we obtain Euler backward and for $k = 1$
we find the trapezoidal formula (both one-step!) (see also (A.6) in
Appendix A).

(ii) Determine the coefficients of σ of the interpolation polynomial for
f at the points $t_{i-k+1}, ..., t_{i+1}$ and integrate this on the interval
(t_i, t_{i+1}). We then find the so-called *Adams-Bashforth formula*

$$x_{i+1}^h - x_i^h = h \sum_{j=1}^{k} \beta_j \, f(t_{i-j+1}, x_{i-j+1}^h) \ .$$

Table VII.2

k	β_1	β_2	β_3	β_4
1	1			
2	$\frac{3}{2}$	$-\frac{1}{2}$		
3	$\frac{23}{12}$	$-\frac{16}{12}$	$\frac{5}{12}$	
4	$\frac{55}{24}$	$-\frac{59}{24}$	$\frac{37}{24}$	$-\frac{9}{24}$

Note that $\alpha_0 = 1$, $\alpha_1 = -1$ and $\alpha_j = 0$, $j = 2, ..., k$. For $k = 1, 2, 3, 4$ the coefficients are given in Table VII.2. (See also (A.5) in Appendix A.)

(iii) Determine the coefficients of ρ by differentiating (for $t = t_{i+1}$) an interpolation polynomial at $t_{i-k+1}, ..., t_{i+1}$. One can also view the resulting method by taking backward differences; we return to this in Chapter VIII. We obtain the so-called *Backward Differences Formula* (BDF)

$$\sum_{j=0}^{k} \alpha_j \, x_{i-j+1}^h = h \, f(t_{i+1}, x_{i+1}^h) \, ,$$

where for $k = 1, 2, 3, 4$ the coefficients α_j are given in Table VII.3. Note that for $k = 1$ we again have Euler backward.

Table VII.3

k	α_0	α_1	α_2	α_3	α_4
1	1	-1			
2	$\frac{3}{2}$	-2	$\frac{1}{2}$		
3	$\frac{11}{6}$	-3	$\frac{3}{2}$	$-\frac{1}{3}$	
4	$\frac{25}{12}$	-4	3	$-\frac{4}{3}$	$\frac{1}{4}$

2. Consistency; local errors

As with one-step methods a basic requirement for an LMM is that the difference formula is consistent with the differential equation. This means that the LMM reproduces the ODE if $h \to 0$. Hence we can introduce a *local discretisation error* δ, being the difference between the exact value of x at t_{i+1} and its approximation, divided by h, assuming exact input data given. In other words, it is the residual as defined in

$$(2.1) \qquad \delta(x(t_{i+1}), h) := \frac{1}{h} \left[\rho(E) \, x(t_{i-k+1}) - h \, \sigma(E) \, f \Big(t_{i-k+1}, x(t_{i-k+1}) \Big) \right] .$$

If $\delta(\cdot) = O(h^p)$, $p \geq 1$, we call the LMM *consistent of order p*.

Property 2.2. *The LMM (1.2) is consistent if and only if $\rho(1) = 0$, $\dot{\rho}(1) = \sigma(1)$.*

Proof: Use Taylor expansions of $x(t_{i-j+1})$ around t_{i-k+1}, i.e.

$$x(t_{i-j+1}) = x(t_{i-k+1}) + (k-j)\,\dot{x}(t_{i-k+1}) + O(h^2)$$

$$h\,\dot{x}(t_{i-j+1}) = h\,\dot{x}(t_{i-k+1}) + O(h^2) .$$

Collecting terms gives

$$\delta(\cdot) = \frac{1}{h}\left\{ \left(\sum_{j=0}^{k} \alpha_j\right) x_{i-k+1} + h\sum_{j=0}^{k} (\alpha_j(k-j) - \beta_j)\,\dot{x}_{i-k+1}\right\} +$$

$$+ O(h^2) .$$

Clearly the first term is zero if and only if $\rho(1) = 0$, whereas the second (zeroth order) term is zero if and only if $\dot{\rho}(1) = \sigma(1)$. $\quad\square$

In §1 we have introduced LMM through polynomial approximation. This then leads to the conclusion that such a method has consistency order p if it is exact for polynomials of degree p or less (i.e. solution $x \in \Pi_p$, see Appendix A). This is shown in the following:

Property 2.3. *The LMM (1.2) is of order p if and only if it is exact for a polynomial solution of degree p or less.*

Proof: Let (1.2) have consistency order p. By Taylor expansions of the local error, around $t = t_0$ and up to order p, we find (for $p \geq 1$, cf. Property 2.2)

$$\delta(\cdot) = \frac{1}{h}\left\{\sum_{q=2}^{p} h^q \sum_{j=0}^{k} [\alpha_j(i-j+1)^q - \beta_j\,h^q(i-j+1)^{q-1}]\cdot \right.$$

$$\left. \cdot \frac{x^{(q)}(t_0)}{q!}\right\} + O(h^{(p)}) .$$

Since the terms between { } must disappear for each q $(2 \leq q \leq p)$ we find

$$\sum_{j=0}^{k} h^q\,\alpha_j(i-j+1)^q = h\sum_{j=0}^{k} h^{q-1}\,\beta_j(i-j+1)^{q-1} .$$

This is precisely the LMM applied to a problem with solution $x(t) = t^q$. If, on the other hand, the LMM is exact for polynomial solutions up to degree p, then the Taylor argument shows that all terms with factors h^q, $q < p$, disappear.$\square$

Since our general LMM has a degree of freedom we may as well scale it such that our definition of consistency includes

$$\frac{1}{h}\sum_{j=0}^{k} \alpha_j\,x(t_{i-j+1}) \to \dot{x}(t_{i+1}) \qquad \text{if } h \to 0 .$$

Hence $\dot{\rho}(1) = 1$, which we shall assume henceforth. Note that the trivial 'ρ-polynomial' for a one-step method, viz. $\rho(\lambda) = \lambda - 1$, automatically satisfies $\dot{\rho}(1) = 1$.

Example VII.2.
In general one may find the local error through Taylor expansions. If we have the so-called *midpoint rule*

$$\tfrac{1}{2} x^h_{i+1} - \tfrac{1}{2} x^h_{i-1} = h f(t_i, x^h_i) \,,$$

then we see that $\rho(x) = \tfrac{1}{2} \lambda^2 - \tfrac{1}{2}$, $\sigma(\lambda) = \lambda$. Clearly we have consistency $(\rho(1) = 0, \dot{\rho}(1) = \sigma(1) = 1)$. More precisely we find by expansion at $t = t_i$

$$x(t_{i+1}) = x(t_i) + h \dot{x}(t_i) + \frac{h^2}{2} \ddot{x}(t_i) + \frac{h^3}{6} \dddot{x}(t_i) + O(h^4)$$

$$x(t_{i-1}) = x(t_i) - h \dot{x}(t_i) + \frac{h^2}{2} \ddot{x}(t_i) - \frac{h^3}{6} \dddot{x}(t_i) + O(h^4) \,.$$

Hence $\delta(x(t_{i+1}), h) = \tfrac{1}{6} h^2 \dddot{x}(t_i) + O(h^3)$.

More generally LMM have a local error of the form

$$(2.4) \qquad \delta\big(x(t_{i+1}), h\big) = c\, h^p\, x^{(p+1)}(t_{i+1}) + O(h^{p+1}) \,.$$

The quantity c is called the *error constant* and depends on the method, not on the solution.

Given the $2k$ parameters for the implicit case and $2k - 1$ for the explicit case one might think that there are $2k - 1$ and $2k - 2$ free parameters, respectively, to be used for obtaining maximal order. Unfortunately this is not the case. We have more requirements, on ρ in particular, as we show in the next section.

3. Root stability

One should realise that LMM are higher order difference equations, needing e.g. more than one initial value (if $k \geq 2$). Applied to a linear ODE this means that they have a k-dimensional linear manifold of solutions, implying that some basis solutions (i.e. satisfying the homogeneous part) are nonphysical; we shall call the latter *parasitic solutions*. A problem not met with one-step methods is the controlling of these parasitic solutions in such a way that they do not affect the computed result in a serious way. As we saw in §V.8 this can be investigated through Lyapunov stability or total stability.

First consider the simple ODE

$$\dot{x} = 0 \,.$$

Clearly the k-step LMM gives

$$\sum_{j=0}^{k} \alpha_j x^h_{i-j+1} = 0 \,.$$

If we write this as a first order system (cf. (I.5.4), (V.8.4))

$$\mathbf{x}_{i+1}^h = \mathbf{A}\,\mathbf{x}_i^h \ ,$$

we conclude that the matrix $\mathbf{A}$ must be stable. We shall show now that this requirement, for the zeros of ρ, is sufficient for the aforementioned stability in more general situations. We thus have the following:

Definition 3.1. *An LMM is called root (or zero-) stable if all roots of ρ are at most 1 in modulus and those which have modulus 1 are simple.*

Note that consistency implies $\rho(1) = 0$. The root 1 is called the *essential root*, the other ones *parasitic roots*.

This root stability imposes further restrictions on the choice of the coefficients $\{\alpha_i\}$ and $\{\beta_i\}$. In fact we have the following result, which we give without proof (cf. [21]).

Theorem 3.2 (Dahlquist barrier). *The maximum order of a k-step LMM is $k+1$ for k odd and $k+2$ for k even.*

Example VII.3.
Consider the implicit two-step based on Simpson's rule:

$$\frac{1}{2}x_{i+1}^h - \frac{1}{2}x_{i-1}^h = \frac{h}{6}\left[f(t_{i+1}, x_{i+1}^h) + h\,f(t_i, x_i^h) + f(t_{i-1}, x_{i-1}^h) \right] \ .$$

Since $\rho(\lambda) = \lambda^2 - 1$, it follows that the essential root equals 1 and the parasitic root equals -1. Simpson's rule is based on second order interpolation, but is exact for all skew symmetric functions, centred around t_i as well, and in particular for $(t - t_i)^3$. As this rule is fourth order, it has optimal order.

We now consider the more general linear ODE

$$(3.3) \quad \begin{cases} \dot{x} = \mu(t)\,x + g(t) \ , \\ x(0) = x_0 \ . \end{cases}$$

In what follows statements can be generalised to nonlinear ODE as well; we omit this for the sake of clarity. The corresponding difference equation then reads

$$(3.4) \quad \sum_{j=0}^{k}\left[\alpha_j - h\,\beta_j\,\mu(t_{i-j+1}) \right] x_{i-j+1}^h = h\sum_{j=0}^{k}\beta_j\,g(t_{i-j+1}) \ .$$

In order to write this as a matrix-vector system we define

$$
(3.5) \qquad \mathbf{x}_i^h := \begin{bmatrix} x_i^h \\ \vdots \\ \vdots \\ x_{i+k-1}^h \end{bmatrix} , \qquad \mathbf{g}_i^h := \begin{bmatrix} 0 \\ \vdots \\ \dfrac{h \sum\limits_{j=0}^{k} \beta_j \, g(t_{i-j+k})}{\alpha_0 - h\,\beta_0\,\mu(t_{i+k})} \end{bmatrix} ,
$$

$$
\mathbf{A}_i^h := \begin{bmatrix} 0 & 1 & & & \\ & 0 & 1 & & \\ & & \ddots & \ddots & \\ & & & 0 & 1 \\ \gamma_i^h & & \cdots & & \gamma_{i+k-1}^h \end{bmatrix} ,
$$

where

$$
\gamma_{i-j+k}^h = \left[h\,\beta_j\,\mu(t_{i-j+k}) - \alpha_j \right] / \left[\alpha_0 - h\,\beta_0\,\mu(t_{i+k}) \right] , \qquad j = 1, \dots, k .
$$

We obtain then

$$
(3.6) \qquad \mathbf{x}_{i+1}^h = \mathbf{A}_i^h\,\mathbf{x}_i^h + \mathbf{g}_i^h , \qquad i \geq 0 .
$$

One should realise that an initial vector $\mathbf{x}_0^h$ is not directly available from the original IVP as only the first coordinate is given in (3.3). For $h = 0$, the matrix $\mathbf{A}_i^h$ becomes the constant companion matrix (cf. (I.4.5b)) with ρ as its characteristic polynomial.

Theorem 3.7. Let $ih \leq T$ for some $T < \infty$ and $h < \hat{h} := \dfrac{|\alpha_0|}{|\beta_0|\,\max\,|\mu(t)|}$. Let $\mathbf{A}^0 := \mathbf{A}_0^0$ $(= \mathbf{A}_i^0$, for all $i > 0)$. Assume that the LMM is root stable, i.e. $\|(\mathbf{A}^0)^i\| \leq K$ for some $K < \infty$. Define the constant M by

$$
M := \max_t \frac{|\mu(t)|}{|\alpha_0|\,(|\alpha_0| - |h\,\mu(t)\,\beta_0|)} \left\{ |\alpha_0| \sum_{j=1}^{k} |\beta_j| + |\beta_0| \sum_{j=1}^{k} |\alpha_j| \right\} .
$$

Then for $i \geq l$

$$
\left\| \prod_{j=l}^{i-1} \mathbf{A}_j^h \right\| \leq K(1 + h\,KM)^{i-l} \leq K\,e^{KM\,h(i-l)} \leq K\,e^{KMT} .
$$

Proof: Write $\mathbf{A}_i^h =: \mathbf{A}^0 + \mathbf{B}_i^h$, then for the last row of $\mathbf{B}_i^h$ (the only one which is nontrivial), say $(\bar{\gamma}_i^h, \dots, \bar{\gamma}_{i+k}^h)$, we have

$$
\bar{\gamma}_{i-j+k}^h = \frac{\alpha_j - h\,\beta_j\,\mu(t_{i-j+k})}{h\,\beta_0\,\mu(t_{i+k})} + \frac{\alpha_j}{\alpha_0} =
$$

$$= h \, \frac{\alpha_j \, \beta_0 \, \mu(t_{i+k}) - \alpha_0 \, \beta_j \, \mu(t_{i-j+k})}{\alpha_0 (h \, \beta_0 \, \mu(t_{i+k}) - \alpha_0)} \, .$$

The product $\mathbf{A}^h_{i-1} \ldots \mathbf{A}^h_l$ is a sum of terms, each of which consists of factors $\mathbf{B}^h_m$ ($l \leq m \leq i-1$) surrounded by factors which are powers of $\mathbf{A}_0$. A typical term with j such factors $\mathbf{B}^h_m$ can be bounded in norm by $K^{j+1} \max_m \|\mathbf{B}^h_m\|$. A combinatorial argument then shows that

$$\|\mathbf{A}^h_{i-1} \ldots \mathbf{A}^h_l\| \leq \sum_{j=0}^{i-l} \binom{i-l}{j} K^{j+1} \max \|\mathbf{B}^h_m\|^j \leq$$

$$\leq K \sum_{j=0}^{i-1} \binom{i-l}{j} (KhM)^j = K(1 + h\,KM)^{i-l} \, . \quad \square$$

From Theorem 3.7 we conclude that the difference equation (3.4) is totally stable for $h < \hat{h}$ (uniform in i, h, as long as $ih \leq T$).

One may expect the upper bound in the estimate in Theorem 3.7 to be sharper in special cases. However, the exponential character cannot be improved upon in general, even for $\mu < 0$; this is shown by the next example.

Example VII.4.
The midpoint rule reads

$$\tfrac{1}{2} x^h_{i+1} - \tfrac{1}{2} x^h_{i-1} = h \, f(t_i, x^h_i) \, .$$

If we let $f = \mu x$ ($\mu < 0$), then we obtain for (3.4)

$$\tfrac{1}{2} x^h_{i+1} - h \, \mu \, x^h_i - \tfrac{1}{2} x^h_{i-1} = 0 \, ,$$

and for the matrix $\mathbf{A}^h_i$ (cf. (3.5))

$$\mathbf{A}^h_i = \begin{bmatrix} 0 & 1 \\ 1 & 2h\,\mu \end{bmatrix} \, .$$

Its eigenvalues are

$$\rho^h_1 = \sqrt{1 + \mu^2 h^2} + \mu\,h \doteq 1 + \mu h + \tfrac{1}{2} \mu^2 h^2$$

$$\rho^h_2 = -\sqrt{1 + \mu^2 h^2} + \mu\,h \doteq -1 + \mu h - \tfrac{1}{2} \mu^2 h^2 \, .$$

Although the essential root ρ^0_1 induces a proper approximation (viz. $x^h_{i+1} = (1 + \mu\,h)\,x^h_i \doteq e^{\mu\,h} x^h_i$) the parasitic root $\rho^0_2 = -1$ induces an unstable mode $(y^h_{i+1} = (-1 + \mu\,h)\,y^h_i \doteq -e^{-\mu\,h} y^h_i)$. Hence $\|\mathbf{A}^h_{i-1} \ldots \mathbf{A}^h_l\| \approx e^{|\mu|(i-j)h}$. $\quad \square$

The previous example shows a phenomenon that may occur more generally for so-called *weakly stable* LMM, viz. for which ρ has more than one root on the unit circle.

Example VII.5.
If we use an interpolation polynomial for f at the points $t_{i-k+1}, ..., t_i$ and integrate this on t_{i-1}, t_{i+1}, we obtain a so-called *Nyström method*

$$\frac{1}{2} x_{i+1}^h - \frac{1}{2} x_{i-1}^h = h \sum_{j=1}^{k} \beta_j f(t_{i-j+1}, x_{i-j+1}^h) \ .$$

Note that $\rho(\lambda)$ is divisible by $\lambda^2 - 1$. Hence the method is weakly stable. Apparently the midpoint rule is the member of this family with $k = 1$.

4. Convergence

As in §III.3 we define the *global error* by

$$(4.1) \qquad e_i^h := x(t_i) - x_i^h \ .$$

In contrast to one-step methods, the interpretation of an LMM applied to an ODE as a perturbation of the latter is not so straightforward, due to the higher dimensionality of the solution manifold. We therefore choose another approach and project the exact solution onto the space where the discrete solution is defined, i.e. we only consider x at the grid points t_i. In fact we know from our definition of local discretisation error that the thus 'projected' exact solution can be viewed to satisfy a perturbed difference equation (viz. the local error). A convergence proof then intuitively follows from the total stability shown in the previous section.

An additional complication is that we need to produce approximations of $x_1^h, ..., x_{k-1}^h$ first. An obvious way to obtain x_1^k etc. from x_0 is through some RK method. For codes employing a family of LMM (of various orders) we start off with the one-step member and gradually increase the order. Clearly the step size choice is a matter of concern in both cases, as we like equispaced points. For a low order starting procedure the initial step size is likely to be quite small.

We now have the following assumption in the remainder of this section:

Assumption 4.2.

(a) Let the solution x exist on $[0,T]$, and let f be Lipschitz continuous on $[0,T] \times \Omega$ with Lipschitz constant L.

(b) Let $|x_j^h - x(t_j)| \leq Q\,h^q$, $Q, q > 0$, for $j = 1, ..., k-1$.

Note that the first requirement generalises the linear case, as treated before, in a standard way.

We obtain the following convergence result:

Theorem 4.3 (Convergence). *Let the LMM be root stable, and let K be bound as in Theorem 3.7 ($\|(\mathbf{A}^0)^i\| \leq K$ for all i), and have consistency order p, i.e. let $|\delta(\cdot)| \leq P\,h^p$ for $P, p > 0$. Define moreover the following constants:*

$$\hat{h} := \frac{\alpha_0}{\beta_0\,L}, \quad M := \frac{L}{|\alpha_0|\,(|\alpha_0| - h_0\,\beta_0\,L)} \sum_{j=1}^{k} |\alpha_0\,\beta_j| + |\beta_0\,\alpha_j|,$$

$$for\ h_0 < \hat{h}\ .$$

Then for $ih \leq T$ we have the global error estimate

$$|e_i^h| \leq e^{KT\,t_i}\left[\frac{P}{M}\,h^p + KQ\,h^q\right].$$

Proof: We only consider the linear case (cf. (3.3)), as the nonlinear case fairly straightforwardly (though tediously) follows through a Lipschitz argument. Let $\|\mathbf{x}\| = \|\mathbf{x}\|_\infty := \max_{j=1,\ldots,h} |x_j|$. For simplicity we shall write δ_i^h instead of $\delta(x(t_{i+1}), h)$. Clearly the exact solution satisfies the perturbed difference equation

$$\sum_{j=1}^{k}[\alpha_j - h\,\beta_j\,\mu(t_{i-j+1})]\,x(t_{i-j+1}) = h\sum_{j=0}^{k}\beta_j\,g(t_{i-j+1}) + h\,\delta_i^h\ .$$

Defining (cf. (3.5))

$$\mathbf{e}_i^h := \begin{bmatrix} e_i^h \\ \vdots \\ e_{i+k-1}^h \end{bmatrix}, \quad \mathbf{d}_0^h := \begin{bmatrix} e_0^h \\ \vdots \\ e_{k-1}^h \end{bmatrix},$$

$$\mathbf{d}_i^h := \begin{bmatrix} 0 \\ \vdots \\ 0 \\ h\,\delta_{i+k}^h/[\alpha_0 - h\,\beta_0\,\mu(t_{i+k})] \end{bmatrix},$$

the associated global error discrete IVP reads

$$\begin{cases} \mathbf{e}_{i+1}^h = \mathbf{A}_i^h + \mathbf{d}_i^h\ , \\ \mathbf{e}_0^h = \mathbf{d}_0^h\ . \end{cases}$$

Hence

$$e_i^h = \sum_{j=0}^{i-1} \prod_{l=j+1}^{i-1} A_l^h \, d_j^h \; .$$

From Theorem 3.7 we deduce that this can be bounded by

$$\|e_i^h\| \leq \sum_{j=0}^{i-1} K \, e^{KM \, h(i-j-1)} \max_j \|d_j\| + K \, e^{KM \, hi} \|d_0^h\| \leq$$

$$\leq K \, \frac{e^{KM \, hi} - 1}{e^{KM \, h} - 1} \, P \, h^{p+1} + K \, e^{KM \, hi} \, Q \, h^q \; .$$

Since $\dfrac{1}{e^{KM \, h} - 1} \leq \dfrac{1}{KM \, h}$ the bound easily follows. $\square$

Remark 4.4. The proof for the nonlinear case should use the Lipschitz constant of the LMM which (in contrast to the one for RK methods) follows fairly directly from L.

Remark 4.5. We directly replaced the 'real local error' (based on the solution $x_i(t)$, see §III.2) by the one obtained from substituting x instead. Anticipating the convergence result, this will give errors of similar order (again as in Chapter III).

We now formally define convergence through the following:

Definition 4.6. *For the class of problems as indicated in Assumption 4.2 the LMM with consistency order p is called convergent if $\lim\limits_{h \to 0} x_i^h = x(t)$, for all $t = ih \leq T$.*

Clearly the *convergence order* equals $\min(p, q)$. Although we may expect starting errors to die out for an asymptotically stable problem (see §V.2), it is obvious that one should try to ensure $q \geq p$, so that the global error equals p throughout the interval. If a low order starting procedure is used this can effectively be achieved by taking small steps. If RK methods are employed they should obviously be of order p at least.

In Chapter III we saw for RK methods that in fact consistency and convergence were equivalent (for a suitable class of IVP). Here we can show an analogue for LMM:

Theorem 4.7. *If Assumption 4.2 holds, then an LMM is convergent if and only if it is root stable and consistent.*

Proof: The 'if' is already shown in Theorem 4.3. So we consider the 'only if': let the LMM be convergent.

(a)   Suppose the LMM is not stable. Then consider the IVP

$$\begin{cases} \dot{x} = 0 \,, \\ x(0) = 0 \,. \end{cases}$$

Apparently there exists a homogeneous solution of the induced difference equation, say $\{y_i\}_{i=0}^{\infty}$, such that $y_i = i\, y_0$ or $y_i = \mu^i\, y_0$, with $|\mu| > 1$. Defining $z_i^h := y_i \sqrt{h}$ we see that $\{z_i^h\}_{i=1}^{\infty}$ is a solution of the difference equation; $\{z_i^h\}$ satisfies Assumption 4.2(b) and the discretised IVP as well. So $\lim\limits_{h \to 0} |z_i^h| = \infty$, which contradicts $|z_i^h - z(ih)| \to 0$, $h \to 0$.

(b$_1$) Next consider the IVP

$$\begin{cases} \dot{x} = 0 \,, \\ x(0) = 1 \end{cases}$$

(with $x(t) \equiv 1$). Choose $x_1^h = \ldots x_{k-1}^h = 1$. Since $x_i^h \to x(ih) = 1$ we have $\rho(1) = \sum\limits_{j=0}^{k} \alpha_j = \lim\limits_{h \to 0} \sum\limits_{j=0}^{k} \alpha_j\, x_{i-j+1}^h = 0.$

(b$_2$) From part (a) it follows that $\dot{\rho}(1) \neq 0$, as this would imply a double root on the unit circle and thus contradicts root stability. Consider the IVP

$$\begin{cases} \dot{x} = 1 \,, \\ x(0) = 0 \end{cases}$$

(with $x(t) = t$). Clearly the approximate solution $\{x_i\}$ satisfies

(*) $\sum\limits_{j=0}^{k} \alpha_j\, x_{i-j+1}^h = h\, \sigma(1) \,.$

If we now define $\{y_i^h\}$ with $y_i^h := ih\, \sigma(1)/\dot{\rho}(1)$, then $\lim\limits_{\substack{h \to 0 \\ j \leq k-1}} y_i^h = 0$ (i.e. $\{y_i^h\}$ satisfies Assumption 4.2(b)). $\{y_i^h\}$ is also a solution of (*), since

$$\sum_{j=0}^{k} \alpha_j \, y_{i-j+1}^h = \frac{h \, \sigma(1)}{\dot{\rho}(1)} \sum_{j=0}^{k} \alpha_j (i - j + 1)$$

$$= \frac{h \, \sigma(1)}{\dot{\rho}(1)} \left\{ (i - k + 1) \sum_{j=0}^{k} \alpha_j + \sum_{j=0}^{k} (k - j) \alpha_j \right\}$$

$$= h \, \sigma(1) \ .$$

So, $y_i^h \to x(ih)$, $h \to 0$, i.e. $\displaystyle\lim_{\substack{h \to 0 \\ ih = t}} ih \frac{\sigma(1)}{\dot{\rho}(1)} = t \Rightarrow \sigma(1) = \dot{\rho}(1).$ □

5. Asymptotic expansion of the global error

As in Chapter III we can give a more precise expression for the global error, not depending on the (often crude) Lipschitz constant, as was the case in Theorem 4.3. We have the following:

Theorem 5.1. *Let the LMM be root stable and have consistency order p. Let Assumption 4.2 hold with $q = p$. Then there exists a function $d \in C^1([0, T] \times \Omega \to I\!\!R)$, the so-called global error function, that satisfies (cf. (2.4))*

$$\begin{cases} \dot{d} = \dfrac{\partial f}{\partial x} (t, x) \, d + c \, x^{(p+1)}(t) \ , \\[2mm] d(0) = 0 \ . \end{cases}$$

This d is such that

$$e_i^h = d\Big(t_i, x(t_i)\Big) \, h^p + O(h^{p+1}) \ .$$

Proof: From Taylor expansion we find

$$f(t_j, x_j^h) - f\Big(t_j, x(t_j)\Big) = \frac{\partial f}{\partial x}\Big(t_j, x(t_j)\Big) + \frac{1}{2} \frac{\partial^2 f}{\partial x^2} (t_j, x_j^*)(e_j^h)^2 \ ,$$

where x_j^* is some intermediate value. Hence

$$\sum_{j=0}^{k} \left[\alpha_j - h \, \beta_j \, \frac{\partial f}{\partial x} \Big(t_{i-j+1}, x(t_{i-j+1})\Big) \right] e_{i-j+1}^h =$$

$$= h^{p+1} \, c \, x^{(p+1)}(t_i) + O(h^{p+2})$$

(here we have used the fact that $e_j^h = O(h^p)$, according to Theorem 4.3). Expanding $x^{(p+1)}(t_i)$ at t_{i-j+1}, $0 \leq j \leq k$, and using $\sum_0^k \beta_j = 1$ results in

$$\sum_{j=0}^{k} \left[\alpha_j - h\,\beta_j \frac{\partial f}{\partial x}\left(t_{i-j+1}, x(t_{i-j+1})\right) \right] e_{i-j+1}^h =$$

$$= \sum_{j=0}^{k} h^{p+1}\,\beta_j\,c\,x^{(p+1)}(t_{i-j+1}) + O(h^{p+2}) \ .$$

Using the *magnified error* $\hat{e}_i^h := e_i^h / h^p$, we obtain

$$\sum_{j=0}^{k} \left[\alpha_j - h_j\,\beta_j \frac{\partial f}{\partial x}\left(t_{i-j+1}, x(t_{i-j+1})\right) \right] \hat{e}_{i-j+1}^h =$$

$$= \sum_{j=0}^{k} h\,\beta_j\,c\,x^{(p+1)}(t_{i-j+1}) + O(h^{2}) \ .$$

This is precisely our LMM applied to the ODE of the global error function with an additional source term $O(h^2)$. With the estimate of Theorem 4.3 and noting that starting errors $\hat{e}_0^h, ..., \hat{e}_{k-1}^h$ are $O(h)$, we then find that $|d(t_i) - \hat{e}_i^h| = O(h)$, from which the desired result follows.   □

We shall not dwell on techniques for estimating the global error or for employing them in an actual procedure. The reasons for this are similar to what we said about the use of global error estimates in Chapter III: it is too expensive, not in the least due to the evolutionary character of the problem, which would require finding approximate solutions on ever increasing intervals consecutively.

6. Predictor-corrector techniques

One of the problems of implicit methods is that they need some iteration procedure, which in turn requires starting data. In this section we consider iteration through successive substitution, and use an explicit method for starting this process. Consider the following implicit LMM:

$$(6.1) \qquad \sum_{j=0}^{k} \bar{\alpha}_j\,x_{i-j+1}^h = h \sum_{j=0}^{k} \bar{\beta}_j\,f(t_{i-j+1}, x_{i-j+1}^h) \qquad (\bar{\beta}_0 \neq 0, \ \bar{\alpha}_0 = 1) \ .$$

Clearly, if $x_{i-k+1}^h, ..., x_i^h$ are known then x_{i+1}^h is an equilibrium point of the equation

(6.2) $y = \varphi(y) + \psi$,

where

$$\varphi(y) := h\,\bar{\beta}_0\, f(t_{i+1}, y) ,$$

$$\psi := -\sum_{j=1}^{k} \left[\bar{\alpha}_j\, x^h_{i-j+1} - h\,\bar{\beta}_j\, f(t_{i-j+1}, x^h_{i-j+1}) \right] .$$

Given an approximate value y_0 we can find a sequence $\{y_l\}_{l \geq 1}$ from the iteration

$$y_{l+1} = \varphi(y_l) + \psi , \quad l \geq 0 .$$

If f is Lipschitz continuous in x, with Lipschitz constant L, we find

$$|y_{l+1} - y| \leq |\varphi(y_l) - \varphi(y)| = h\,|\bar{\beta}_0|\,|f(t_{i+1}, y_l) - f(t_{i+1}, y)| \leq$$

$$\leq h\,|\bar{\beta}_0|\, L\,|y_l - y| .$$

Hence, if only $h\,|\bar{\beta}_0|\, L < 1$ and $|y_0 - y|$ is small enough, $\{y_l\}$ will converge to the desired solution of (6.2), viz. x^h_{i+1}. Consequently, if $|\partial f/\partial x|$ is only moderate, this convergence will be assured for not unduly small h (hopefully commensurate with accuracy requirements).

An approximate value for x^h_{i+1} is found by applying an *explicit* LMM, say

(6.3) $$\sum_{j=0}^{k} \hat{\alpha}_j\, x^h_{i-j+1} = h \sum_{j=1}^{h} \hat{\beta}_j\, f(t_{i-j+1}, x^h_{i-j+1}) , \qquad \hat{\alpha}_0 = 1 .$$

The method (6.3) is called the *predictor* and (6.1) the *corrector*.

There exist two predictor-corrector strategies basically: their main distinction is whether at time $t = t_{i+1}$ the f-values are found from the predictor or from the corrector.

(i)   *PEC algorithm.*
      P: *predict*, i.e. determine $x^h_{i+1}(\text{pred})$ from

$$x^h_{i+1}(\text{pred}) := -\sum_{j=1}^{k} \hat{\alpha}_j\, x^h_{i-j+1}(\text{cor}) +$$

$$+ h \sum_{j=1}^{k} \hat{\beta}_j\, f\!\left(t_{i-j+1}, x^h_{i-j+1}(\text{pred}) \right) .$$

      E: *evaluate* $f\!\left(t_{i+1}, x^h_{i+1}(\text{pred}) \right)$.
      C: *correct*, i.e. determine $x^h_{i+1}(\text{cor})$ from

$$x_{i+1}^h(\text{cor}) := -\sum_{j=1}^{k} \bar{\alpha}_j\, x_{i-j+1}^h(\text{cor}) +$$

$$+ h \sum_{j=0}^{k} \bar{\beta}_j\, f\left(t_{i-j+1}, x_{i-j+1}^h(\text{pred})\right).$$

Note that only 'corrected' x_{i+1}^h are found (and not $f(t_{i+1}, x_{i+1}^h)$-values).

(ii) *PECE algorithm.*
P: *predict*, i.e. determine $x_{i+1}^h(\text{pred})$ from

$$x_{i+1}^h(\text{pred}) := -\sum_{j=1}^{k} \hat{\alpha}_j\, x_{i-j+1}^h(\text{cor}) +$$

$$+ h \sum_{j=1}^{k} \hat{\beta}_j\, f\left(t_{i-j+1}, x_{i-j+1}^h(\text{cor})\right).$$

E: *evaluate* $f\left(t_{i+1}, x_{i+1}^h(\text{pred})\right)$.
C: *correct*, i.e. determine $x_{i+1}^h(\text{cor})$ from

$$x_{i+1}^h(\text{cor}) := -\sum_{j=1}^{k} \bar{\alpha}_j\, x_{i-j+1}^h(\text{cor}) +$$

$$+ h \sum_{j=1}^{k} \bar{\beta}_j\, f\left(t_{i-j+1}, x_{i-j+1}^h(\text{cor})\right) +$$

$$+ h\, \bar{\beta}_0\, f\left(t_{i+1}, x_{i+1}^h(\text{pred})\right).$$

E: *evaluate* $f\left(t_{i+1}, x_{i+1}^h(\text{cor})\right)$.
Note that both 'corrected' x_{i+1}^h and $f(t_{i+1}, x_{i+1}^h)$ are found.

Remark 6.4. One can perform more correction steps. For m steps we obtain $P(EC)_m$ and $P(EC)_m E$ analogues to **(i)** and **(ii)** respectively.

Example VII.6.
Consider the trapezoidal rule

$$x_{i+1}^h = x_i^h + \frac{h}{2}\left[f(t_i, x_i^h) + f(t_{i+1}, x_{i+1}^h)\right],$$

and Euler forward

$$x_{i+1}^h = x_i^h + h\, f(t_i, x_i^h)\,.$$

The latter can be used as predictor for the former. In a PECE mode, with $\bar{x}_i^h := x_i^h(\text{cor})$, we then find

$$\bar{x}_{i+1}^h = \bar{x}_i^h + \frac{h}{2}\left[f(t_i, \bar{x}_i^h) + f\Big(t_{i+1}, \bar{x}_i^h + h\, f(t_i, \bar{x}_i^h)\Big)\right]\,.$$

This example happens to give rise to *Heun's second order method* (which we already met in Example III.1(iv).   □

Although the corrector method would require an infinite number of iterations to obtain the theoretical limit (fixed point), it turns out that a finite number suffices to obtain an approximation within a tolerance set by the discretisation error. Actually the situation is much better, so that already a PEC or PECE algorithm (i.e. one correction step) does a good job, see Property 6.5. For clarity we shall denote the local discretisation error of the predictor and the corrector by $\delta_i^h(\text{pred})$ and $\delta_i^h(\text{cor})$ respectively.

Property 6.5. Let $\delta_i^h(\text{pred}) = O(h^q)$ be the local discretisation error of the predictor and $\delta_i^h(\text{cor})$ the local discretisation error of the corrector. Then we have for the local discretisation error of the PECE mode, say $\delta_i^h(\text{PECE})$,

$$(6.5a)\qquad \delta_i^h(\text{PECE}) := \delta_i^h(\text{cor}) - \bar{\beta}_0\, h\, \frac{\partial f}{\partial x}\Big(t_{i+1}, x(t_{i+1})\Big)\,\delta_i^h(\text{pred}) + O(h^{q+2})\,.$$

Proof: We substitute the exact solution into the PECE formula and determine the residual (we assume properly normed coefficients so that $\sigma(1) = 1$). We obtain:

$$\delta_i^h(\text{PECE}) = \delta_i^h(\text{cor}) + \bar{\beta}_0\left[f\Big(t_{i+1}, x(t_{i+1}) - h\,\delta_i^h(\text{pred})\Big)\right.$$

$$\left. - f\Big(t_{i+1}, x(t_{i+1})\Big)\right] =$$

$$= \delta_i^h(\text{cor}) - h\,\bar{\beta}_0\, \frac{\partial f}{\partial x}\Big(t_{i+1}, x(t_{i+1})\Big)\,\delta_i^h(\text{pred}) + O(h^{q+2})\,.\qquad □$$

We thus obtain the following important result:

Corollary 6.6. If $\delta_i^h(\text{cor}) = O(h^q)$ and $q \geq p - 1$ then $\delta_i^h(\text{PECE}) = O(h^p)$. For example, we find that the PECE mode in Example VII.6 results in a second order method. An interesting class of multisteps where this predictor-corrector technique can be used are Adams methods, of which Table VII.1 and VII.2

contain the simplest members. Here one uses Adams-Bashforth as predictor
and Adams-Moulton as corrector. A practical implementation of this family is
considered in §7.

We conclude this section by giving a practical method to estimate the local
errors (which are of the same importance in practice here as in the one-step
case). Suppose the predictor and the corrector have the same consistency order
p. Then we see from Property 6.5 that the error constant (i.e. the coefficient
of the principal term) of a PECE algorithm equals that of the corrector, so in
first order $\delta_i^h(\text{PECE}) = \delta_i^h(\text{cor})$. Let

$$(6.7) \quad \begin{cases} \delta_i^h(\text{pred}) &= c_p\, h^p\, x_i^{(p+1)}(t_i) + O(h^{p+1}) \\ \delta_i^h(\text{cor}) &= c_c\, h^p\, x_i^{(p+1)}(t_i) + O(h^{p+1}) , \end{cases}$$

then

$$(6.8) \quad x_{i+1}^h(\text{pred}) - x_{i+1}^h(\text{cor}) = (c_p - c_c)\, h^{p+1}\, x_i^{(p+1)}(t_i) + O(h^{p+2}) ,$$

which then gives us

$$(6.9) \quad \delta_i^h(\text{PECE}) = \frac{c_c}{h(c_p - c_c)}\, [x_{i+1}^h(\text{pred}) - x_{i+1}^h(\text{cor})] + O(h^{p+1}) .$$

This so-called *Milne's device* (cf. [53]) is an analogue of a local extrapolation
method.

Of course we also may use embedding. Suppose we also have a $(p+1)$-st order
corrector, which gives an approximation, $x_{i+1}^n(\text{cor2})$ say, and leaves a local
discretisation error, say $\delta_i(\text{cor2})$, with

$$(6.10) \quad \delta_i^h(\text{cor2}) = c_{c2}\, h^{p+1}\, x_i^{(p+2)}(t_i) + O(h^{p+2}) .$$

Then we obtain

$$(6.11) \quad \delta_i^h(\text{cor}) = \delta_i^h(\text{cor}) - \delta_i^h(\text{cor2}) + O(h^{p+1}) =$$

$$= [x_{i+1}^h(\text{cor2}) - x_{i+1}(\text{cor})]/h + O(h^{p+1}) .$$

By using a family of methods (notably Adams methods) one can compute this
$x_{i+1}^h(\text{cor2})$ rather inexpensively. Of course analogues with PECE modes (cf.
(6.5a)) are straight forward.

7. Variable step, variable order algorithms

Although much of the foregoing analysis applies to LMM in general, the most
important family for nonstiff problems (see Chapter VIII) is given by Adams
methods. We recall that a k-step (explicit) Adams-Bashforth formula and a
k-step (implicit) Adams-Moulton formula pair is given by

$$(7.1) \qquad x_{i+1}^h = x_i^h + h \sum_{j=1}^k \hat{\beta}_j \, f(t_{i-j+1}, \bar{x}_{i-j+1}^h)$$

and

$$(7.2) \qquad \bar{x}_{i+1}^h = x_i^h + h \sum_{j=0}^k \bar{\beta}_j \, f(t_{i-j+1}, \bar{x}_{i-j+1}^h)$$

respectively. Note that the former is of order k and the latter of order $k + 1$; for some values of $\{\hat{\beta}_j\}$ and $\{\bar{\beta}_j\}$ see Table VII.2 and Table VII.1 respectively.

Adams methods are trivially seen to be root stable (apart from the essential root 1, all roots are 0), have maximum consistency order and, as it turns out, also possess some interesting properties that make them very suitable for practical implementation in a so-called *variable step-variable order algorithm*: changing a step is less trivial for LMM than for one-step methods. Given a method it is therefore sometimes more convenient to change the order. Also, error estimation has to be done using formulae with different orders.

In order to fully benefit from their relationship it is useful to reformulate the Adams formulae in terms of backward differences (cf. (I.5.8c)). Define

$$(7.3a) \qquad \nabla^0 x_i := x_i \; ,$$

$$(7.3b) \qquad \nabla^j x_i := \nabla^{j-1} x_i - \nabla^{j-1} x_{i-1} \; .$$

This yields

$$\nabla x_i = x_i - x_{i-1}$$

$$\nabla^2 x_i = x_i - x_{i-1} - (x_{i-1} - x_{i-2}) = x_i - 2x_{i-1} + x_{i-2}$$

$$\vdots$$

$$\nabla^k x_i = \sum_{j=0}^k (-1)^j \binom{k}{j} x_j \; .$$

This backward difference notation can be applied to $x(t_i)$ and $f(t_i, x_i^h)$ in a similar (obvious) way. Denoting the uncorrected function value by

$$f_i := f(t_i, x_i^h) \; ,$$

and the corrected function value by

$$\bar{f}_i := f(t_i, \bar{x}_i^h) \; ,$$

we see, e.g., that $\nabla^k f_{i+1}$ (the k-th backward difference) is a linear combination of $f_{i+1}, \ldots, f_{i-k+1}$; clearly $\nabla^j f_{i+1}$, for $j < k$, can also be thought of as a linear combination of this set. Hence, we can rewrite (7.1) and (7.2) as

(7.4) $\qquad x_{i+1}^h = \bar{x}_i^h + h \sum_{j=0}^{k-1} \hat{\gamma}_j \nabla^j \bar{f}_i$

and

(7.5) $\qquad \bar{x}_{i+1}^h = \bar{x}_i^h + h \sum_{j=0}^{k} \bar{\gamma}_j \nabla^j \tilde{f}_{i+1}$,

where the tilde in $\nabla^j \tilde{f}_{i+1}$ is there to remind the reader that the differences are employing the uncorrected f_{i+1}-values. Actually, in Appendix A it is shown that $\{\hat{\gamma}_j\}$ and $\{\bar{\gamma}_j\}$ are independent of k.

If we use a $(k+1)$-step (so $(k+2)$nd order) Adams-Moulton method, say

(7.6) $\qquad \bar{\bar{x}}_{i+1}^h = \bar{x}_i^h + h \sum_{j=0}^{k+1} \bar{\gamma}_j \nabla^j \tilde{f}_{i+1}$,

then we can conclude from (6.11) that the local discretisation of the k-step corrector, δ_i^h (k-step) is estimated by

(7.7) $\qquad \delta_i^h (k-\text{step}) \doteq [\bar{\bar{x}}_{i+1}^h - x_{i+1}^h]/h = \hat{\gamma}_{k+1} \nabla^{k+1} \tilde{f}_{i+1} =: \text{EST}_i$.

Though this gives an easy criterion for estimating the error, the actual procedure for employing this is a (new) problem by itself, as a possibly new step size requires values of x and/or of f at unknown grid points. A practical way is to employ interpolation on f-values as follows: we write

(7.8) $\qquad f(t_i + (1-j)\omega h) := f\left(t_i + (1-j)\omega h, x_i^h((1-j)\omega h)\right)$,

where $x_i^h((1-j)\omega h)$ is a (numerical) approximation of $t_i + (1-j)\omega h$. The quantity ω is the factor by which the step size should be decreased or increased. If we use k-th order interpolation, we only make errors consistent with the (k-th order) method.

An attractive procedure to store appropriate information and change it in a simple fashion uses the so-called *Nordsieck vector* $\mathbf{V}_i^h$ (cf. [57]),

(7.9) $\qquad \mathbf{V}_i^h := \left(x(t_i), h\,\dot{x}(t_i), \frac{h^2}{2}\,\ddot{x}(t_i), ..., \frac{h^k}{k!}\,\frac{d^h}{dt^h}\,x(t_i) \right)^T$.

The coordinate values of $\mathbf{V}_i^h$ are found from differentiations of the interpolation polynomial after multiplication by factor $h^l/l!$, $l = 0, ..., k$; this interpolation polynomial in turn can be written in Lagrangian form (with polynomials in t as coefficients of f_{i-j}). Since Adams formulae are based on integration of these interpolation polynomials, we can make use of sums of components of the Nordsieck vector.

Example VII.7.
Consider the second degree interpolation polynomial at t_{i-2}, t_{i-1}, t_i:

$$p(t) := \frac{(t - t_i)(t - t_{i-1})}{2h^2} f_{i-2} + \frac{(t - t_i)(t - t_{i-2})}{-h^2} f_{i-1} +$$

$$+ \frac{(t - t_i)(t - t_{i-2})}{2h^2} f_i \ .$$

We obtain

$$\dot{p}(t) = \frac{1}{2h} f_{i-2} - \frac{2}{h} f_{i-1} + \frac{3}{2} \frac{1}{h} f_i \ ,$$

$$\ddot{p}(t_i) = \frac{1}{h^2} f_{i-2} - \frac{2}{h^2} f_{i-1} + \frac{1}{h^2} f_i \ .$$

Hence $\mathbf{V}_i^h$ is given by $(f_i := f(t_i, x(t_i))$

$$\mathbf{V}_i^h = \left(x(t_i), h\, f_i, \frac{h}{4} \, [3f_i - 4f_{i-1} + f_{i-2}], \right.$$

$$\left. \frac{h}{6} \, [f_i - 2f_{i-1} + f_{i-2}] \right)^T \ .$$

The two-step Adams-Bashforth is precisely the sum of the coefficients:

$$x(t_{i+1}) = x(t_i) + h \left[\tfrac{23}{12} f_i - \tfrac{16}{12} f_{i-1} + \tfrac{5}{12} f_{i-2} \right] \ .$$

Given the vector $\mathbf{V}_i^h$ it is clearly simple to change the step length from h to ωh.

As we remarked above one may also wish to change the order of the method. Suppose we have estimates of the $(k-1)$st, the k-th and the $(k+1)$st order method (cf. (7.7)), i.e.

(7.10) $\mathrm{EST}_i^{(j)} := \bar{\gamma}_j \, \nabla^j \, \tilde{f}_{i+1} \ , \qquad j = k-1, k, k+1 \ .$

If we use e.g. the EPUS criterion (see III.6(ii)) and have a tolerance TOL, we investigate which of the three step size increment estimates

$$\left| \frac{\mathrm{TOL}}{\mathrm{EST}_i^{(j)}} \right|^{\frac{1}{j}} , \qquad j = k-1, k, k+1$$

is the largest. For this j we estimate a new step size h_{new} via

(7.11) $h_{\mathrm{new}} := h_i \left| \dfrac{\mathrm{TOL}}{\mathrm{EST}_i^{(j)}} \right|^{\frac{1}{j}} \ .$

In this book we shall use an implementation of this idea in the PASCAL procedure Adams (cf. [43], [68]).

Example VII.8.
As in Chapter III we consider the model problem

$$\begin{cases} \dot{x} = \lambda x \, , \\ x(0) = 1 \, , \end{cases}$$

for a positive and a negative value of λ and for absolute and relative error control. Ideally EST_i should behave like $\lambda^j e^{\lambda t_i}$ for a $(j+1)$st order method, i.e. $|\text{TOL}/\text{EST}_i|^{(1/j)} \approx |\text{TOL}/e^{\lambda t_i}|^{\frac{1}{j}}/\lambda$. If λt_i is such that $|\text{TOL}/\text{EST}_i| > 1$ a lower order will be chosen. A relative control will not influence this choice qualitatively. We shall demonstrate this order below, where the examples have $\lambda = \pm 2$ and ABSTOL and RELTOL are 10^{-2}.

(i) In Table VII.4 we see that for $\lambda = 2$ and ABSTOL $= 10^{-2}$ the highest order is indeed preferred. Note that the order nicely increases (starting from $k = 1$ of course) and that the step size is doubled each time, until it becomes constant. In Example III.5(i) the step size was gradually decreasing. The total number of steps for ABSTOL becomes 103, as compared to 56 for Fehlberg (see last column of Table VII.6). However, for Fehlberg one has to perform 6 function evaluations per step, so that Adams is more efficient nevertheless here.

(ii) In Table VII.5 we have printed similar values to those in Table VII.4 but now for RELTOL $= 10^{-2}$. Qualitatively the results are the same as under (i), cf. also Example III.5(ii).

(iii) For the practically more relevant case $\lambda = -2$ we see in Table VII.6 that for ABSTOL $= 10^{-2}$ the order first increases and then decreases; this can be explained from the growth of the quantity $|\text{TOL}/e^{-2t}|$.

(iv) Finally for RELTOL $= 10^{-2}$ and $\lambda = -2$ we have qualitatively similar results to those under (ii) but in a somewhat weaker appearance, because the error is also divided by a factor $e^{-\lambda t_i}$; see Table VII.7.

Table VII.4 $\lambda = 2$, ABSTOL $= 10^{-2}$.

i	t_i	h_i	e_i^h	$e_i^h/x(t_i)$	x_i^h	k	nfe
0					1.00E+00		
1	1.768E-02	1.768E-02	-7.43E-06	-7.17E-06	1.04E+00	1	3
2	5.303E-02	3.536E-02	-1.03E-05	-9.28E-06	1.11E+00	2	5
3	1.237E-01	7.071E-02	-1.38E-05	-1.07E-05	1.28E+00	3	7
4	2.652E-01	1.414E-01	-3.35E-05	-1.97E-05	1.70E+00	4	9
5	5.480E-01	2.828E-01	3.85E-05	1.29E-05	2.99E+00	5	11
6	1.114E+00	5.657E-01	-2.05E-02	-2.21E-03	9.26E+00	6	13
7	1.299E+00	1.853E-01	-3.36E-02	-2.50E-03	1.34E+01	5	16
8	1.484E+00	1.853E-01	-5.11E-02	-2.63E-03	1.94E+01	5	18
9	1.670E+00	1.853E-01	-7.61E-02	-2.70E-03	2.81E+01	5	20
10	1.855E+00	1.853E-01	-1.12E-01	-2.75E-03	4.07E+01	5	22
11	2.040E+00	1.853E-01	-1.65E-01	-2.79E-03	5.90E+01	5	24
12	2.225E+00	1.853E-01	-2.42E-01	-2.83E-03	8.55E+01	5	26
13	2.411E+00	1.853E-01	-3.53E-01	-2.84E-03	1.24E+02	6	28
14	2.596E+00	1.853E-01	-5.12E-01	-2.85E-03	1.79E+02	6	30
15	2.781E+00	1.853E-01	-7.44E-01	-2.86E-03	2.60E+02	7	32
16	2.967E+00	1.853E-01	-1.08E+00	-2.86E-03	3.76E+02	7	34
17	3.152E+00	1.853E-01	-1.56E+00	-2.86E-03	5.45E+02	8	36
18	3.337E+00	1.853E-01	-2.27E+00	-2.86E-03	7.90E+02	8	38
19	3.522E+00	1.853E-01	-3.29E+00	-2.87E-03	1.14E+03	9	40
20	3.708E+00	1.853E-01	-4.76E+00	-2.87E-03	1.66E+03	9	42
21	3.893E+00	1.853E-01	-6.90E+00	-2.87E-03	2.40E+03	10	44
22	4.078E+00	1.853E-01	-1.00E+01	-2.87E-03	3.48E+03	10	46
23	4.264E+00	1.853E-01	-1.45E+01	-2.87E-03	5.04E+03	11	48
24	4.449E+00	1.853E-01	-2.10E+01	-2.87E-03	7.29E+03	11	50
25	4.634E+00	1.853E-01	-3.05E+01	-2.87E-03	1.06E+04	10	52
26	4.819E+00	1.853E-01	-4.41E+01	-2.88E-03	1.53E+04	11	54
27	5.000E+00	1.853E-01	-6.34E+01	-2.88E-03	2.20E+04	12	56

Table VII.5 $\lambda = 2$, RELTOL $= 10^{-2}$.

i	t_i	h_i	e_i^h	$e_i^h/x(t_i)$	x_i^h	k	nfe
0					1.00E+00		
1	1.768E-02	1.768E-02	-7.43E-06	-7.17E-06	1.04E+00	1	3
2	5.303E-02	3.536E-02	-1.03E-05	-9.28E-06	1.11E+00	2	5
3	1.237E-01	7.071E-02	-1.38E-05	-1.07E-05	1.28E+00	3	7
4	2.652E-01	1.414E-01	-3.35E-05	-1.97E-05	1.70E+00	4	9
5	5.480E-01	2.828E-01	3.85E-05	1.29E-05	2.99E+00	5	11
6	1.114E+00	5.657E-01	-2.05E-02	-2.21E-03	9.26E+00	6	13
7	1.559E+00	4.448E-01	-1.40E-01	-6.22E-03	2.24E+01	5	15
8	2.003E+00	4.448E-01	-6.23E-01	-1.13E-02	5.43E+01	5	17
9	2.448E+00	4.448E-01	-2.26E+00	-1.69E-02	1.32E+02	5	19
10	2.893E+00	4.448E-01	-7.31E+00	-2.24E-02	3.18E+02	5	21
11	3.338E+00	4.448E-01	-2.20E+01	-2.78E-02	7.71E+02	5	23
12	3.783E+00	4.448E-01	-6.39E+01	-3.31E-02	1.87E+03	5	25
13	4.228E+00	4.448E-01	-1.71E+02	-3.65E-02	4.53E+03	6	27
14	4.672E+00	4.448E-01	-4.51E+02	-3.94E-02	1.10E+04	6	29
15	5.000E+00	4.448E-01	-9.16E+02	-4.16E-02	2.11E+04	7	31

Table VII.6 $\lambda = -2$, ABSTOL $= 10^{-2}$.

i	t_i	h_i	e_i^h	$e_i^h/x(t_i)$	x_i^h	k	nfe
0					1.00E+00		
1	1.768E-02	1.768E-02	7.30E-06	7.56E-06	9.65E-01	1	3
2	5.303E-02	3.536E-02	4.63E-06	5.15E-06	8.99E-01	2	5
3	1.237E-01	7.071E-02	6.51E-06	8.34E-06	7.81E-01	3	7
4	2.652E-01	1.414E-01	-5.07E-06	-8.61E-06	5.88E-01	4	9
5	5.480E-01	2.828E-01	3.00E-05	8.98E-05	3.34E-01	5	11
6	8.309E-01	2.828E-01	3.05E-04	1.61E-03	1.90E-01	5	14
7	1.114E+00	2.828E-01	5.90E-04	5.47E-03	1.08E-01	5	16
8	1.397E+00	2.828E-01	7.05E-04	1.15E-02	6.19E-02	5	18
9	1.679E+00	2.828E-01	6.56E-04	1.89E-02	3.54E-02	5	20
10	1.962E+00	2.828E-01	5.21E-04	2.64E-02	2.03E-02	5	22
11	2.528E+00	5.657E-01	-1.01E-03	-1.58E-01	5.36E-03	6	24
12	3.094E+00	5.657E-01	1.34E-03	6.54E-01	3.40E-03	5	26
13	3.659E+00	5.657E-01	4.06E-03	6.13E+00	4.73E-03	5	28
14	4.225E+00	5.657E-01	3.73E-04	1.74E+00	5.87E-04	4	30
15	5.356E+00	1.131E+00	-2.40E-02	-1.08E+03	-2.40E-02	4	32
16	6.233E+00	8.765E-01	-2.38E-02	-6.16E+03	-2.38E-02	3	34
17	7.022E+00	7.888E-01	-4.03E-03	-5.06E+03	-4.03E-03	3	36
18	7.811E+00	7.888E-01	1.00E-02	6.09E+04	1.00E-02	3	38
19	8.599E+00	7.888E-01	5.51E-03	1.62E+05	5.51E-03	3	40
20	9.388E+00	7.888E-01	-5.25E-03	-7.50E+05	-5.25E-03	3	42
21	1.000E+01	7.888E-01	-2.87E-03	-1.39E+06	-2.87E-03	4	44

Table VII.7 $\lambda = -2$, RELTOL $= 10^{-2}$.

i	t_i	h_i	e_i^h	$e_i^h/x(t_i)$	x_i^h	k	nfe
0					1.00E+00		
1	1.768E-02	1.768E-02	7.30E-06	7.56E-06	9.65E-01	1	3
2	5.303E-02	3.536E-02	4.63E-06	5.15E-06	8.99E-01	2	5
3	1.237E-01	7.071E-02	6.51E-06	8.34E-06	7.81E-01	3	7
4	2.652E-01	1.414E-01	-5.07E-06	-8.61E-06	5.88E-01	4	9
5	5.480E-01	2.828E-01	3.00E-05	8.98E-05	3.34E-01	5	11
6	8.309E-01	2.828E-01	3.05E-04	1.61E-03	1.90E-01	5	14
7	1.114E+00	2.828E-01	5.90E-04	5.47E-03	1.08E-01	5	16
8	1.397E+00	2.828E-01	7.05E-04	1.15E-02	6.19E-02	5	18
9	1.679E+00	2.828E-01	6.56E-04	1.89E-02	3.54E-02	5	20
10	1.962E+00	2.828E-01	5.21E-04	2.64E-02	2.03E-02	5	22
11	2.245E+00	2.828E-01	2.25E-04	2.01E-02	1.14E-02	6	24
12	2.528E+00	2.828E-01	9.65E-05	1.51E-02	6.47E-03	6	26
13	2.811E+00	2.828E-01	9.45E-05	2.61E-02	3.71E-03	7	28
14	3.094E+00	2.828E-01	1.26E-05	6.13E-03	2.07E-03	7	30
15	3.376E+00	2.828E-01	7.97E-06	6.83E-03	1.18E-03	6	32
16	3.659E+00	2.828E-01	2.12E-05	3.20E-02	6.84E-04	5	34
17	3.942E+00	2.828E-01	5.34E-06	1.42E-02	3.82E-04	4	36
18	4.225E+00	2.828E-01	5.11E-06	2.39E-02	2.19E-04	5	38
19	4.366E+00	1.414E-01	4.17E-06	2.59E-02	1.65E-04	5	41
20	4.508E+00	1.414E-01	3.08E-06	2.54E-02	1.25E-04	5	43
21	4.649E+00	1.414E-01	2.35E-06	2.57E-02	9.39E-05	5	45
22	4.791E+00	1.414E-01	1.77E-06	2.57E-02	7.08E-05	5	47
23	5.000E+00	2.828E-01	1.13E-06	2.48E-02	4.65E-05	5	49

Exercises Chapter VII

1. Determine the local discretisation error of the method (cf. §1)

$$\tfrac{3}{2}\, x_{i+1}^h - 2x_i^h + \tfrac{1}{2}\, x_{i-1}^h = h\, f(t_{i+1}, x_{i+1}^h) \ .$$

2. a) Find the linear interpolation polynomial $p(t)$ of $g(t)$ with $p(t_i) = g(t_i)$, $p(t_{i-1}) = g(t_{i-1})$, and use this to determine the ρ and σ polynomial of the two-step Adams-Bashforth formula.

    b) Determine the local discretisation error.

3. a) Find the quadratic interpolation polynomial $p(t)$ of $g(t)$ with $p(t_j) = g(t_j)$, $j = i - 1, i, i + 1$, and use this to determine the ρ- and σ-polynomial of the two-step Adams-Moulton formula.

    b) Determine the local discretisation error.

4. Show that the trapezoidal rule is a one-step Adams-Moulton method. Determine the order.

5. a) Determine the coefficients of the two-step method with the highest consistency order.

    b) The same as a) but now with the requirement that it is root stable.

6. Simpson's method reads:

$$x_{i+1}^h = x_{i-1}^h + \frac{h}{3} \left[f(t_{i-1}, x_{i-1}^h) + 4 f(t_i, x_i^h) + f(t_{i+1}, x_{i+1}^h) \right] .$$

Show that it has 'maximum order'.

7. Consider the following two-step family

$$x_{i+1}^h - (1+a)\, x_i^h + a\, x_{i-1}^h =$$
$$= \tfrac{1}{2} h \left[(3-a) f(t_i, x_i^h) - (1+a) f(t_{i-1}, x_{i-1}^h) \right] .$$

    a) Show that these methods are consistent.

    b) For which values of a do we have root stability?

8. Consider the family of three-step methods

$$x_{i+1}^h + a\, x_i^h - a\, x_{i-1}^h - x_{i-2}^h =$$
$$= \tfrac{1}{2} h \left[(3+a) f(t_i, x_i^h) + (3+a) f(t_{i-1}, x_{i-1}^h) \right] .$$

    a) For which values of a do we have root stability?
        Hint: Realise that 1 should always be a root.

    b) Show that the order cannot exceed 2 for a root-stable member.

9. a) Apply the trapezoidal rule to $\dot{x} = \lambda x$.

    b) Determine the global error at t for a step size h.

10. Consider the following predictor-corrector pair:

$$P : x_{i+1}^h + 4x_i^h - 3x_{i-1}^h = h \left[f(t_i, x_i^h) + f(t_{i-1}, x_{i-1}^h) \right]$$

$$C : x_{i+1}^h - x_{i-1}^h = \frac{h}{3} \left[f(t_{i-1}, x_{i-1}^h) + 4f(t_i, x_i^h) + f(t_{i-1}, x_{i-1}^h) \right] .$$

    a) Show that P is not root stable and C only weakly root stable (i.e. the ρ-polynomial has two roots on the unit circle).

    b) Why is the lack of stability of the predictor not important in a PECE mode?

c) Determine the recursion we obtain when applying a PECE algorithm to $\dot{x} = \lambda x$. Can you show that the zero solution of this Δ-equation is asymptotically stable for $|h\lambda|$ small, $\lambda < 0$?

11. If the Δ-equation

$$x_{i+1} = a_0\, x_i + \ldots + a_k\, x_{i-k+1}$$

is exact for polynomials of degree p $(p \leq k)$, i.e. $x_i = i^p$, then its characteristic polynomial $\rho(\lambda)$ is divisible by a factor $(\lambda - 1)^p$. Show this. Hint: Recall the proof of Property 2.3.

12. Recall the k-th order Adams pair:

$$x_{i+1}^h = x_i^h + h \sum_{j=1}^{k} \hat{\beta}_j\, f(t_{i-j+1}, x_{i-j+1}^h) \qquad \text{(Bashforth)}$$

$$x_{i+1}^h = x_i^h + h \sum_{j=0}^{k-1} \bar{\beta}_j\, f(t_{i-j+1}, x_{i-j+1}^h) \qquad \text{(Moulton)} .$$

Show that

$$\hat{\beta}_k = (-1)^{k-1}\, \bar{\beta}_0$$

$$\hat{\beta}_j = \bar{\beta}_j + (-1)^j \begin{pmatrix} k \\ j \end{pmatrix} \bar{\beta}_0 , \qquad j = 2, \ldots, k-1 .$$

Hint: Both methods are exact for polynomials of degree $\leq k - 1$. If $x_i = (ih)^p$, $p \leq k$, then subtraction of the formulae would still be an exact expression. Now use Exercise 11.

VIII

Singular Perturbations and Stiff Differential Equations

In §1 we define singularly perturbed ODE, i.e. ODE where the highest order derivative has a coefficient depending on a small parameter ε, and analyse the solution behaviour when ε approaches zero. In §2 we introduce the method of matched asymptotic expansions, which gives approximations for the solution in terms of (truncated) power series in this parameter ε. In §3 it is shown that treating such problems numerically may impose a severe restriction on the step size for certain methods including all explicit ones). This so-called stiffness problem necessitates the introduction of new numerical stability notions, such as A-stability, as is done in §5, after a more precise scrutiny of the increment per step has been carried out in §4. The most important class of LMM for use in stiffness problems, viz. of backward difference formulae, is described in §6, along with some practical implementation aspects, including the step size control.

1. Singular perturbations

Consider a scalar higher order linear ODE. If we let the coefficient of the highest order term go to zero we obtain an equation of one order less; the latter equation also needs one initial condition less than the former. Now if this coefficient is very small but nonzero, we may expect solutions of this equation to resemble (in some way) appropriate solutions of the equation without the highest order term; but the limit case must have a *singular* behaviour in the sense that the solution of this problem cannot satisfy all initial conditions of the former. If we have, for example,

$$(1.1) \qquad \varepsilon \ddot{x} + \alpha \dot{x} + \beta x = 0 \,, \qquad |\varepsilon| \ll |\alpha|, |\beta| \,,$$

and $|\ddot{x}|$ is not unduly large, the solution x should resemble a solution of $\alpha \dot{x} + \beta x = 0$. The latter equation is referred to as the *reduced equation* corresponding to (1.1). In domains where $|\varepsilon \ddot{x}|$ is not negligible we have a

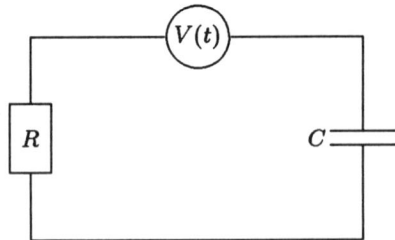

Figure VIII.1   Example of an electric circuit, which may give rise to a stiff ODE.

so-called *layer*; this is a transient where the solution of (1.1) rapidly moves to a solution of the reduced equation; the term $\varepsilon\ddot{x}$ is the *singular perturbation* of this reduced ODE. Such problems occur in many areas, e.g. in electric circuits.

Example VIII.1.
Consider a simple circuit with source $V(t)$, constant resistor R and constant capacitor C, see Fig. VIII.1. If we denote the current in the circuit by I and the voltage over the capacitor by U_C, we obtain $I = C\,dU_C/dt$. For the resistor we have $U_R = IR$. From Kirchhoff's second law (sum of all voltages in the circuit vanishes) we obtain

$$\frac{dV}{dt} = R\frac{dI}{dt} + \frac{I}{C}\ .$$

Suppose we have an alternating current $V(t) = \sin t$. If we define $\varepsilon := RC$ we find

$$\varepsilon\frac{dI}{dt} = -I + C\cos t\ .$$

In order to establish whether the ε term in this equation is small, the equation should be brought into dimensionless form. Because the technique of non-dimensionalisation is not yet dealt with, we only give the results: we may read the equation as a non-dimensional one, with $C = 1$, $\varepsilon \ll 1$, and I being of order 1. So in the following we take $C \equiv 1$. We shall provide I with a superscript ε to indicate its dependence on ε. Hence, if we further prescribe $I^\varepsilon(0) = 0$, the solution is given by (cf. (IV.1.3b))

$$I^\varepsilon(t) = \frac{1}{\varepsilon}e^{-t/\varepsilon}\int_0^t e^{s/\varepsilon}\cos s\,ds =$$

$$= \frac{1}{1+\varepsilon^2}\cos t + \frac{\varepsilon}{1+\varepsilon^2}\sin t - \frac{1}{1+\varepsilon^2}e^{-t/\varepsilon}\ .$$

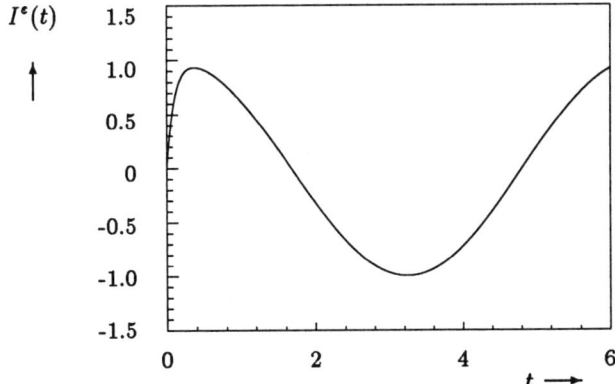

Figure VIII.2   Current $I^\varepsilon(t)$ as a function of time t in the circuit in **Fig. VIII.1**, for $\varepsilon \ll 1$.

For $t > \varepsilon$ we clearly see that $I^{(\varepsilon)}(t) \approx \cos t$ because $\varepsilon \ll 1$. Note that the first order ODE results in an algebraic equation (zeroth order ODE) after putting $\varepsilon = 0$, with precisely the solution $I^0(t) = \cos t$. It should be noted that $I^0(0) \neq 0$. Therefore the third term, i.e. $-(1/(1+\varepsilon^2))e^{-t/\varepsilon}$, is needed to describe the *transient* part, where I^ε rapidly goes from its initial value to the solution of the reduced problem. Clearly the term $(\varepsilon/(1+\varepsilon^2))\sin t$ does not matter very much for $\varepsilon \ll 1$. See Fig. VIII.2. We conclude that for all $t > 0$

$$\lim_{\varepsilon \to 0} I^\varepsilon(t) = I^0(t) \ .$$

However, for every $\varepsilon > 0$ we have

$$\lim_{t \downarrow 0} I^\varepsilon(t) = 0 \ .$$

The example explains the terminology *singular perturbation*: the limits $\varepsilon \to 0$ and $t \downarrow 0$ may not be interchanged.

One important conclusion from the foregoing example is that singular perturbation problems exhibit *more time scales*: the slowly varying solution corresponds to the slow time scale and the fast boundary layer solution to the fast time scale. It may be deceptive to identify the latter with a small parameter in front of the derivative exclusively. Indeed when replacing t/ε by τ we obtain the ODE

$$\frac{dI(\tau)}{d\tau} = I(\tau) + \cos(\tau/\varepsilon) \ .$$

Note, however, that the slow time scale, i.e. the one that properly describes the 'slow solution', is of course still a factor $1/\varepsilon$ larger than the fast time scale. In many practical problems one can often find dimensionless parameters which indicate the presence of multiple time scales and therefore require a careful treatment (which, as we shall see, is needed even more so numerically).

We may write the ODE in Example VIII.1 as

$$(1.2) \qquad \varepsilon \dot{x}^\varepsilon = -x^\varepsilon + b(t) \ , \qquad x^\varepsilon(0) = x_0 \ ,$$

with $0 < \varepsilon \ll 1$ and $b \in C^1$ (i.e. the space of continuously differentiable functions). This singularly perturbed ODE in fact shows all features of interest. If we take $\varepsilon = 0$ we obtain the solution of the *reduced equation*

$$x^0(t) = b(t) \ .$$

If $b(t) = O(1)$, then $|\varepsilon \dot{x}^0(t)| = \varepsilon |\dot{b}(t)| = O(\varepsilon)$. It does not satisfy the initial value at $t = 0$ in general. Hence we have a small region, the *initial (or boundary) layer* where $|\varepsilon \dot{x}^\varepsilon(t)| = O(1)$. The width of this layer can be found easily. If we have a small interval, $(0, \varphi(\varepsilon))$ say, then $b(t)$ is nearly constant there, i.e. $b(t) \doteq b(0)$. So we have approximately

$$(1.3) \qquad \begin{cases} \varepsilon \dot{x}^\varepsilon \doteq x^\varepsilon + b(0) \ , \\ x^\varepsilon(0) = x_0 \ . \end{cases}$$

This gives the solution

$$(1.4) \qquad x^\varepsilon(t) \doteq e^{-t/\varepsilon} x_0 + b(0)(1 - e^{-t/\varepsilon}) \ .$$

Clearly for $t \gtrsim \varepsilon$ the first term becomes less important than the second. Hence we may take $\varphi(\varepsilon) := \varepsilon$. In the layer we have a so-called *boundary layer correction term* $e^{-t/\varepsilon}$.

The observation above can be generalised. Consider the scalar autonomous singularly perturbed IVP where f is Lipschitz continuous:

$$(1.5) \qquad \begin{cases} \varepsilon \dot{x}^\varepsilon = f(x^\varepsilon) \ , & 0 < \varepsilon \ll 1 \ , \\ x^\varepsilon(0) = x(0) \ . \end{cases}$$

The reduced problem can now have more than one solution, to be found from solving $f(x^0) = 0$. We are interested in the stable root, i.e. we require that at $x = x^0$

$$(1.6) \qquad \frac{\partial f}{\partial x} < 0 \ .$$

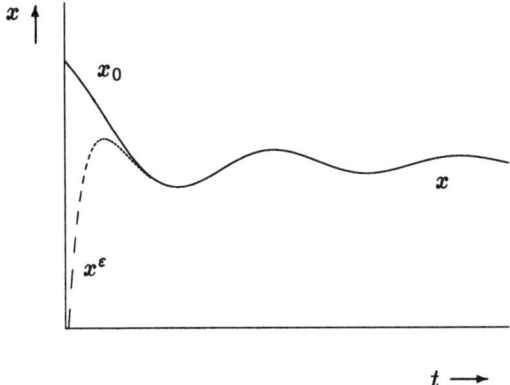

Figure VIII.3

Theorem 1.7. *Suppose the equation* $f(x) = 0$ *(cf. (1.5)) has at least one stable root* φ. *Denote the immediately preceding and following roots as* φ_1 *and* φ_2 *respectively, i.e.* $\varphi_1 < \varphi < \varphi_2$ *(if* φ_1 *or* φ_2 *does not exist we take* $\varphi_1 = -\infty$ *or* $\varphi_2 = \infty$). *If* $\varphi_1 < x_0 \le \varphi_2$ *then the solution* x^ε *of (1.5) exists for* $t \ge 0$, *whereas* $\lim_{\varepsilon \downarrow 0} x^\varepsilon(t) = 0$, $t > 0$.

Proof: Suppose $\varphi_1 < x_0 < \varphi$. For $\varphi < x_0 < \varphi_2$ the proof is similar. Since $\dot{x} = f(x)/\varepsilon > 0$ for $x_0 < x < \varphi$, we conclude that x^ε is increasing; at the same time x^ε must be bounded above by the equilibrium solution $x \equiv \varphi$. Hence x^ε exists according to Corollary II.3.5. Also $\lim_{t \to \infty} x^\varepsilon(t) = \varphi$, because we can find a strong Lyapunov function $V(x) := (x - \varphi)^2$ for which

$$D_f V = 2(x - \varphi)\, f(x)/\varepsilon < 0\,, \qquad \varphi_1 < x < \varphi_2 \quad (x \ne \varphi,\ \varepsilon > 0)\,;$$

hence φ is asymptotically stable in an attraction area (φ_1, φ_2).

To prove the asymptotics in ε we employ the monotonicity of $x^\varepsilon(t)$ for $\varphi_1 < x^\varepsilon(t) < \varphi$. Consider the IVP

$$(*) \qquad \begin{cases} \dfrac{dt^\varepsilon}{dx} = \varepsilon/f(x)\,, \\[2mm] t^\varepsilon(x_0) = 0\,. \end{cases}$$

This IVP is a 'regularly' perturbed problem. The unperturbed problem is $t^0 \equiv 0$. Because of monotonicity of x^ε the solution $t^\varepsilon(x)$ of $(*)$ exists for all $x_0 \le x \le \varphi - \mu$, μ arbitrarily small. Because of the perturbation Theorem II.5.2 the distance $|t^\varepsilon(x) - t^0|$ will become arbitrarily small for $\varepsilon \downarrow 0$ uniformly on $x_0 \le x \le \varphi - \mu$. $\qquad \square$

For higher order systems the time scales sometimes arise quite naturally in 'double deck' systems like

(1.8a)
$$\begin{cases} \dfrac{dy^\varepsilon}{dt} = f(t, y^\varepsilon, z^\varepsilon) \,, \\[2mm] \varepsilon \dfrac{dz^\varepsilon}{dt} = g(t, y^\varepsilon, z^\varepsilon) \,, \end{cases}$$

with initial conditions

(1.8b)
$$\begin{cases} y^\varepsilon(0) = y_0 \,, \\[2mm] z^\varepsilon(0) = z_0 \,. \end{cases}$$

Example VIII.2.
Consider a simple chemical reaction with a catalyst. Let A be the initial species and D be the expected end product, B a catalyst and C an intermediate product. Then the reactions may read as follows:

$$A + B \; \underset{k_2}{\overset{k_1}{\rightleftarrows}} \; C \; \overset{k_3}{\longrightarrow} \; B + D \,,$$

where k_1, k_2 and k_3 are reaction constants. Usually the concentration of B is much smaller than that of A and $k_2 \gg k_3$. Let the concentrations at $t = 0$ be A_0, B_0, and $C_0 = D_0 = 0$. Note that the sum of the concentrations of B and C is constant, and given by B_0. Using this the relevant reaction equations are

$$\frac{d}{dt} A = -k_1 (B_0 - C) A + k_2 C$$

$$\frac{d}{dt} C = k_1 (B_0 - C) A - (k_2 + k_3) C \,.$$

We first make these equations dimensionless (see §VIII.2) and introduce

$$\begin{aligned} y(t) &:= A(t)/A_0 \\ z(t) &:= C(t)/B_0 \\ \varepsilon &:= B_0/A_0 \\ p &:= (k_2 + k_3)/(k_1 A_0) \\ q &:= k_2/(k_1 A_0) \\ s &:= t k_1 B_0 \,. \end{aligned}$$

This leads to the following IVP:

$$\begin{cases} \dfrac{dy}{ds} = -(1-z)\,y + qz \;, \\[2mm] \varepsilon\,\dfrac{dz}{ds} = (1-z)\,y - pz \;, \qquad 0 < \varepsilon \ll 1 \;, \\[2mm] y(0) = 1 \;, \\[2mm] z(0) = 0 \;. \end{cases}$$

☐

For the double deck IVP (1.8) we can give a result akin to Theorem 1.7. First note that the reduced equation reads as

(1.9a) $g(t, y^0, z^0) = 0$.

We assume that we can write the solution of (1.9a) as

(1.9b) $z^0 = \varphi(t, y^0)$,

and that potentially multiple solutions are isolated in a relevant domain. A solution $z^0 = \varphi(t, y^0)$ is called *stable* on domain G, if

(1.10) $\dfrac{\partial g}{\partial z}(t, \varphi(t, y^0), y^0) < 0$, $(t, y^0) \in G$.

Theorem 1.11 (Tichonov). *Let there be constants $b > 0$, $d > 0$ and $T > 0$ and assume that*

(i)   *the partial derivatives with respect to y^ε and z^ε of f and g in (1.8a) are continuous in a domain*

$$H := \{(t, y) \in G, \; z < d\} \;,$$

where G is defined as

$$G := \{0 \le t \le T, \; |y| \le b\} \;.$$

(ii)   *the reduced solution z^0 (cf. (1.9b)) and $\partial z^0/\partial y^0$ are continuous on G.*

(iii)   *$\varphi(t, y^0)$ is a stable root.*

(iv)   *the solution y^0 of*

$$\begin{cases} \dfrac{dy^0}{dt} = f(t, y_0, \varphi(t, y^0)) \;, \\[2mm] y^0(0) = y_0 \end{cases}$$

is defined for $0 \le t \le T$ and stays in G.

(v) φ_1, φ_2 are defined as in Theorem 1.7, such that $\varphi_1 < z_0 < \varphi_2$.

Then $(y^\varepsilon, z^\varepsilon)^T$ exists on $[0, T]$ and

$$\lim_{\varepsilon \downarrow 0} y^\varepsilon(t) = y^0(t) , \qquad\qquad 0 \leq t \leq T$$

$$\lim_{\varepsilon \downarrow 0} z^\varepsilon(t) = \varphi(t, y^0(t)) , \qquad\qquad 0 < t < T . \qquad\qquad \square$$

The proof is rather technical but similar to that of Theorem 1.7. We emphasise that convergence of y^ε to y^0 is *uniformly* in ε but that of z^ε to z^0 is *not!*

2. Matched asymptotic expansions

The different time scales require different approximations. Here we shall investigate how this should be done when power series expansions in ε are employed. The idea is that we use different expansions inside and outside the layer regions and match the approximate solutions appropriately. Before going into detail we first dwell on the concept of asymptotic expansion.

Example VIII.3.
Consider the *exponential integral*

$$E(x) := \int_x^\infty \frac{e^{-t}}{t} \, dt , \qquad x > 0 .$$

Partial integration results in

$$E(x) = \frac{e^{-x}}{x} \left(1 - \frac{1}{x} + \frac{2!}{x^2} - \frac{3!}{x^3} + \dots + (-1)^{n-1} \frac{(n-1)!}{x^{n-1}} \right) +$$

$$+ R_n(x) ,$$

where the remainder term R_n is given by

$$R_n(x) := (-1)^{n-1} (n-1)! \int_x^\infty \frac{e^{-t}}{t^n} \, dt .$$

Note that we cannot let $n \to \infty$ since this series is *not converging*. It is still a useful expression, e.g. if we want to use it for larger x. So defining $\varepsilon := x^{-1}$ and $E^\varepsilon := \exp(1/\varepsilon) \, E(1/\varepsilon)$ we find

$$E^\varepsilon = \varepsilon - \varepsilon^2 + 2! \, \varepsilon^3 - 3! \, \varepsilon^4 + \dots + \bar{R}_n(\varepsilon) ,$$

where

$$\bar{R}_n(\varepsilon) := (-1)^n (n-1)! \exp(1/\varepsilon) \int\limits_{1/\varepsilon}^{\infty} \frac{e^{-t}}{t^n} \, dt \ .$$

Note that for $\bar{R}_n$ we have

$$|\bar{R}_n(\varepsilon)| \le (n-1)! \, \varepsilon^n \ .$$

Hence, if we keep n fixed we conclude that $\bar{R}_n(\varepsilon)$ exponentially converges to zero for $\varepsilon \downarrow 0$. □

The example above induces the following definition of an asymptotic expansion. Let $\mathbf{x}^\varepsilon \in \mathbb{R}^n$ be a quantity we would like to approximate, and let $\mathbf{x}_j, \ j = 0, 1, 2, ...$, be such that for each $N \in \mathbb{N}$

$$\left\| \frac{1}{\varepsilon^{N+1}} \left[\mathbf{x}^\varepsilon - \sum_{j=0}^{N} \varepsilon^j \, \mathbf{x}_j \right] \right\| \qquad \text{is bounded for } \varepsilon \to 0 \ ,$$

then $\sum\limits_{j=0}^{\infty} \varepsilon^j \, \mathbf{x}_j$ is called an *asymptotic expansion* for $\mathbf{x}^\varepsilon$. As a consequence we have for each $N \in \mathbb{N}$

$$\mathbf{x}^\varepsilon = \sum_{j=0}^{N} \varepsilon^j \, \mathbf{x}_j + O(\varepsilon^{N+1}) \ .$$

One may consider various other power expansions, including rational powers of ε for example, but the one considered above is sufficient for our purpose. We remark that all quantities may depend on another parameter such as t, which of course will affect the global domain of validity of the expansion.

The method of *matched asymptotic expansions* is a collective terminology for a number of techniques with a variety of ad hoc ideas (cf. [59, 75]). We shall illustrate the idea by treating the problem considered in §1.

Consider the (model) problem (1.2). We call the solution outside the layer the *outer solution*, denoted by $X^\varepsilon(t)$, and assume an asymptotic expansion to exist:

$$(2.1) \qquad X^\varepsilon(t) = \sum_{j=0}^{\infty} \varepsilon^j \, X_j(t) \ .$$

Substitution in (1.2) results in

$$(2.2) \qquad \frac{d}{dt} \sum_{j=0}^{\infty} \varepsilon^{j+1} X_j(t) = - \sum_{j=0}^{\infty} \varepsilon^j \, X_j(t) + b(t) \ .$$

From a limit argument it follows that equal powers on the left-hand side and on the right-hand side should have equal coefficients. From zeroth order terms we find

(2.3a) $X_0(t) = b(t)$.

Note that this is precisely the solution of the reduced problem. From higher order terms we obtain

(2.3b) $\dfrac{d}{dt} X_{j-1}(t) = -X_j(t)$.

Assuming sufficient smoothness of b this then results in

(2.4) $X^j(t) = (-1)^j \, b^{(j)}(t)$.

Next we seek a boundary layer correction in the initial layer, a so-called *inner solution*, using a transformation of the variable, $t = \varphi(\varepsilon)\,\tau$, to map the layer $[0, \varphi(\varepsilon)]$ onto $[0, 1]$. Usually $\varphi(\varepsilon) = \varepsilon^\alpha$, with α some positive number. Using this so-called *stretched variable* τ, we obtain for (1.2), with $\xi^\varepsilon(\tau) := x(t)$,

(2.5) $\dfrac{\varepsilon}{\varphi(\varepsilon)} \dfrac{d}{d\tau} \xi^\varepsilon = -\xi^\varepsilon + b(\varphi(\varepsilon)\,\tau)$.

We shall assume that ξ^ε can be written as an asymptotic series

(2.6) $\xi^\varepsilon(\tau) = \displaystyle\sum_{j=0}^{\infty} \varepsilon^j \, \xi_j(\tau)$,

and substitute this in (2.5). It is clear now that the only meaningful, consistent choice of $\varphi(\varepsilon)$ equals ε. Note that this is in agreement with the analysis in §1. This then leads to the conclusion that the layer has a width of $O(\varepsilon)$ and we obtain

(2.7) $\dfrac{d}{d\tau} \displaystyle\sum_{j=0}^{\infty} \varepsilon^j \, \xi_j(t) = - \sum_{j=0}^{\infty} \varepsilon^j \, \xi_j(t) + b(\varepsilon\tau)$.

Equating

(2.8a) $\dfrac{d}{d\tau} \xi_0(\tau) = -\xi_0(\tau) + b(\varepsilon\tau)$.

We now use the initial condition x_0, take

(2.8b) $\xi_0(0) = x_0$,

and solve for ξ_0. For higher powers in ε we obtain

(2.9) $\dfrac{d}{d\tau} \xi_j(\tau) = -\xi_j(\tau)$,

i.e. $\xi_j(\tau) = c_j e^{-\tau}$, $c_j \in \mathbb{R}$. The original problem does not provide initial conditions for (2.9), but we can now require the solution to *match* with the outer solution at $t \approx \varepsilon$. For this we have to formulate the inner solution in terms of t. First note that the latter is given by (cf. (2.7), (2.8), (2.9))

$$(2.10) \qquad \xi^\varepsilon(\tau) = e^{-\tau} x_0 + \int_0^\tau e^{s-\tau} b(\varepsilon s)\, ds + \sum_{j=1}^\infty \varepsilon^j c_j e^{-\tau} .$$

So for $t = \varepsilon$ we obtain the approximation

$$(2.11) \qquad x^\varepsilon(t) = e^{-t/\varepsilon} x_0 + \int_0^{t/\varepsilon} e^{(s-t)/\varepsilon} b(\varepsilon s)\, ds + \sum_{j=1}^\infty \varepsilon^j c_j e^{-t/\varepsilon} ,$$

which should be compared to the outer solution, which is given by

$$(2.12) \qquad x^\varepsilon(t) = b(t) + \sum_{j=1}^\infty (-1)^j \varepsilon^j b^{(j)}(t) .$$

Substituting $t = \varepsilon$ we compare powers of ε again. This requires an expansion of the integral in (2.11):

$$(2.13) \qquad \int_0^1 e^s b(\varepsilon s)\, ds = \frac{1}{\varepsilon} \int_0^\varepsilon e^{p/\varepsilon} b(p)\, dp = \sum_{j=0}^\infty (-\varepsilon)^j \left[e\, b^{(j)}(\varepsilon) - b^{(j)}(0) \right] .$$

By substituting this in (2.11) and comparing (2.11) and (2.12) for $t = \varepsilon$ we immediately find

$$(2.14) \qquad c_j = b^{(j)}(0) .$$

Although this example shows the principle of matching, it is deceptively simple, because (1.2) is linear and first order. In general expressions are recursive in nature, requiring lower order terms to be known before proceeding to a higher order term. Often these higher order terms involve ODE (or differential-algebraic equations, see Chapter IX) of at least as complex a character as the original problem to be solved!

As our second illustration we will apply the matching method to (1.8). We start with expansions for the outer solution:

$$(2.15) \qquad \begin{cases} y^\varepsilon(t) = \displaystyle\sum_{j=0}^\infty \varepsilon^j y_j(t) , \\[2mm] z^\varepsilon(t) = \displaystyle\sum_{j=0}^\infty \varepsilon^j z_j(t) . \end{cases}$$

Thus $y^0(t) = y_0(t)$, $z^0(t) = z_0(t)$. Substitution in (1.8) yields

(2.16a) $\displaystyle\sum_{j=0}^{\infty} \varepsilon^j \frac{d}{dt} y_j = f(t, y^\varepsilon, z^\varepsilon) = f(t, y^0, z^0) + O(\varepsilon)$,

(2.16b) $\displaystyle\sum_{j=0}^{\infty} \varepsilon^{j+1} \frac{d}{dt} z_j = g(t, y^\varepsilon, z^\varepsilon) = g(t, y^0, z^0) + O(\varepsilon)$.

From zeroth order terms we obtain

(2.17a) $\dfrac{d}{dt} y_0 = f(t, y_0, z_0)$,

(2.17b) $g(t, y_0, z_0) = 0$.

We assume that we can solve for z_0 from (2.17b), so

(2.18) $\quad z_0 = \varphi(t, y_0)$,

and require $y_0(0) = y_0$.

  Next we look for the inner solution, using a stretching $\tau := t/\varepsilon$. By writing

(2.19) $\quad y^\varepsilon(\tau) = y(\varepsilon\tau)$, $\qquad \zeta^\varepsilon(\tau) = z(\varepsilon\tau)$,

we may try the expansions

(2.20) $\qquad \begin{cases} \eta^\varepsilon(\tau) = \displaystyle\sum_{j=0}^{\infty} \varepsilon^j\, \eta_j(\tau) , \\[2mm] \zeta^\varepsilon(\tau) = \displaystyle\sum_{j=0}^{\infty} \varepsilon^j\, \zeta_j(\tau) . \end{cases}$

Substitution in the equation gives

(2.21a) $\displaystyle\sum_{j=0}^{\infty} \varepsilon^j \frac{d}{d\tau} \eta_j = \varepsilon\, f(\varepsilon\tau, \eta^\varepsilon, \zeta^\varepsilon) = \varepsilon\, f(0, \eta_0, \zeta_0) + O(\varepsilon)$,

(2.21b) $\displaystyle\sum_{j=0}^{\infty} \varepsilon^j \frac{d}{d\tau} \zeta_j = g(\varepsilon\tau, \eta^\varepsilon, \zeta^\varepsilon) = g(0, \eta_0, \zeta_0) + O(\varepsilon)$.

From (2.21a) we obtain for $j = 0$:

$$\frac{d}{d\tau} \eta_0 = 0 .$$

Hence we conclude for the inner solution

(2.22) $\quad \eta_0 = y(0)$.

Note that the inner solution and the outer solution already match in their zeroth order terms. From (2.21b) we conclude that

(2.23a) $\quad \dfrac{d}{d\tau}\zeta_0 = g(0, y_0, \zeta_0)$.

Since we use the initial condition

(2.23b) $\quad \zeta_0(0) = z_0$,

we can solve for ζ_0 in principle. The complexity of all this depends on f and g. This will be demonstrated by the next example.

Example VIII.4.
Consider again the IVP in Example VIII.2. It has the form (1.8a), with

$$f(y, z) = -(1 - z)y + qz ,$$

$$g(y, z) = (1 - z)y - pz ,$$

and initial values $y(0) = 1$, $z(0) = 0$.
Solving the reduced equation, which gives zeroth order approximations of the outer solution, results in

$$z_0 = \frac{y_0}{y_0 + p} .$$

This can be used to find the other zeroth order component of the outer solution, which satisfies (cf. (2.17a))

$$\begin{cases} \dfrac{d}{dt}y_0 = -\dfrac{y_0}{y_0 + p}(p - q) , \\ y_0(0) = 1 . \end{cases}$$

Integrating via separation of variables and using the initial conditions gives

$$y_0(t) + p \ln y_0(t) + (p - q)t = 1 .$$

Next, for the inner solution we find (cf. (2.23))

$$\begin{cases} \dfrac{d}{d\tau}\zeta_0 = 1 - (1 + p)\zeta_0 , \\ \zeta_0(0) = 0 . \end{cases}$$

Solving this IVP gives

$$\zeta_0(\tau) = \frac{1}{1 + p}[1 - \exp(-(1 + p)\tau)] .$$

Computing higher order terms is much more complicated. In Fig. VIII.10a we show some numerical solutions of this problem.

The importance – and power – of these asymptotic methods is that they give insights into the qualitative behaviour of the solution. Some experience is crucial, e.g. to find out about the 'size' of the layer. They also provide useful knowledge for the numerical computations, which will be dealt with next. For more theory see e.g. [25, 59].

3. Stiff differential equations

Let us consider the model equation (1.2) once more, now omitting the superscript ε:

(3.1) $\varepsilon \, \dot{x} = -x + b(t) \,,$ $\varepsilon > 0 \,.$

We can put this ODE into standard form by multiplying both sides by $\lambda := -\dfrac{1}{\varepsilon}$, so

(3.2) $\dot{x} = \lambda \Big(x - b(t) \Big) \,,$ $\lambda < 0 \,.$

Note that $|\lambda|$ is large now.

If we discretise (3.2) by Euler forward with step size h, we have

(3.3) $x_{i+1}^{h} = (1 + h \, \lambda) \, x_i - h \, \lambda \, b(t_i) \,.$

The next example shows what may happen then.

Example VIII.5.
Take e.g. $\lambda = -100$ and $b(t) = 0.99 \, e^{-t}$ in (3.2). If we choose $x(0) = 1$ we obtain $x(t) = e^{-t}$. Incidentally, from a singular perturbation point of view, we see that this solution is close to the solution of the reduced equation (viz. $x(t) = b(t)$); it only differs by $O(\varepsilon)$, i.e. at most 1%!

In Table VIII.1 we have given some numerical results for $h = 0.05$, which would seem to be a reasonable step for having errors of a few percent at most. Yet the results are disastrous.

Table VIII.1   Euler forward, $h = 0.05$.

t_i	x_i^h	$x(t_i)$	$|e_i^h|$	$|e_i^h/x(t_i)|$
0.000	1.00000E+00	1.00000E+00	0.0E+00	0.0E+00
0.100	9.08586E-01	9.04837E-01	3.7E-03	4.1E-03
0.200	8.82094E-01	8.18731E-01	6.3E-02	7.7E-02
0.300	1.75770E+00	7.40818E-01	1.0E+00	1.4E+00
0.400	1.69432E+01	6.70320E-01	1.6E+01	2.4E+01
0.500	2.60975E+02	6.06531E-01	2.6E+02	4.3E+02
0.600	4.16645E+03	5.48812E-01	4.2E+03	7.6E+03
0.700	6.66549E+04	4.96585E-01	6.7E+04	1.3E+05
0.800	1.06647E+06	4.49329E-01	1.1E+06	2.4E+06
0.900	1.70635E+07	4.06570E-01	1.7E+07	4.2E+07
1.000	2.73016E+08	3.67879E-01	2.7E+08	7.4E+08

If we take $h = 0.00625$, i.e. a factor 8 smaller, then we get what we expect, see Table VIII.2.

Table VIII.2   Euler forward, $h = 0.00625$.

t_i	x_i^h	$x(t_i)$	$|e_i^h|$	$|e_i^h/x(t_i)|$
0.000	1.00000E+00	1.00000E+00	0.0E+00	0.0E+00
0.100	9.04809E-01	9.04837E-01	2.9E-05	3.1E-05
0.200	8.18705E-01	8.18731E-01	2.6E-05	3.1E-05
0.300	7.40795E-01	7.40818E-01	2.3E-05	3.1E-05
0.400	6.70299E-01	6.70320E-01	2.1E-05	3.1E-05
0.500	6.06512E-01	6.06531E-01	1.9E-05	3.1E-05
0.600	5.48794E-01	5.48812E-01	1.7E-05	3.1E-05
0.700	4.96570E-01	4.96585E-01	1.6E-05	3.1E-05
0.800	4.49315E-01	4.49329E-01	1.4E-05	3.1E-05
0.900	4.06557E-01	4.06570E-01	1.3E-05	3.1E-05
1.000	3.67868E-01	3.67879E-01	1.2E-05	3.1E-05

The results in Table VIII.1 are not so strange on second thoughts. Indeed, by noticing that the homogeneous part of (3.3) generates solutions growing geometrically like $(1 - 100h)^i$, it follows that the *recursion* (3.3) is *numerically unstable* for $h > 0.02$. We wish to emphasise that this is not at all contradicting the convergence theory of Chapter III, since there we essentially let h approach zero, and set

$$e^{h\lambda} \doteq 1 + h\lambda , \qquad h \to 0 .$$

Euler forward is also not a proper method for singularly perturbed problems more generally, as the considerations above simply generalise to more complex ODE. The essential point here is the *existence of at least two time scales*. If one is interested in approximating outer solutions, i.e. the ones with larger time scales, Euler forward still has to comply with a stability requirement imposed by the smaller time scale, and thus needs unduly many grid points. This undesired condition on the step size is not present for Euler backward, where we have after rewriting the Δ-equation in explicit form, which is possible due to the linearity,

$$(3.4) \qquad x_{i+1}^h = \frac{x_i^h}{1 - h\lambda} - \frac{h\lambda}{1 - h\lambda} g(t_i) .$$

Example VIII.6.
In Table VIII.3 an Table VIII.4 we have displayed results for Euler backward analogous to those as for the problem in Example VIII.5 for $h = 0.05$ and 0.00625. Note that also the large step size gives satisfactory results. Of course the factor $1/(1 + 100h)$ (cf. (3.4)) is always smaller than 1, implying numerical stability.

Table VIII.3   Euler backward, $h = 0.05$.

t_i	x_i^h	$x(t_i)$	$|e_i^h|$	$|e_i^h/x(t_i)|$
0.000	1.00000E+00	1.00000E+00	0.0E+00	0.0E+00
0.100	9.05063E-01	9.04837E-01	2.3E-04	2.5E-04
0.200	8.18941E-01	8.18731E-01	2.1E-04	2.6E-04
0.300	7.41008E-01	7.40818E-01	1.9E-04	2.6E-04
0.400	6.70492E-01	6.70320E-01	1.7E-04	2.6E-04
0.500	6.06686E-01	6.06531E-01	1.6E-04	2.6E-04
0.600	5.48953E-01	5.48812E-01	1.4E-04	2.6E-04
0.700	4.96713E-01	4.96585E-01	1.3E-04	2.6E-04
0.800	4.49444E-01	4.49329E-01	1.2E-04	2.6E-04
0.900	4.06674E-01	4.06570E-01	1.0E-04	2.6E-04
1.000	3.67974E-01	3.67879E-01	9.4E-05	2.6E-04

Table VIII.4   Euler backward, $h = 0.00625$.

t_i	x_i^h	$x(t_i)$	$|e_i^h|$	$|e_i^h/x(t_i)|$
0.000	1.00000E+00	1.00000E+00	0.0E+00	0.0E+00
0.100	9.04866E-01	9.04837E-01	2.9E-05	3.2E-05
0.200	8.18757E-01	8.18731E-01	2.6E-05	3.2E-05
0.300	7.40842E-01	7.40818E-01	2.3E-05	3.2E-05
0.400	6.70341E-01	6.70320E-01	2.1E-05	3.2E-05
0.500	6.06550E-01	6.06531E-01	1.9E-05	3.2E-05
0.600	5.48829E-01	5.48812E-01	1.7E-05	3.2E-05
0.700	4.96601E-01	4.96585E-01	1.6E-05	3.2E-05
0.800	4.49343E-01	4.49329E-01	1.4E-05	3.2E-05
0.900	4.06583E-01	4.06570E-01	1.3E-05	3.2E-05
1.000	3.67891E-01	3.67879E-01	1.2E-05	3.2E-05

The numerical terminology for the phenomenon above is the celebrated notion of *stiffness*. Originally used in elasticity, it is in fact quite closely related to the multiple time scale problem met in singularly perturbed problems. We like to use it in the sense that the dominant time scale is different from a faster time scale, present in the problem. For linear problems we may say it slightly more precisely. Consider

$$(3.5) \qquad \dot{\mathbf{x}} = \mathbf{A}\,\mathbf{x} + \mathbf{r}(t) \ .$$

Let $\rho(\mathbf{A})$ denote the spectral radius of $\mathbf{A}$, see Appendix B. Then the ODE (3.5) is *stiff* on an interval (T_1, T_2) if $|\rho(\mathbf{A})(T_2 - T_1)| \gg 1$. This definition reflects the delicate relation between the relevant length of the interval and the time scale. In an obvious way one may generalise the notion of stiffness for nonlinear problems by considering Jacobian matrices.

Remark 3.6. Some people like to link stiffness to the existence of absolutely small and large eigenvalues of $\mathbf{A}$, or alternatively, to moderate and very large negative real parts. Although this may induce different time scales, it is not general enough to cover e.g. our simple model problem (3.1).

Remark 3.7. We shall not deal with problems with highly oscillatory modes, although they may genuinely be called stiff as well, when different time scales are present.

4. The increment function

In the previous section we saw that the (in)stability was related to the way the homogeneous part of the ODE was represented in the difference equation. This then induces the notion of *increment function* ψ for a difference equation, being a discretisation of the homogeneous form of (3.2). Since the time step times the coefficient λ turns out to be an invariant quantity (i.e. they always appear jointly), we may write the difference equation as

$$(4.1) \qquad x_{i+1}^h = \psi(h\,\lambda)\,x_i^h \ .$$

For Euler forward and Euler backward we thus have $\psi(h\,\lambda) = (1 + h\,\lambda)$ and $\psi(h\,\lambda) = 1/(1 - h\,\lambda)$ respectively.

For general RK formulae we have the following interesting theorem (cf. [22]):

Theorem 4.2. *Let* $\begin{bmatrix} \rho & \mathbf{\Gamma} \\ & \beta^T \end{bmatrix}$ *be the Butcher matrix of an RK formula. Then*

$$\psi(h\,\lambda) = 1 + \beta^T\, h\,\lambda(\mathbf{I} - h\,\lambda\,\mathbf{\Gamma})^{-1}\,\mathbf{e} \ , \qquad where\ \mathbf{e} := (1, 1, ..., 1)^T \ .$$

Proof: Recall the general RK formula (III.1.3). Substituting $f(t, x) = \lambda\,x$ we find

$$k_j = \lambda\Big(x_i^h + h \sum_{j=1}^{m} \gamma_{jl}\,k_l\Big) \ ,$$

whence, writing $\mathbf{k} := (k_1, ..., k_m)^T$,

$$\mathbf{k} = \lambda\,x_i^h\,\mathbf{e} + h\,\lambda\,\mathbf{\Gamma}\,\mathbf{k} \ ,$$

so

$$x_{i+1}^h = (1 + \beta^T\, h\,\lambda(\mathbf{I} - h\,\lambda\,\mathbf{\Gamma})^{-1}\,\mathbf{e})\,x_i^h \ . \qquad\qquad \square$$

Corollary 4.3. *If the RK method is explicit, so that* Γ *is nilpotent (i.e.* $\Gamma^m = 0$), *then we have*

$$\psi(h\,\lambda) = 1 + \beta^T h\,\lambda \sum_{j=0}^{m-1} (h\,\lambda\,\Gamma)^j \, \mathbf{e} \, ,$$

which is a polynomial in $h\,\lambda$ *of degree m.*

For a k-step LMM the homogeneous difference equation generates k basis solutions, each having its own growth behaviour as a function of $h\,\lambda$. Given $h\,\lambda$ we can represent them all, once we have the roots $\omega_j(h\,\lambda)$, $j = 1, ..., n$, of the characteristic equation

$$(4.4) \qquad \sum_{j=0}^{k} (\alpha_j - h\,\lambda\,\beta_j)\,\omega^{k-j} = 0 \, .$$

A straightforward generalisation of (4.1) is then to define the increment function of an LMM as

$$(4.5) \qquad \psi(h\,\lambda) := \max_j |\omega_j(h\,\lambda)| \, .$$

It is noteworthy that $\psi(h\,\lambda)$ can be viewed as an approximation of $e^{h\lambda}$. For LMM we should then omit the modulus on the right-hand side, strictly speaking; however, $|h\,\lambda|$ must be small for it to be called an approximation anyway. The Euler forward approximation e.g. then consists of the first two terms of a Taylor expansion. In general we have a rational function in $h\,\lambda$ in the case of implicit RK, or a polynomial in $h\,\lambda$ in the case of explicit RK or LMM, as an approximation for $e^{h\lambda}$. Consistency clearly relates to this kind of approximation. The accuracy can be indicated as $O((h\,\lambda)^p)$ for suitable p. However, a polynomial increment function grows unbounded, as $h\,\lambda \to \infty$, whereas a rational increment function may remain bounded in suitable cases. This is precisely what we wish to achieve by the notions considered in the next section.

5. A-stable, $A(\alpha)$-stable methods

The quantity ψ in the previous section can now be used to determine whether or not a method will be acceptable. Since the problem (3.2) may be a model for all kinds of ODE, including ones with Jacobian matrices having complex eigenvalues, we allow λ to be complex. Rather than of $h\,\lambda$ we shall consider ψ as a function of a single complex variable, z say. From the definition of stability in §V.8 we see that a discretisation formula induces a so-called *stability domain* S, where

(5.1) $S := \{z \in C \mid |\psi(z)| \le 1\}$.

For explicit RK formulae of order p, $p = 1, 2, ...$, one can thus indicate domains as shown in Fig. VIII.4.

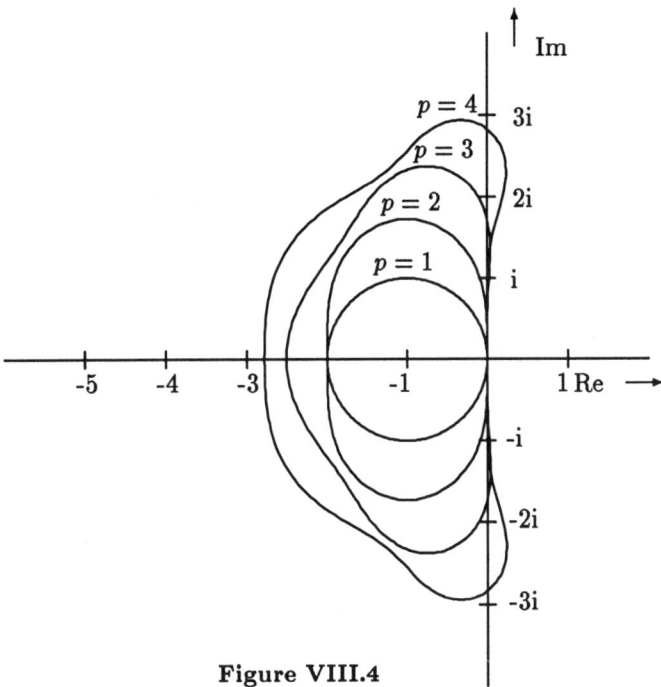

Figure VIII.4

One can construct implicit RK formulae which show so-called A-stability. The definition of this form of stability is given below. Some can be derived using Gaussian quadrature formulae (cf. Appendix A) in particular. Fully implicit formulae are not so easy to handle. However, for solving boundary value problems they are quite natural on the other hand (see Chapter X). Fortunately, one can also construct semi-implicit formulae, where the strictly upper triangular part of the Butcher matrix is zero. We shall not dwell on implicit RK here, however.

We next consider LMM. Note that implicit 'RK' formulae like the trapezoidal rule or Euler backward can also be viewed as LMM. For a general k-step LMM we have the following:

Theorem 5.2. *If $\psi(z) \leq 1$, for all $z \in \mathbb{R}$, then the LMM is implicit.*

Proof: Write $h\lambda = \rho(\omega)/\sigma(\omega)$. For a k-step LMM, ρ is of degree k and σ of degree $\leq k$. If the LMM is explicit, σ is of degree $\leq k-1$. Hence for $\omega \to -\infty$ we find $\rho(\omega)/\sigma(\omega) \to (\alpha_0/\beta_1)\,\omega$. Consequently, for $|h\lambda|$ large enough and $\lambda < 0$, one would always find some root $\omega_j(h\lambda)$ with $|\omega_j(h\lambda)| > 1$ which rules out explicit formulae. $\square$

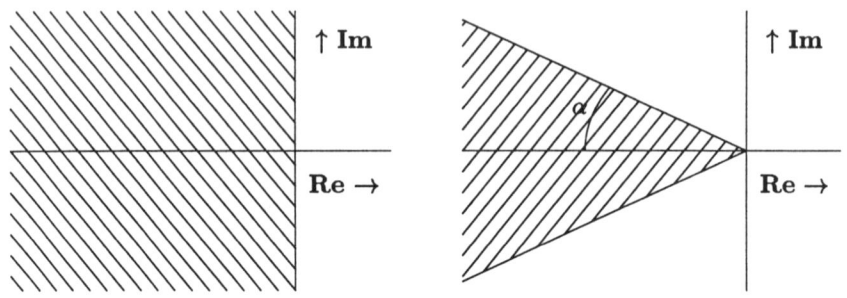

Figure VIII.5   Figure VIII.6

Thinking of singular perturbation problems, one would like to employ a discretisation method which is at least numerically stable for values of $h\lambda \in \mathbb{R}_-$, but preferably for a larger part of the negative complex half-plane. This induces the following special cases. A method is called *A-stable* if

$$C_- := \{z \in \mathbb{C} \mid \operatorname{Re} z < 0\} \subset S$$

and *A(α)-stable* if

$$C(\alpha) := \{z \in \mathbb{C} \mid -\alpha < \pi - \arg(z) < \alpha,\ 0 < \alpha < \pi/2\} \subset S.$$

In Fig. VIII.5 and Fig. VIII.6 we have shaded the area of C_- and $C(\alpha)$ respectively.

Without proof we give the following famous barrier theorem (cf [21]).

Theorem 5.3 (Dahlquist barrier). *The order of an A-stable LMM cannot exceed 2. Moreover, of all second order A-stable LMM the trapezoidal rule has the smallest error constant (cf. (VII.2.4)).*

The situation for implicit RK methods is more promising as there also exist higher order *A*-stable methods (see [28]).

Example VIII.7.

(i) *Euler forward.* This is actually an RK method of order 1. From Fig. VIII.7 we see that the stability region is a disc of radius 1 centred at $z = -1$. Being an explicit method the stability region cannot even contain $\mathbb{R}_-$.

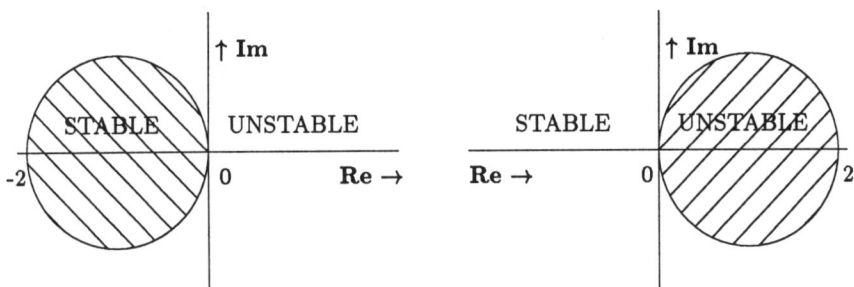

Figure VIII.7   Figure VIII.8

(ii) *Euler backward.* The stability region is the area outside a disc of radius 1 centred at $z = 1$ (see Fig. VIII.8). Hence this method is clearly A-stable.

Sometimes one is interested in more than (absolute) stability and likes the homogeneous modes to decay geometrically; for an LMM this means that all $|\omega_j(hz)| < 1$. This is not a trivial matter. For the trapezoidal rule we have

(5.4) $\qquad \psi(z) = \dfrac{1 + \frac{1}{2} z}{1 - \frac{1}{2} z}$.

Hence $\psi(z) \to -1$ if $z \to -\infty$.

In order to have criteria for monitoring the stability in more general situations one has invented many other stability notions. We only wish to show here the limitations of the A-stability concept for a non-autonomous, but still simple, linear ODE:

(5.5) $\qquad \dot{x} = \lambda(t)\, x(t) , \qquad \lambda(t) < 0 , \quad t > 0 .$

Euler backward yields

$$x^h_{i+1} = \frac{1}{1 - h\, \lambda(t_{i+1})}\, x^h_i ,$$

i.e. we still have stability, and even geometrical decay. On the other hand, if we employ the trapezoidal rule we find

(5.6) $x_{i+1}^h = \dfrac{1 + \frac{1}{2} h \lambda(t_i)}{1 - \frac{1}{2} h \lambda(t_{i+1})} \, x_i^h$.

As one may easily check $\dfrac{1 + \frac{1}{2} h \lambda(t_i)}{1 - \frac{1}{2} h \lambda(t_{i+1})} < -1$ if $\lambda(t_{i+1}) - \lambda(t_i) > \dfrac{4}{h}$. Hence if $\dfrac{d\lambda}{dt} > \dfrac{4}{h^2}$ the increment in (5.6) is absolutely larger than 1!

Remark 5.7. Rather than using the trapezoidal rule we may employ the so-called *one-leg* variant (cf. Exercise 5), being the midpoint rule, see Example VII.2. This gives

$$x_{i+1}^h = \dfrac{1 + \frac{1}{2} h \lambda(t_{i+\frac{1}{2}})}{1 - \frac{1}{2} h \lambda(t_{1+\frac{1}{2}})} \, x_i^h \, ,$$

which clearly gives asymptotic stability.

For stability in nonlinear (and also in non-autonomous) situations it is sufficient to prove *contractivity* for the Δ-equation. To this end consider the ODE

(5.8) $\dfrac{d\mathbf{x}}{dt} = \mathbf{f}(t, \mathbf{x})$,

and, say, the one-step method

(5.9) $\mathbf{x}_{i+1}^h = \mathbf{x}_i^h + h \, \Phi(t_i, \mathbf{x}_i^h, \mathbf{x}_{i+1}^h, h)$.

This Δ-equation is called *contractive* on some domain Ω if for each two solutions $\{\mathbf{x}_i^h\}$, $\{\mathbf{y}_i^h\}$, with $\mathbf{x}_i^h, \mathbf{y}_i^h \in \Omega$ for all i, there exists a $\kappa < 1$ such that

$$\|\mathbf{x}_{i+1}^h - \mathbf{y}_{i+1}^h\| \leq \kappa \, \|\mathbf{x}_i^h - \mathbf{y}_i^h\| \, .$$

In particular for Euler backward e.g. we have the following property:

Property 5.10. *Let (5.8) be contractive on a domain* Ω, *i.e. there exists a negative function* $\nu(t)$ *such that*

$$\Big((\mathbf{f}(t, \mathbf{x}) - \mathbf{f}(t, \mathbf{y})) \cdot \big(\mathbf{x}(t) - \mathbf{y}(t) \big) \Big) < \nu(t) \, \|\mathbf{x}(t) - \mathbf{y}(t)\|_2^2 < 0$$

for all solutions $\mathbf{x}$, $\mathbf{y}$ *as long as* $\mathbf{x}(t), \mathbf{y}(t) \in \Omega$. *Then Euler backward is contractive on* Ω.

Proof: Clearly

$$\mathbf{y}^h_{i+1} - \mathbf{x}^h_{i+1} - (\mathbf{y}^h_i - \mathbf{x}^h_i) = h\,\mathbf{f}(t_{i+1}, \mathbf{y}^h_{i+1}) - h\,\mathbf{f}(t_{i+1}, \mathbf{x}^h_{i+1}) ,$$

so

$$\|\mathbf{y}^h_{i+1} - \mathbf{x}^h_{i+1}\|^2_2 - (\mathbf{y}^h_i - \mathbf{x}^h_i) \cdot (\mathbf{y}^h_{i+1} - \mathbf{x}^h_{i+1}) \le$$

$$\le h\,\nu(t_{i+1})\,\|\mathbf{y}^h_{i+1} - \mathbf{x}^h_{i+1}\|^2_2$$

whence

$$\|\mathbf{y}^h_{i+1} - \mathbf{x}^h_{i+1}\|_2 \le \|\mathbf{y}^h_i - \mathbf{x}^h_i\|_2 + h\,\nu(t_{i+1})\,\|\mathbf{y}^h_{i+1} - x^h_{i+1}\|_2 .$$

In other words, we may take $\kappa := \dfrac{1}{1 - h\,\nu}$. $\square$

In Exercise 13 we consider contractivity for the midpoint rule.

6. BDF methods and their implementation

From Theorem 5.2 we deduce that LMM, to be used for stiff problems, should be implicit. Unfortunately the Adams-Bashforth family is not $A(\alpha)$-stable, except for its first $(k = 1)$ member. The BDF family (Backward Differences Formula, cf. Example VII.1(iii)), of which again Euler backward is the lowest order member, has more favourable stability properties. They have the form

$$(6.1) \qquad \sum_{j=0}^{k} \alpha_j\, x^h_{i-j+1} = h\,\beta_0\, f(t_{i+1}, x^h_{i+1}) .$$

Recalling the operator ∇ as defined in (I.5.8c), then

$$\nabla^0\, x_{i+1} = x_{i+1} ,$$

$$\nabla^1\, x_{i+1} = \nabla^0\, x_{i+1} - \nabla^0\, x_i ,$$

$$\vdots$$

$$\nabla^k\, x_{i+1} = \nabla^{k-1}\, x_{i+1} - \nabla^k\, x_i .$$

We can write linear combinations of $x_{i+1}, ..., x_{i-k+1}$ in terms of these backward differences. In particular, there exist $\gamma_0, ..., \gamma_k$ such that (6.1) is equivalent to

$$(6.2) \qquad \sum_{j=0}^{k} \gamma_j\, \nabla^j\, x^h_{i+1} = h\,\beta_0\, f(t_{i+1}, x^h_{i+1}) .$$

We can use either formulation to derive the formulae by requiring optimal consistency order, although they would also directly follow from writing the interpolation formula at $t = t_{i+1}, ..., t_{i+1-k}$ and differentiating this once.

Example VIII.8.

(i)   We trivially have for Euler backward

$$x_{i+1}^h - x_i^h = h\, f(t_{i+1}, x_{i+1}^h)\,,$$

so

$$\nabla^1 x_{i+1}^h = h\, f(t_{i+1}, x_{i+1}^h)\,.$$

(ii) BDF with $k = 2$:

$$\gamma_0\, x_{i+1}^h + \gamma_1 (x_{i+1}^h - x_i^h) + \gamma_2 (x_{i+1}^h - 2x_i^h + x_{i-1}^h) = h\, \beta_0 (t_{i+1}, x_{i+1}^h)\,.$$

Consistency implies (expansion around t_{i+1}):

$$(\gamma_0 + \gamma_1 + \gamma_2) - (\gamma_1 + 2\gamma_2) + \gamma_2 = 0$$
$$2(\gamma_0 + \gamma_1 + \gamma_2) - (\gamma_1 + 2\gamma_2) = \beta_0\,.$$

Scaling yields $\beta_0 = 1$.
One degree of freedom is left. We may thus require second order
consistency, giving:

$$\gamma_1 + 2\gamma_2 - 4\gamma_2 = 0\,.$$

We thus find

$$\nabla^1 x_{i+1} + \tfrac{1}{2}\nabla^2 x_{i+1}^h = h\,(t_{i+1}, x_{i+1}^h)\,,$$

or

$$\tfrac{3}{2} x_{i+1}^h - 2x_i^h + \tfrac{1}{2} x_{i-1}^h = h\, f(t_{i+1}, x_{i+1}^h)\,.$$

The coefficients γ_j in (6.2) have a nice form in general, for

(6.3) $$\gamma_j = \frac{1}{j}\,,\quad j \ge 1\,,\quad \gamma_0 = 0\,.$$

Only for $k \le 6$ are the BDF $A(\alpha)$-stable; for $k = 1$ or 2 the method is even
A-stable. We have displayed their stability plots in Fig. VIII.9.

Although the power of BDF methods was recognised by Curtis and
Hirschfelder ([20]) as far back as 1952, their popularity only came in 1971,
when Gear published an ODE book containing a code DIFSUB (cf. [28]). Our
PASCAL procedure Gear is based on the latter.

A particular problem is to find an appropriate predictor, since the
BDF formula is implicit. Indeed, as an explicit LMM may give very bad
approximations, one uses extrapolation of approximate values of x at previous
time points as a predictor. In a layer where the step sizes are small anyway,

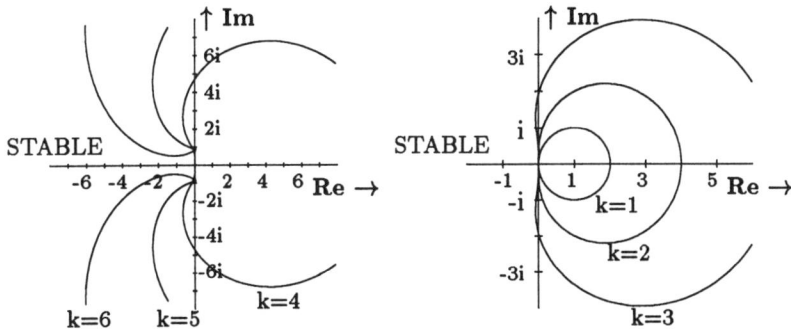

Figure VIII.9   Stability regions of BDF methods.

this is certainly accurate, whereas this method virtually circumvents stiffness in a smooth region. So, given an extrapolated value $x_{i+1}^h(\text{pred})$, we might compute a corrector value $x_{i+1}^h(\text{corr})$ through

$$(6.4) \qquad \sum_{j=1}^{k} \frac{1}{j} \nabla^j x_{i+1}^h(\text{corr}) = h\, f\!\left(t_{i+1}, x_{i+1}^h(\text{pred})\right) .$$

As in §§VII.6, 7 one can see that the local error is estimated by

$$(6.5) \qquad \text{EST}_i := \frac{1}{h}\,[x_{i+1}^h(\text{corr}) - x_{i+1}^h(\text{pred})] .$$

A variable step, variable order method then follows essentially the same pattern as outlined in §VII.7.

It is important to realise that straightforward successive substitution is not appropriate for stiff problems. Recall from §VII.6 that convergence of the latter process requires $|h\, \partial f/\partial x| < 1$. However, for stiff problems $|\partial f/\partial x|$ tends to be rather large, so that h then should be comparatively small (see notion of stiffness, §3). So the fully implicit problem reads as (cf. (6.4))

$$(6.4)' \qquad \sum_{j=1}^{k} \frac{1}{j} \nabla^j x_{i+1}^h(\text{corr}) = h\, f\!\left(t_{i+1}, x_{i+1}^h(\text{corr})\right) .$$

This can be solved by Newton's method, using $x_{i+1}^h(\text{pred})$ as a starting value. Error estimation and step/order strategy remain the same.

In actual problems the ODE is often a vector-valued function, which requires extra care for an efficient Newton iteration, in particular as far as the Jacobian matrices are concerned. We shall describe a way to compute the latter numerically.

Suppose we have the equation

(6.6) $\mathbf{w}(\mathbf{y}) = 0$.

We typically arrive at this equation (6.4′) if $\mathbf{x}_{i+1}^h(\text{corr})$ is a vector. Then the Newton iteration runs as follows:

(6.7)
$$\begin{cases} \mathbf{y}^0 \text{ given, i.e. } \mathbf{y}^0 := \mathbf{x}_{i+1}^h(\text{pred}) \\ \mathbf{y}^{j+1} = \mathbf{y}^j - \mathbf{J}^{-1}(\mathbf{y}^j)\, \mathbf{w}(\mathbf{y}^j) \,, \end{cases}$$

where $\mathbf{J}(\mathbf{y}) := \partial \mathbf{w}/\partial \mathbf{y}$, $\mathbf{w} := (w_1, ..., w_n)^T$, $\mathbf{y} := (y_1, ..., y_n)^T$.

Now let $e_1, ..., e_n$ be the standard basis with $e_1 := (1, 0, ..., 0)^T$ etc. and $\delta_1, ..., \delta_n$ small numbers, approximately equal to $\sqrt{\varepsilon_M}$, ε_M being the machine accuracy. Then the j-th column of the Jacobian matrix is reasonably well approximated by

(6.8) $\dfrac{1}{\delta_j} \left[\mathbf{w}(\mathbf{y} + \delta_j\, e_j) - \mathbf{w}(\mathbf{y}) \right]$.

Note that the choice for the magnitude of δ_j is expected to give optimal accuracy under normal circumstances. One immediately sees that such an approximation of $\mathbf{J}$ costs $(n + 1)$ vector function evaluations. In general, when the Newton process is slow, one often freezes the Jacobian matrix a few iteration steps. This is known as the 'modified Newton' method.

In our particular algorithm, one retains the Jacobian matrix often for several time steps as long as no convergence difficulties are met. Only after even a new Jacobian matrix has not been able to settle this convergence problem is a new step size looked for.

The algorithm as described above is implemented in our procedure Gear by which the next examples have been computed.

Example VIII.9.
Consider the model problem

$$\begin{cases} \dot{x} = -1000(x - \cos t) \,, & 0 \le t \le 5 \,, \\ x(0) = 0 \,. \end{cases}$$

Table VIII.5

TOL	10^{-2}	10^{-3}	10^{-4}	10^{-5}	10^{-6}	10^{-7}	10^{-8}
Gear	95	163	217	333	395	427	470
Adams	6768	6955	7483	7670	8280	8433	8550

If we use the predictor-corrector code Adams, we need quite a few function evaluations, see Table VIII.5. If we use Gear then the result is clearly more efficient. The overkill in Adams is entirely due to stiffness problems. The problem contains a layer $[0, 10^{-3}]$.

Example VIII.10.
Recall the chemical reaction in Example VIII.2:

$$\begin{cases} \dot{y} = -(1-z)y + qz , \\ \varepsilon \dot{z} = (1-z)y - pz , \end{cases}$$

with initial conditions

$$\begin{cases} y(0) = 1 , \\ z(0) = 0 . \end{cases}$$

Table VIII.6

i	t_i	h_i	y_i^h	z_i^h
0	0.000E+00		1.000E+00	0.000E+00
1	9.642E-04	9.642E-04	9.992E-01	8.143E-02
2	1.928E-03	1.615E-03	9.985E-01	1.502E-01
3	3.543E-03	1.615E-03	9.974E-01	2.507E-01
4	5.159E-03	1.615E-03	9.966E-01	3.251E-01
5	6.774E-03	4.493E-03	9.960E-01	3.797E-01
6	1.127E-02	4.493E-03	9.951E-01	4.594E-01
7	1.477E-02	3.500E-03	9.946E-01	4.861E-01
8	1.827E-02	3.500E-03	9.943E-01	5.015E-01
9	2.177E-02	3.500E-03	9.940E-01	5.114E-01
10	2.527E-02	6.814E-03	9.938E-01	5.175E-01
11	3.208E-02	6.814E-03	9.934E-01	5.231E-01
12	3.889E-02	6.814E-03	9.930E-01	5.248E-01
13	4.571E-02	1.894E-02	9.926E-01	5.249E-01
14	6.465E-02	1.894E-02	9.916E-01	5.244E-01
15	8.359E-02	1.412E-01	9.907E-01	5.241E-01
16	2.248E-01	1.412E-01	9.833E-01	5.222E-01
17	3.660E-01	3.131E+00	9.760E-01	5.203E-01
18	3.497E+00	3.131E+00	8.265E-01	4.788E-01
19	6.628E+00	4.880E+00	6.909E-01	4.344E-01
20	1.151E+01	4.880E+00	4.993E-01	3.569E-01

For $\varepsilon = 10^{-2}$, $q = 0.8$, $p = 0.9$ and ABSTOL $= 10^{-2}$ we obtain results as given in Table VIII.6. Note the considerable number of steps in the boundary layer. In Fig. VIII.10a we have made a plot of y and z as a function of t. In order to see the layer behaviour of z we have also made an enlargement of the layer in Fig. VIII.10b.

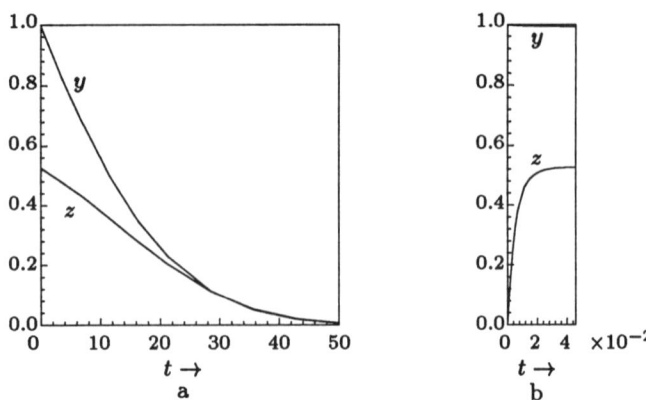

Figure VIII.10   (a) Behaviour of the concentrations in the chemical reaction of Example VIII.10. (b) Enlargement of the stiffness layer.

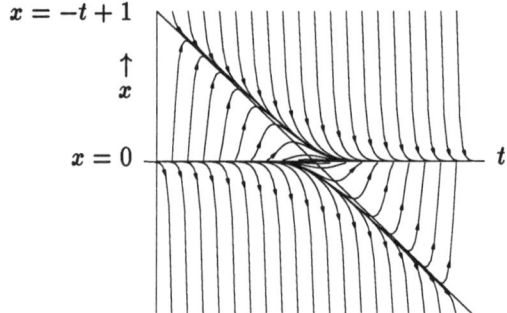

Figure VIII.11   Solution curves of the ODE in Example VIII.11.

Example VIII.11.
Consider the IVP

$$\begin{cases} \varepsilon\, \dot{x} = -x(x + t - 1)\,, & 0 < \varepsilon \ll 1\,, \\ x(0) = 1\,. \end{cases}$$

In Fig. VIII.11 we have given a qualitative overview of solution curves. Note that the reduced solutions $x(t) = 0$ and $x(t) = -t + 1$ attract or repel solutions.

In Table VIII.7 we have shown numerical results for $\varepsilon = 10^{-2}$, ABSTOL $= 10^{-2}$ and $t \in [0, 13]$, which are as expected. In Table VIII.8 we have shown the result, now for $\varepsilon = 10^{-4}$ and ABSTOL $= 10^{-2}$. One should observe that the approximation goes astray for $t > 1$. The explanation is not so complicated. For BDF methods have a growth factor $\psi(h\lambda)$ well bounded above by 1, not only for negative values of $h\lambda$, but also for large positive values! For example, Euler backward has $\psi(h\lambda) = 1/(1 - h\lambda)$. Once h is fairly large, the integrator does not notice that it may jump over a layer region. Indeed, convergence beyond $t = 1$ on the 'wrong' branch is still guaranteed then. This phenomenon is sometimes called *superstability*.

	Table VIII.7			Table VIII.8	
i	t_i	x_i^h	i	t_i	x_i^h
0	0.000E+00	1.00E+00	0	0.000E+00	1.00E+00
1	1.000E-02	9.95E-01	1	1.000E-02	9.90E-01
2	2.000E-02	9.88E-01	2	2.000E-02	9.80E-01
3	4.370E-02	9.66E-01	3	1.369E-01	8.63E-01
4	6.741E-02	9.43E-01	4	2.538E-01	7.46E-01
5	1.485E-01	8.63E-01	5	6.866E+00	-5.87E+00
6	2.296E-01	7.83E-01	6	1.348E+01	-1.25E+01
7	4.911E-01	5.27E-01			
8	7.525E-01	2.81E-01			
9	8.334E-01	2.09E-01			
10	9.143E-01	1.43E-01			
11	1.075E+00	4.23E-02			
12	1.236E+00	2.47E-03			
13	1.397E+00	-2.05E-03			
14	1.557E+00	-5.03E-04			
15	1.718E+00	2.61E-06			
16	1.879E+00	1.67E-05			
17	2.040E+00	1.96E-06			
18	2.201E+00	-2.13E-07			
19	2.361E+00	-6.00E-08			
20	2.522E+00	-5.12E-10			
21	2.683E+00	1.01E-09			
22	2.844E+00	7.33E-11			
23	3.004E+00	-1.07E-11			
24	1.300E+01	-8.52E-16			

Exercises Chapter VIII

1. Apply matched asymptotic expansions to the following IVP:

    a) $\varepsilon \dot{x} + x^2 - 1 = 0$, $\qquad x(0) = 0$

    b) $\varepsilon \dot{x} + tx - 1 = 0$, $\qquad x(0) = 1$

    c) $\varepsilon \ddot{x} + \dot{x} + x = 0$, $\qquad x(0) = 1$, $\dot{x}(0) = 0$

    d) $\ddot{x} + \varepsilon \dot{x} + x = 1$, $\qquad x(0) = 0$, $\dot{x}(0) = 1$

    e) $\dot{x} + \varepsilon x + \sin t = 0$, $\qquad x(0) = 0$.

2. a) Compute the increment function $\psi(h\lambda)$ for the trapezoidal rule

    $$x_{i+1}^h = x_i^h + \frac{h}{2} \left[f(t_i, x_i^h) + f(t_{i+1}, x_{i+1}^h) \right] .$$

    b) Compute $\psi(h\lambda)$ for an 'optimal' explicit Runge-Kutta method of order 2.

3. Show that if we apply a one-step method to $\dot{x} = \lambda x$, then we have for the local error: $h\delta(x(t_{i+1}), h) = e^{\lambda h} - \psi(h\lambda)$.

4. Show that a one-step (explicit) method is consistent if and only if the increment function is asymptotically equal to $1 + h\lambda + O(h^2\lambda^2)$. Can you give an analogous property for LMM ?

5. Instead of the 'classical' trapezoidal rule one can also use

    $$x_{i+1}^h = x_i^h + h f\left(\frac{t_i}{2} + \frac{t_{i+1}}{2}, \frac{x_i^h}{2} + \frac{x_{i+1}^h}{2}\right) .$$

    Show that this method is consistent and stable for all λ with $\operatorname{Re}\lambda < 0$.
    Remark: The latter method is a so-called *one-leg* variant. For an arbitrary LMM one can define a one-leg variant by

    $$\sum_{j=0}^{k} \alpha_j x_{i-j+1}^h = h f\left(\sum_{j=0}^{k} \beta_j t_{i-j+1}, \sum_{j=0}^{k} \beta_j x_{i-j+1}^h\right) .$$

6. Weakly stable LMM are in fact inherently unstable. As an example we consider the midpoint rule

    $$x_{i+1}^h = x_{i-1}^h + 2h f(t_i, x_i^h) .$$

    Apply this to the ODE $\dot{x} = \lambda x$ and show that the roots of the characteristic polynomial of the recursion with step h are given by $\pm(1 + \frac{1}{2}(h\lambda)^2) - h\lambda$. Formulate your conclusion.

7. Determine the three-step BDF method.

8. Determine the error constant of the two-step BDF method and check Theorem 5.2.

9. Show that for BDF methods $\beta_0 = (-1)^{k+1} k \, \alpha_k$.

10. Consider the model problem $\dot{x} = \lambda x + g(t)$. Indicate why from a family of multisteps the lowest order method should be chosen in the layer. For simplicity we assume that the error constants are the same (cf. Example III.5(iii)).

11. Consider the IVP

$$\begin{cases} \dot{x} = -x + 2x^2 \, e^{-t} \; , \\ x(0) = \frac{1}{3} \; . \end{cases}$$

If we use Fehlberg with ABSTOL = RELTOL = 10^{-4} we never get a step size exceeding 1. Explain this, after having argued why the problem can be considered stiff for larger values of t.

12. A way to determine solutions of IVP with large Lipschitz constants is so-called *exponential fitting*. For a suitable family of methods one chooses the free parameters such that the ODE $\dot{x} = \lambda x$ for fixed h is solved exactly at the time points $t_i = ih$. As an example consider the ϑ-methods:

$$x_{i+1}^h = x_i^h + h \, \vartheta \, f(t_i, x_i^h) + h(1 - \vartheta) \, f(t_{i+1}, x_{i+1}^h)$$

Determine ϑ such that $\psi(h\lambda) = e^{\lambda h}$.
Show now that the global error when solving

$$\begin{cases} \dot{x} = \lambda x + g(t) \; , \\ x(0) = x_0 \; , \end{cases}$$

is uniformly bounded in t, if $\lambda < 0$ and g bounded.

13. Let the ODE (5.8) be contractive on Ω. Show that the one-leg variant of the trapezoidal rule (cf. Exercise 5)

$$x_{i+1}^h = x_i^h + \frac{1}{2} f\Big((t_i + t_{i+1})/2, (x_i^h + x_{i+1}^h)/2\Big)$$

is contractive.

IX

Differential-Algebraic Equations

In §1 we give a general introduction to differential-algebraic equations (DAE), i.e. ODE where the solution is subject to an (algebraic) constraint. It can be seen as a limiting case of certain singularly perturbed ODE. In §2 a theory for linear DAE is developed. Here the notion of the matrix pencil turns out to be crucial. It induces an index concept, which relates to a degree of complexity of the problem. Next, more general DAE are introduced with a corresponding notion of (differential) index in §3. An important class of DAE, viz. semi-explicit DAE, is the subject of §4. As far as numerical methods are concerned, this chapter mainly deals with (implicit) multisteps. Therefore, we consider BDF methods for DAE in §5. Since higher index problems cause (numerical) difficulties, one is often interested in lowering the index, which can always be achieved by differentiation. However, this may cause so-called drift, and that is why regularization methods for alleviating this problem are considered in §6.

1. Introduction

The type of equations we have discussed so far always contained an explicit derivative, i.e. could be written as

(1.1) $\qquad \dfrac{d\mathbf{x}}{dt} - \mathbf{f}(t, \mathbf{x}) = 0 \ .$

We may consider this as a special case of the *implicit* ODE

(1.2) $\qquad \mathbf{k}(t, \mathbf{x}, \dot{\mathbf{x}}) = 0 \ .$

Only if $\partial \mathbf{k}/\partial \mathbf{x}$ is nonsingular may we rewrite (1.2) as an *explicit* ODE. If this is not the case, we have an essentially more complicated problem. In order to understand this, consider the following set of equations:

(1.3a) $\qquad \dot{\mathbf{x}}_1 = \mathbf{f}(t, \mathbf{x}_1, \mathbf{x}_2) \ , \qquad \mathbf{x}_1(t) \in I\!\!R^p \ , \quad \mathbf{x}_2(t) \in I\!\!R^q \ ,$

(1.3b) $\qquad 0 = \mathbf{g}(t, \mathbf{x}_1, \mathbf{x}_2) \ , \qquad \mathbf{g}(t, \mathbf{x}_1, \mathbf{x}_2) \in I\!\!R^q \ .$

Clearly (1.3a) looks like an ODE for x_1, albeit with a parameter-like, time dependent additional variable x_2, whereas (1.3b) is a constraint equation. Hence we call (1.3) a differential-algebraic equation (DAE). Such equations arise in many applications; two important areas are electrical networks and mechanical systems. We shall discuss the latter in Chapters XI and XII in more detail. The former are demonstrated below.

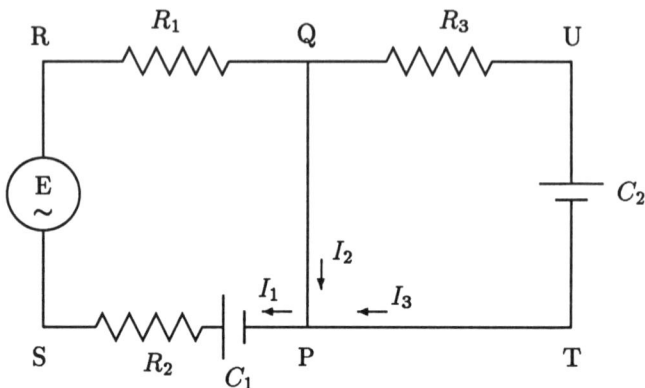

Figure IX.1   Electrical network with resistors R_1, R_2, R_3, capacitors C_1, C_2, currents I_1, I_2, I_3, voltage drops V_1, V_2, V_3, V_4, V_5 respectively, and voltage source E.

Example IX.1.
Consider an electrical network with resistors (R), capacitors (C), and a voltage source (E) as shown in Fig. IX.1. A point like P is called a *node* and $SPQR$, $PTUQ$ and $STUR$ are called *loops* in the circuit. Then we have *Kirchhoff's laws*:

(i)   *The algebraic sum of all currents meeting at any node in a circuit is zero.*

For instance, given the currents I_1, I_2 and I_3 of Fig. IX.1, we have at P: $I_1 - I_2 - I_3 = 0$. Here we use the convention that a minus sign is taken if the current flows towards the point under consideration.

(ii)   *Around any closed loop in a circuit the sum of the voltage drops is equal to the sum of the voltage sources in that loop.*

If the voltage drop at R_1, R_2 and C_1 in Fig. IX.1 equals V_1, V_2 and V_4, respectively, we thus have

$$E = V_1 + V_2 + V_4 \ .$$

Voltages and currents are related through $V = IR$ for a resistor (Ohm's law) and $I = C\,dV/dt$ for a capacitor. So we have, e.g.,

$$V_1 = I_1 R_1 , \qquad I_3 = C_2 \frac{dV_5}{dt} .$$

Summarising we see that Kirchhoff's laws give algebraic relations for the currents and voltages, as does Ohm's law, whereas we obtain a differential relation for them at a capacitor. For the 8 unknowns in the circuit of Fig. IX.1 (3 currents plus 5 voltages) we have 8 equations. Note that the current law applied at Q would yield the same result as at P. Applying the voltage law at the circuit $PTUQ$ gives for the voltages V_3 and V_5 at R_3 and C_2, respectively,

$$0 = V_3 + V_5 .$$

The circuit $STUR$ would give redundant information now. It is easy to check that we thus obtain the DAE

$$\frac{d}{dt} \begin{bmatrix} V_4 \\ V_5 \end{bmatrix} = \begin{bmatrix} C_1^{-1} & I_1 \\ C_2^{-1} & I_3 \end{bmatrix} ,$$

$$0 = \begin{bmatrix} 1 & 1 & 0 & 1 & 0 & 0 & 0 & 0 \\ 0 & 0 & 1 & 0 & 1 & 0 & 0 & 0 \\ 0 & 0 & 0 & 0 & 0 & 1 & -1 & -1 \\ 1 & 0 & 0 & 0 & 0 & -R_1 & 0 & 0 \\ 0 & 1 & 0 & 0 & 0 & 0 & -R_2 & 0 \\ 0 & 0 & 1 & 0 & 0 & 0 & 0 & -R_3 \end{bmatrix} \begin{bmatrix} V_1 \\ V_2 \\ V_3 \\ V_4 \\ V_5 \\ I_1 \\ I_2 \\ I_3 \end{bmatrix} - \begin{bmatrix} E \\ 0 \\ 0 \\ 0 \\ 0 \\ 0 \\ 0 \\ 0 \end{bmatrix} .$$

The DAE (1.3) can be thought of as a limiting case of the singularly perturbed ODE dealt with in §VIII.1, i.e.

(1.4a) $\qquad \dot{x}_1 = f(t, x_1, x_2) ,$

(1.4b) $\qquad \varepsilon \dot{x}_2 = g(t, x_1, x_2) ,$

where $\partial g/\partial x_2$ is assumed to be nonsingular. Recall that the limiting case $\varepsilon = 0$ of (1.4) is called the *reduced problem*, i.e. (1.3). We see that we can only describe p conditions for x_1 in (1.3a), while x_2 should be determined from (1.3b). We see that the initial conditions induce a p-dimensional manifold as solution space. If $\partial g/\partial x_2$ is nonsingular, we can differentiate (1.3b) to obtain

(1.5) $\qquad 0 = \dfrac{\partial g}{\partial t} + \dfrac{\partial g}{\partial x_1} \dfrac{dx_1}{dt} + \dfrac{\partial g}{\partial x_2} \dfrac{dx_2}{dt} .$

Substituting (1.3a), we thus find

(1.3b)' $\dfrac{d\mathbf{x}_2}{dt} = -\left[\dfrac{\partial \mathbf{g}}{\partial \mathbf{x}_2}\right]^{-1}\left[\dfrac{\partial \mathbf{g}}{\partial \mathbf{x}_1}\mathbf{f} + \dfrac{\partial \mathbf{g}}{\partial t}\right]$.

Hence (1.3a), (1.3b)' constitute an explicit ODE, for which q additional initial conditions can be found by computing $\mathbf{x}_2(0)$ from (1.3b). Yet, practical problems are to be expected. First of all, the equation (1.3b)' may be cumbersome to obtain. But secondly, one should realise that (1.5) is *not* completely equivalent to (1.3b), and so one may expect, at least numerically, deviations from the original constraint when using (1.3b)'. We return to this so-called *drift problem* later.

2. Linear DAE

If the DAE (1.2) is linear both in $\mathbf{x}$ and $\dot{\mathbf{x}}$, we can analyse it in more detail. So consider

(2.1) $\mathbf{E}\,\dot{\mathbf{x}}(t) + \mathbf{A}\,\mathbf{x}(t) = \mathbf{q}(t)$,

where, for the moment, $\mathbf{E}$ and $\mathbf{A}$ are constant $n \times n$ matrices. If $\mathbf{E}$ is nonsingular, we can easily obtain an explicit ODE as remarked in §1. Such an ODE admits solutions of the form

$$\mathbf{x}(t) = e^{\lambda t}\,\mathbf{x}_0 , \qquad \mathbf{x}_0 \in I\!\!R^n .$$

With a similar ansatz inserted in (2.1) we find for λ the eigenvalue equation

(2.2) $\det(\mathbf{E}\lambda + \mathbf{A}) = 0$.

The quantity $\mathbf{E}\lambda + \mathbf{A}$ is called the *matrix pencil* (cf. [29]), and is said to be *singular* if it is singular for all values of λ; otherwise it is called *regular*. If we make (2.1) an IVP by requiring

(2.3) $\mathbf{x}(0) = \mathbf{x}_0$,

then we call $\mathbf{x}_0$ a *consistent* initial value for (2.1), if (2.1) with (2.3) possesses at least one solution. This concept also applies to general DAE.

What one really likes is to have a manifold of solutions which are uniquely prescribed by the initial values. If such a manifold exists, one calls the DAE (2.1) *solvable* or *tractable*, (cf.[36, 15]). For the linear constant coefficient case we have the following:

Theorem 2.4. *The DAE (2.1) is solvable if and only if the matrix pencil* $\lambda\mathbf{E} + \mathbf{A}$ *is regular.*

Proof: First we assume (2.1) to be solvable. Let $x(0) = 0$ and assume by contradiction that the pencil is singular. Choose a set of λ_j and $v_j \neq 0$ such that $(\lambda_j E + A) v_j = 0$, $j = 1, ..., n + 1$. Since $\{v_j\}_{j=1}^{n+1}$ must be a dependent set, there exist scalars $\alpha_1, ..., \alpha_{n+1}$ such that

$$\sum_{j=1}^{n+1} \alpha_j v_j = 0,$$

whereas

$$y(t) := \sum_{j=1}^{n+1} \alpha_j e^{\lambda_j t} v_j \not\equiv 0.$$

Hence (2.1) allows for at least two independent solutions, viz. y and $x \equiv 0$, corresponding to the initial value $y(0) = x(0) = 0$. So we conclude that solvability implies regularity.

Assume now that the pencil is regular. Then the polynomial $\det(\lambda E + A)$, which has at most degree n, has at most $k \leq n$ zeros; they induce a solution space as in the ODE case, from which uniqueness follows trivially.   □

In what follows we shall assume $\lambda E + A$ to be regular now. It turns out that we can give a useful characterisation of DAE through the so-called *Weierstrass-Kronecker canonical form*, which is introduced in the following theorem:

Theorem 2.5. *If the pencil $\lambda E + A$ is regular, there exist nonsingular matrices P and Q such that*

$$(2.5a) \quad PEQ = \begin{bmatrix} I & 0 \\ 0 & N \end{bmatrix} ; \quad PAQ = \begin{bmatrix} C & 0 \\ 0 & I_l \end{bmatrix}, \quad for\ some\ l,$$

where the matrix N consists of nilpotent Jordan blocks N_i, in other words $N := \mathrm{diag}(N_1, ..., N_k)$, for some k, with $N_i, i = 1, ..., k$ given by

$$N_i := \begin{bmatrix} 0 & 1 & & & \\ & 0 & 1 & & \\ & & \ddots & \ddots & \\ & & & 0 & 1 \end{bmatrix} \quad (or\ possibly\ N_i = [0]),$$

and C consists of Jordan blocks with nonzero eigenvalues.

Proof: The general proof can be found in [29]. Here we consider the case that A is nonsingular only. Let TJT^{-1} be a Jordan canonical form of $A^{-1}E$ (cf. Appendix C), such that the blocks with nonzero eigenvalues appear first in $\mathbf{J}$. Clearly we can then take $\mathbf{P} := \mathbf{T}\mathbf{A}^{-1}$ and $\mathbf{Q} := \mathbf{T}^{-1}$. In this case we have $\mathbf{C} = \mathbf{I}_{n-l}$.   □

This Weierstrass-Kronecker form induces new (partitioned) variables through

$$(2.6a) \qquad \mathbf{x} =: \mathbf{Q}\begin{bmatrix} \mathbf{u} \\ \mathbf{v} \end{bmatrix}, \qquad \text{where } \mathbf{v} \in I\!R^l .$$

By also introducing

$$(2.6b) \qquad \mathbf{P}\mathbf{q} =: \begin{bmatrix} \mathbf{r} \\ \mathbf{s} \end{bmatrix}, \qquad \mathbf{s} \in I\!R^l ,$$

we can thus rewrite (2.1) as

$$(2.7a) \qquad \mathbf{u}^{(1)} + \mathbf{C}\mathbf{u} = \mathbf{r} ,$$

$$(2.7b) \qquad \mathbf{N}\mathbf{v}^{(1)} + \mathbf{v} = \mathbf{s} .$$

Note that (2.7a) is an ODE, but (2.7b) is *not*. By differentiating the latter, we can eliminate $\dot{\mathbf{v}}$ and obtain

$$(2.8a) \qquad \mathbf{N}\mathbf{v}^{(2)} + \mathbf{v}^{(1)} = \mathbf{s}^{(1)} \quad \Rightarrow \quad \mathbf{v} = \mathbf{s} - \mathbf{N}\mathbf{s}^{(1)} - \mathbf{N}^2\mathbf{v}^{(2)} .$$

If we differentiate (2.7b) twice and eliminate the second derivative via (2.8a), we find

$$(2.8b) \qquad \mathbf{v} = \mathbf{s} - \mathbf{N}\mathbf{s}^{(1)} + \mathbf{N}^2\mathbf{s}^{(2)} - \mathbf{N}^{(3)}\mathbf{v}^{(3)} .$$

In general we thus obtain

$$(2.8c) \qquad \mathbf{v} = \mathbf{s} - \mathbf{N}\mathbf{s}^{(1)} + \mathbf{N}^2\mathbf{s}^{(2)} + ...(-1)^{m-1}\mathbf{N}^{m-1}\mathbf{s}^{(m-1)} + $$
$$+ (-1)^m\mathbf{N}^m\mathbf{v}^{(m)} .$$

Since all Jordan blocks are nilpotent, i.e. $(\mathbf{N}_i)^{m_i} = 0$ for some m_i, there must be an integer ν such that $\mathbf{N}^\nu = 0$, where $\mathbf{N}^{\nu-1} \neq 0$. We shall call ν the (*nilpotency*) *index*. If we choose m in (2.8c) equal to ν, we find

$$(2.9) \qquad \mathbf{v}(t) = \sum_{i=0}^{m-1} (-1)^i \mathbf{N}^i \mathbf{s}^{(i)}(t) .$$

So we conclude that (2.7) allows for solutions $(\mathbf{u}^T, \mathbf{v}^T)^T$ of which only the first $(n - l)$ components, viz. $\mathbf{u}(0)$, should be prescribed as initial values. Hence the set of initial values of the solution manifold is given by $\mathbf{Q}^{-1}\begin{bmatrix} I\!R^{n-l} \\ 0 \end{bmatrix}$.

Below we now summarise some properties of DAE (2.1):

- The index ν is at most equal to l, the number of zero eigenvalues of $\mathbf{E}$.

- The solution lies on an $(n - l)$-dimensional manifold.

- The solution depends on derivatives of the source function $\mathbf{q}$, if $\nu \geq 1$. It has thus a lesser degree of smoothness as compared to solutions of ODE, because the latter only depend on source functions as such.

The next example shows that it is not directly clear how the index of a general DAE should be defined.

Example IX.2.
Consider the DAE

$$\begin{cases} \dot{x}_2 + x_1 = q_1 \ , \\ \dot{x}_3 + x_2 = q_2 \ , \\ x_3 = q_3 \ . \end{cases}$$

Only the last relation looks like a constraint equation. However, straightforward substitution reveals that

$$\begin{cases} x_1 = q_1 - \dot{q}_2 + \ddot{q}_3 \ , \\ x_2 = q_2 - \dot{q}_3 \ , \\ x_3 = q_3 \ . \end{cases}$$

So apparently this is a completely algebraic 'equation'. Rewriting it in matrix vector form as in (2.1) we see that

$$\mathbf{E} := \begin{bmatrix} 0 & 1 & 0 \\ 0 & 0 & 1 \\ 0 & 0 & 0 \end{bmatrix}, \qquad \mathbf{A} := \begin{bmatrix} 1 & 0 & 0 \\ 0 & 1 & 0 \\ 0 & 0 & 1 \end{bmatrix}.$$

Due to the simple form of $\mathbf{A}$ we see that $\mathbf{P} = \mathbf{Q} = \mathbf{I}$ in the Weierstrass-Kronecker form and actually $\mathbf{N}$ consists merely of one order-3 Jordan block, so $\mathbf{N}^3 = \mathbf{0}$. Hence we conclude that this DAE has index 3.

Remark 2.10. A constant coefficient ODE is a DAE of (nilpotency) index 0.

3. General DAE

If we want to understand DAE, being more general than (2.1), we clearly need to reformulate the index concept. The following example shows that the (algebraic) nilpotency index has no meaning even for a *time dependent, linear system*.

Example IX.3.
Consider the DAE

$$\mathbf{E}(t)\,\dot{\mathbf{x}} + \mathbf{A}\mathbf{x} = \mathbf{q}\,,$$

where

$$\mathbf{E}(t) := \begin{bmatrix} 1 & t \\ 0 & 0 \end{bmatrix},\ \mathbf{A}(t) := \begin{bmatrix} 0 & 0 \\ 1 & t \end{bmatrix},\ \mathbf{q} := \begin{bmatrix} q_1 \\ q_2 \end{bmatrix},\ \mathbf{x} := \begin{bmatrix} x_1 \\ x_2 \end{bmatrix}.$$

Then $\det(\mathbf{E}\lambda + \mathbf{A}) = 0$ for all λ. It is equivalent to

$$\begin{cases} x_1 + tx_2 = q_2\,, \\ \dot{x}_1 + t\dot{x}_2 = q_1\,. \end{cases}$$

Fortunately we can solve this system. Differentiation of the first equation gives

$$\dot{x}_1 + t\dot{x}_2 + x_2 = \dot{q}_2\,,$$

and so we directly obtain

$$x_2 = \dot{q}_2 - q_1 \quad \Rightarrow \quad x_1 = q_2 - t(\dot{q}_2 - q_1)\,.$$

Clearly there are no initial conditions to be specified.

One suitable generalisation of the index concept is inspired by the way we derived (2.9) from (2.7b), viz. repeated differentiation. This leads to the concept of *(differential) index*. We say that (1.2) has *(differential) index* ν if ν is the minimal value for which the system

(3.1a)
$$\begin{cases} \mathbf{k}(t,\mathbf{x},\dot{\mathbf{x}}) = 0\,, \\[1mm] \dfrac{d}{dt}\,\mathbf{k}(t,\mathbf{x},\dot{\mathbf{x}}) = 0\,, \\[1mm] \quad\vdots \\[1mm] \dfrac{d^\nu}{dt^\nu}\,\mathbf{k}(t,\mathbf{x},\dot{\mathbf{x}}) = 0\,, \end{cases}$$

constitutes an explicit ODE, say

(3.1b) $\dot{\mathbf{x}} = \tilde{\mathbf{f}}(t,\mathbf{x})\,,$

the so-called *underlying ODE*. Recall from the previous section that we have to perform several substitutions (or eliminations) to obtain (3.1b), which may be tedious.

Remark 3.2. A general ODE has (differential) index 0.

Example IX.4.
For the DAE in Example IX.3 we obtain from differentiating once

$$\begin{bmatrix} 0 & 0 \\ 0 & 1 \end{bmatrix} \mathbf{x} + \begin{bmatrix} 0 & 1 \\ 1 & t \end{bmatrix} \mathbf{x}^{(1)} + \begin{bmatrix} 1 & t \\ 0 & 0 \end{bmatrix} \mathbf{x}^{(2)} = \mathbf{q}^{(1)} .$$

Differentiating once more yields

$$\begin{bmatrix} 0 & 0 \\ 0 & 2 \end{bmatrix} \mathbf{x}^{(1)} + \begin{bmatrix} 0 & 2 \\ 1 & t \end{bmatrix} \mathbf{x}^{(2)} + \begin{bmatrix} 1 & t \\ 0 & 0 \end{bmatrix} \mathbf{x}^{(3)} = \mathbf{q}^{(2)} .$$

From this relation we can express $\mathbf{x}^{(2)}$ in the rest and thus eliminate this quantity in the first relation:

$$\begin{bmatrix} 0 & 0 \\ 0 & 1 \end{bmatrix} \mathbf{x} + \begin{bmatrix} 0 & 1 \\ 1 & t \end{bmatrix} \mathbf{x}^{(1)} + \begin{bmatrix} 1 & t \\ 0 & 0 \end{bmatrix} \begin{bmatrix} 0 & 2 \\ 1 & t \end{bmatrix}^{-1} .$$

$$\cdot \left\{ \mathbf{q}^{(2)} - \begin{bmatrix} 0 & 0 \\ 0 & 2 \end{bmatrix} \mathbf{x}^{(1)} + \begin{bmatrix} 1 & t \\ 0 & 0 \end{bmatrix} \mathbf{x}^{(3)} \right\} = \mathbf{q}^{(1)} .$$

Since $\begin{bmatrix} 1 & t \\ 0 & 0 \end{bmatrix} \begin{bmatrix} 0 & 2 \\ 1 & t \end{bmatrix}^{-1} = \begin{bmatrix} 0 & 1 \\ 0 & 0 \end{bmatrix}$, we thus find

$$\begin{bmatrix} 0 & -1 \\ 1 & t \end{bmatrix} \dot{\mathbf{x}} + \begin{bmatrix} 0 & 0 \\ 0 & 1 \end{bmatrix} \mathbf{x} = \dot{\mathbf{q}} - \begin{bmatrix} 0 & 1 \\ 0 & 0 \end{bmatrix} \ddot{\mathbf{q}} ,$$

which can be written as an explicit ODE for all t. We thus conclude that the index is 2.

Remark 3.3. For linear constant coefficient DAE, the differential index coincides with the nilpotency one.

One of the properties of a higher index linear constant coefficient system is the dependence of the solution on higher derivatives of the source term. If we add a perturbation function $\mathbf{d}(t)$ to the right-hand side of (1.2) we obtain

(3.4a) $\mathbf{k}(t, \mathbf{y}, \dot{\mathbf{y}}) = \mathbf{d}(t)$.

It follows from (3.1a) that the underlying ODE has the form

(3.4b) $\dot{\mathbf{y}} = \tilde{\mathbf{f}}(t, \mathbf{y}) + \mathbf{e}(t)$,

where
$$\|\mathbf{e}(t)\| \le C \left[\|\mathbf{d}(t)\| + \|\mathbf{d}^{(1)}(t)\| + \dots + \|\mathbf{d}^{(\nu)}(t)\| \right], \quad C \ge 0,$$

and thus by a standard argument (cf. §II.5)

$$(3.5) \qquad \|\mathbf{x}(t) - \mathbf{y}(t)\| \leq \hat{C} \left[\|\mathbf{x}(0) - \mathbf{y}(0)\| + \max_{s \leq t} \|e(s)\| \right], \ \hat{C} \geq 0 \ .$$

Often one has dependence on derivatives up to $(\nu - 1)$ only, as the next example shows.

Example IX.5.
Consider the linear DAE

$$\begin{cases} \dot{\mathbf{x}}_1 = \mathbf{A}\,\mathbf{x}_1 + \mathbf{B}\,\mathbf{x}_2 + \mathbf{q}_1 \ , \\ 0 = \mathbf{C}\,\mathbf{x}_1 + \mathbf{D}\,\mathbf{x}_2 + \mathbf{q}_2 \ , \qquad \mathbf{D} \text{ nonsingular} \ . \end{cases}$$

This is an index-1 DAE. Let us assume $\mathbf{x}_1(0)$ is given. Solving for $\mathbf{x}_2$ from the algebraic part and substituting this in the differential part gives

$$\dot{\mathbf{x}}_1 = (\mathbf{A} - \mathbf{B}\mathbf{D}^{-1}\mathbf{C})\,\mathbf{x}_1 + \mathbf{q}_1 + \mathbf{B}\mathbf{D}^{-1}\mathbf{q}_2 \ .$$

Hence for $0 < t \leq T < \infty$ there exists a constant C such that

$$\|\mathbf{x}_1(t)\| \leq C \left[\|\mathbf{x}_1(0)\| + \max_{0 \leq s \leq t} \{\|\mathbf{q}_1\| + \|\mathbf{q}_2\|\} \right] \ .$$

Trivially $\mathbf{x}_2$ does not depend on $\dot{\mathbf{q}}_1$ or $\dot{\mathbf{q}}_2$ either.

4. Semi-explicit DAE

In §1 we saw that DAE may be more accessible if the algebraic (constraint) equation is decoupled. One often calls (1.3) a *semi-explicit* DAE. Note that if $\partial g/\partial \mathbf{x}_2$ is nonsingular, we have an index-1 problem. Equations (1.3a) and (1.3b)' form the underlying ODE. For higher index problems semi-explicit forms often appear in so-called *Hessenberg form* of size ν, viz.

$$(4.1) \qquad \begin{cases} \dot{\mathbf{x}}_1 = \mathbf{f}_1(t, \mathbf{x}_1, ..., \mathbf{x}_\nu) \ , \\ \dot{\mathbf{x}}_2 = \mathbf{f}_2(t, \mathbf{x}_1, ..., \mathbf{x}_{\nu-1}) \ , \\ \qquad \vdots \\ \dot{\mathbf{x}}_{\nu-1} = \mathbf{f}_{\nu-1}(t, \mathbf{x}_{\nu-2}, \mathbf{x}_{\nu-1}) \ , \\ 0 = \mathbf{f}_\nu(t, \mathbf{x}_{\nu-1}) \ . \end{cases}$$

One can show that (4.1) has index ν if the matrix

$$\left[\frac{\partial \mathbf{f}_\nu}{\partial \mathbf{x}_{\nu-1}} \cdots \frac{\partial \mathbf{f}_2}{\partial \mathbf{x}_1} \right] \cdot \frac{\partial \mathbf{f}_1}{\partial \mathbf{x}_\nu}$$

is nonsingular. We shall prove this for $\nu = 2$ and $\nu = 3$ below, where we assume the system to be autonomous, for the sake of simplicity. The autonomous size-2 Hessenberg form looks like

(4.2a) $\dot{\mathbf{x}}_1 = \mathbf{f}_1(\mathbf{x}_1, \mathbf{x}_2)$,

(4.2b) $0 = \mathbf{f}_2(\mathbf{x}_1)$.

Differentiating the constraint equation (4.2b) gives

(4.2b)' $0 = \dfrac{\partial \mathbf{f}_2}{\partial \mathbf{x}_1} \dot{\mathbf{x}}_1 = \dfrac{\partial \mathbf{f}_2}{\partial \mathbf{x}_1} \mathbf{f}_1$.

Now (4.2a), (4.2b)' constitute an explicit index-1 system, since

$$\frac{\partial}{\partial \mathbf{x}_2} \left\{ \frac{\partial \mathbf{f}_2}{\partial \mathbf{x}_1} \mathbf{f}_1 \right\} = \frac{\partial \mathbf{f}_2}{\partial \mathbf{x}_1} \frac{\partial \mathbf{f}_1}{\partial \mathbf{x}_2}$$

is nonsingular. Consequently (4.2) must be index-2.

The autonomous size-3 Hessenberg form is given by

(4.3a) $\dot{\mathbf{x}}_1 = \mathbf{f}_1(\mathbf{x}_1, \mathbf{x}_2, \mathbf{x}_3)$,

(4.3b) $\dot{\mathbf{x}}_2 = \mathbf{f}_2(\mathbf{x}_1, \mathbf{x}_2)$,

(4.3c) $0 = \mathbf{f}_3(\mathbf{x}_2)$.

Differentiating (4.3c) gives

(4.3c)' $0 = \dfrac{\partial \mathbf{f}_3}{\partial \mathbf{x}_2} \dot{\mathbf{x}}_2 = \dfrac{\partial \mathbf{f}_3}{\partial \mathbf{x}_2} \mathbf{f}_2$.

The equations (4.3a), (4.3b), (4.3c)' constitute an explicit index-2 system, which can be seen by writing

(4.4) $\mathcal{X}_1 := \begin{bmatrix} \mathbf{x}_1 \\ \mathbf{x}_2 \end{bmatrix}$, $\mathcal{X}_2 := \mathbf{x}_3$, $\mathcal{F}_1 := \begin{bmatrix} \mathbf{f}_1 \\ \mathbf{f}_2 \end{bmatrix}$, $\mathcal{F}_2 := \dfrac{\partial \mathbf{f}_3}{\partial \mathbf{x}_2} \mathbf{f}_2$.

Then we see that

(4.5) $\dfrac{\partial \mathcal{F}_2}{\partial \mathcal{X}_1} \dfrac{\partial \mathcal{F}_1}{\partial \mathcal{X}_2} = \begin{bmatrix} \dfrac{\partial \mathbf{f}_3}{\partial \mathbf{x}_2} \dfrac{\partial \mathbf{f}_2}{\partial \mathbf{x}_1} & \vdots & \dfrac{\partial^2 \mathbf{f}_3}{\partial \mathbf{x}_2^2} \mathbf{f}_2 + \dfrac{\partial \mathbf{f}_3}{\partial \mathbf{x}_2} \dfrac{\partial \mathbf{f}_2}{\partial \mathbf{x}_2} \end{bmatrix} \cdot \begin{bmatrix} \dfrac{\partial \mathbf{f}_1}{\partial \mathbf{x}_3} \\ \text{-----} \\ 0 \end{bmatrix} =$

$$= \frac{\partial \mathbf{f}_3}{\partial \mathbf{x}_2} \frac{\partial \mathbf{f}_2}{\partial \mathbf{x}_1} \frac{\partial \mathbf{f}_1}{\partial \mathbf{x}_3} ,$$

which is nonsingular by assumption. Since we thus have effectively reduced (4.3) to a DAE like (4.2) we conclude that (4.3) has index 3 and can be written as

$$(4.6) \qquad \begin{cases} \dot{\mathcal{X}}_1 = \mathcal{F}(\mathcal{X}_1, \mathcal{X}_2) \,, \\ \\ 0 = \mathcal{F}_2(\mathcal{X}_1) \,. \end{cases}$$

In an IVP setting, say with $t_0 = 0$, one needs *consistent initial values*, i.e. which satisfy the constraints. It is important to realise that higher index problems involve 'hidden constraints'. First consider the index-1 DAE (1.3), with $\partial \mathbf{g}/\partial \mathbf{x}_2$ nonsingular; it is then obvious that we should prescribe only $\mathbf{x}_1(0)$; for all t the value of $\mathbf{x}_2(t)$ follows from $\mathbf{x}_1(t)$ through (1.3b). For index-2 problems like (4.2), we note that (4.2b)$'$ constitutes a hidden constraint, which enables us to determine $\mathbf{x}_2(t)$ from $\mathbf{x}_1(t)$. Now let $\mathbf{x}_1(0) \in \mathbb{R}^p$ and $\mathbf{x}_2(0) \in \mathbb{R}^q$; we then see from (4.2b) that q conditions for $\mathbf{x}_1(0)$ are prescribed already. Since $p \geq q$ (see Exercise 7), we have to prescribe only $p - q$ additional initial conditions for $\mathbf{x}_1(0)$.

For index-3 problems like (4.3), we have two additional hidden constraints, viz. (4.3c) and (cf. (4.2b)$'$)

$$0 = \frac{\partial \mathcal{F}_2}{\partial \mathcal{X}_1} \cdot \mathcal{F}_1 \,.$$

Suppose $\mathbf{x}_1(0) \in \mathbb{R}^p$, $\mathbf{x}_2(0) \in \mathbb{R}^q$ and $\mathbf{x}_3(0) \in \mathbb{R}^r$, then $p \geq q \geq r$ (see Exercise 7). Altogether we thus need $p + q - 2r$ additional initial conditions for $\mathbf{x}_1(0)$ and $\mathbf{x}_2(0)$.

Example IX.6.
Consider the DAE

$$\begin{cases} \dot{x}_1 = x_3 =: f_1(x_3) \,, \\ \\ \dot{x}_2 = x_1 =: f_2(x_1) \,, \\ \\ 0 = x_2 - g =: f_3(x_2) \,, \end{cases}$$

where g is a function of t only. Obviously, $\dfrac{\partial f_3}{\partial x_2} \dfrac{\partial f_2}{\partial x_1} \dfrac{\partial f_1}{\partial x_3} \neq 0$. So this is an index-3 semi-explicit (Hessenberg form) DAE, cf. (4.5). The solution is given by $x_1 = \dot{g}$, $x_2 = g$, $x_3 = \ddot{g}$. So *no* initial conditions are needed. Indeed, $p = q = r = 1 \Rightarrow p + q - 2r = 0$.

Finally, we remark that for certain purposes it is sometimes useful to introduce an extra variable and associate a semi-explicit system with an implicit DAE. Suppose we have the DAE

$$(4.7) \qquad 0 = \mathbf{k}(t, \mathbf{x}, \dot{\mathbf{x}}) \,.$$

By defining

$$(4.8a) \qquad \mathbf{z} := \dot{\mathbf{x}} \,,$$

we have

(4.8b) $0 = \mathbf{k}(t, \mathbf{x}, \mathbf{z})$.

Obviously (4.8) is semi-explicit. Moreover, if (4.7) has index ν, we see that (4.8) has index $\nu + 1$. The latter property can be a disadvantage though.

5. BDF methods for DAE

We now turn to the numerical treatment of DAE. Given the relationship between DAE and singularly perturbed ODE, as already mentioned in §1, it may not be surprising that BDF methods (see Chapter VII) have a good potential. Before we dwell on this, we first illustrate why we *need* implicit methods for general (implicit) DAE.

Example IX.7.

Consider $\mathbf{E}\dot{\mathbf{x}} + \mathbf{A}\mathbf{x} = \mathbf{f}$, where $\mathbf{E} := \begin{bmatrix} 1 & 1 \\ 1 & 1 \end{bmatrix}$, $\mathbf{A} := \begin{bmatrix} -1 & -1 \\ 2 & 0 \end{bmatrix}$.

Euler forward then yields

$$\mathbf{E}\mathbf{x}_{i+1}^{h} - \mathbf{E}\mathbf{x}_{i}^{h} + h\mathbf{A}\mathbf{x}_{i}^{h} = h\mathbf{f}(t_i) .$$

Since $\mathbf{E}$ is singular, we cannot solve for $\mathbf{x}_{i+1}^{h}$! If we were to apply Euler backward, we would find

$$\mathbf{E}\mathbf{x}_{i+1}^{h} - \mathbf{E}\mathbf{x}_{i}^{h} + h\mathbf{A}\mathbf{x}_{i+1}^{h} = h\mathbf{f}(t_{i+1}) .$$

The pencil,

$$\begin{bmatrix} \lambda - 1 & \lambda - 1 \\ \lambda + 2 & \lambda \end{bmatrix} ,$$

of the problem above is regular (i.e. only for $\lambda = 1$ does the determinant vanish). We can thus uniquely solve for $\mathbf{x}_{i+1}^{h}$, unless $h = 1$.

If we have a general multistep method, cf. (VII.1.1), and apply this to the constant coefficient DAE (2.1), we obtain

$$(5.1) \qquad \sum_{j=0}^{k} \alpha_j \mathbf{E}\mathbf{x}_{i-j+1}^{h} - h \sum_{j=0}^{k} \beta_j \mathbf{A}\mathbf{x}_{i-j+1}^{h} = h \sum_{j=0}^{k} \beta_j \mathbf{q}(t_{i-j+1}) .$$

Since we can write this in explicit form, we see that $\alpha_0 \mathbf{E} - h\beta_0 \mathbf{A}$ needs to be nonsingular. Because $\mathbf{E}$ is singular, a necessary condition is $\beta_0 \neq 0$, i.e. the method has to be implicit. However, not all implicit methods can be used (cf. Exercise 8). We can prove the following:

Theorem 5.2. *A k-step BDF with constant step size, applied to the constant coefficient DAE (2.1) with index ν, is convergent with order k after $(\nu - 1)\,k + 1$ steps.*

Proof: Rather than proving the general result we shall restrict ourselves to $k = 1$ (Euler backward) and moreover to $\nu = 1, 2$ only. It is not restrictive to assume that (2.1) is already in Weierstrass-Kronecker form, i.e.

$$\begin{bmatrix} \mathbf{I} & 0 \\ 0 & \mathbf{N} \end{bmatrix} \dot{\mathbf{x}} + \begin{bmatrix} \mathbf{C} & 0 \\ 0 & \mathbf{I} \end{bmatrix} \mathbf{x} = \mathbf{f} \, .$$

In the following let vectors $\mathbf{x}$ be partitioned like $(\mathbf{y}^T, \mathbf{z}^T)^T$ and $\mathbf{f}$ like $(\mathbf{g}^T, \mathbf{h}^T)^T$, where $\mathbf{y}$ and $\mathbf{g}$ have the same order as $\mathbf{C}$. Euler backward gives

$$\begin{bmatrix} \mathbf{I} + h\mathbf{C} & 0 \\ 0 & \mathbf{N} + h\mathbf{I} \end{bmatrix} \mathbf{x}_{i+1} = \begin{bmatrix} \mathbf{I} & 0 \\ 0 & \mathbf{N} \end{bmatrix} \mathbf{x}_i + h\,\mathbf{f}_{i+1} \, .$$

Here we adopt the notation $\mathbf{f}_i := \mathbf{f}(t_{i+1})$. If $\nu = 1$ then $\mathbf{N} = 0$, so

$$(\mathbf{I} + h\mathbf{C})\,\mathbf{y}_{i+1}^1 = \mathbf{x}_i^1 + h\,\mathbf{g}_{i+1} \, ,$$

which is simply Euler's scheme applied to $\dot{\mathbf{x}}^1 + \mathbf{C}\mathbf{x}^1 = \mathbf{f}^1$. Note that the invertibility of $\mathbf{C}$ implies that the error amplification is bounded by $\|\exp(-\mathbf{C}t)\|$ on an interval $(0, t)$. Since we have first order local errors and apparently also stability on bounded intervals, we must have global first order convergence. As for $\mathbf{z}_i$, we see that the second part of the Δ-equation yields $\mathbf{z}_{i+1} = \mathbf{h}_{i+1}$, $i \geq 0$. This result is independent of the choice of $\mathbf{z}_0$!.

For $\nu = 2$ the result for $\mathbf{y}_i$ remains the same. For $\mathbf{z}_i$ we work out the recursion:

$$(\mathbf{N} + h\mathbf{I})\,\mathbf{z}_{i+1} = \mathbf{N}\,\mathbf{z}_i + h\,\mathbf{h}_{i+1} \, .$$

Since $(\mathbf{N} + h\mathbf{I})$ is nonsingular (see (2.5a)) we can write

$$(\mathbf{N} + h\mathbf{I})^{-1} = h^{-1}(h^{-1}\mathbf{N} + \mathbf{I})^{-1} = h^{-1}(\mathbf{I} - h^{-1}\mathbf{N}) \, ,$$

since $\mathbf{N}^2 = 0$. Hence

$$\mathbf{z}_{i+1} = h^{-1}\mathbf{N}\,\mathbf{z}_i + (\mathbf{I} - h^{-1}\mathbf{N})\,\mathbf{h}_{i+1} \, .$$

After two time steps we thus find

$$\mathbf{z}_2 = h^{-2}\mathbf{N}^2\,\mathbf{z}_0 + h^{-1}\mathbf{N}(\mathbf{I} - h^{-1}\mathbf{N})\,\mathbf{h}_1 + (\mathbf{I} - h^{-1}\mathbf{N})\,\mathbf{h}_2 =$$
$$= h^{-1}\mathbf{N}\,\mathbf{h}_1 + (\mathbf{I} - h^{-1}\mathbf{N})\,\mathbf{h}_2 \, .$$

Note that the influence of the initial value $\mathbf{z}_0$ has disappeared after the second step. In general we obtain

$$\mathbf{z}_i = h^{-1} \mathbf{N} \mathbf{h}_{i-1} + (\mathbf{I} - h^{-1} \mathbf{N}) \mathbf{h}_i =$$

$$= \mathbf{N} \frac{1}{h} [\mathbf{h}_{i-1} - \mathbf{h}_i] + \mathbf{h}_i \ ,$$

which is apparently an $O(h)$ approximation of the true solution

$$\mathbf{z}(t) = -\mathbf{N} \dot{\mathbf{h}}(t) + \mathbf{h}(t) \ .$$

It is simple to generalise this proof for $\nu > 2$. For higher order BDF the proof will be similar. $\qquad\qquad\qquad\qquad\qquad\qquad\qquad\qquad\qquad\qquad\qquad\qquad\qquad$ □

Remark 5.3. As said before we do not need to give a full initial condition for (2.1). From the proof of Theorem 5.2 we see that even an arbitrary initial condition automatically gives a consistent solution after ν steps.

For variable coefficient linear DAE the situation may be more complicated, at least for higher order index problems. We demonstrate this for a simple, even semi-explicit index-3 problem and Euler backward.

Example IX.8.
For given smooth function $\alpha(t) \neq 0$ consider the DAE

$$\begin{cases} \dot{x} = \alpha(t) y \ , \\ \dot{y} = \alpha(t) z^3 \ , \\ 0 = x - q(t) \ . \end{cases}$$

Because α can be absorbed in the time variable, we find that solving this problem by Euler backward is equivalent to solving the constant coefficient DAE

$$\begin{cases} \dot{x} = y \ , \\ \dot{y} = z \ , \\ 0 = x - q(t) \end{cases}$$

by variable step (we have identified the "old" and "new" t in $q(t)$ for simplicity). Clearly the solution is $x = q(t)$, $y = \dot{q}(t)$, $z = \ddot{q}(t)$. Discretising with Euler backward we find

$$\begin{cases} x_{i+1}^1 - x^1 = h_i \, y_{i+1} \ , \\ y_{i+1} - y_i = h_i \, z_{i+1}^3 \ , \\ 0 = x_{i+1}^1 - q_{i+1} \ , \end{cases}$$

where $q_{i+1} := q(t_{i+1})$. This leads to

$$z_{i+1} = \frac{1}{h_i} \left[\frac{1}{h_i} (q_{i+1} - q_i) - \frac{1}{h_{i-1}} (q_i - q_{i-1}) \right] =$$

$$= \frac{1}{2} \left(1 - \frac{h_{i-1}}{h_i} \right) \ddot{q}(t_i) + O\left(\max(h_i, h_{i-1}) \right) .$$

We conclude that we obtain an $O(h_i)$ approximation only if $h_{i-1} = h_i(1 + O(h_i))$. Otherwise we have an $O(1)$ error!

Fortunately many problems are of index 1 or 2, or can at least be brought into lower index form by differentiation (cf. §3). If they are of semi-implicit form, an obvious formulation for BDF seems to be to treat the constraint (algebraic) part separately. So for the index-1 DAE

(5.4) $\qquad \begin{cases} \dot{\mathbf{x}} = \mathbf{f}_1(t, \mathbf{x}, \mathbf{y}) , \\ 0 = \mathbf{f}_2(t, \mathbf{x}, \mathbf{y}) . \end{cases}$

We use

(5.5) $\qquad \begin{cases} \displaystyle\sum_{j=0}^{k} \alpha_j \mathbf{x}_{i-j+1}^h = h\, \beta_0\, \mathbf{f}(t_{i+1}, \mathbf{x}_{i+1}^h, \mathbf{y}_{i+1}^h) , \\ 0 = \mathbf{f}_2(t_{i+1}, \mathbf{x}_{i+1}^h, \mathbf{y}_{i+1}^h) . \end{cases}$

For the linear case the DAE (5.4) reads

(5.6) $\qquad \begin{cases} \dot{\mathbf{x}} = \mathbf{A}\,\mathbf{x} + \mathbf{B}\,\mathbf{y} + \mathbf{q} , \\ 0 = \mathbf{C}\,\mathbf{x} + \mathbf{D}\,\mathbf{y} + \mathbf{r} . \end{cases}$

This is index-1 if $\mathbf{D}$ is nonsingular for all (relevant) t. Applying (5.5) to (5.6) results in a Δ-equation for $\{\mathbf{x}_{i+1}^h\}$ and $\{\mathbf{y}_{i+1}^h\}$ of the form

(5.7a) $\qquad \mathbf{J} \begin{bmatrix} \mathbf{x}_{i+1}^h \\ \mathbf{y}_{i+1}^h \end{bmatrix} = \begin{bmatrix} \mathbf{x}_i^h \\ 0 \end{bmatrix} + \begin{bmatrix} -\displaystyle\sum_{j=1}^{k} \alpha_j \mathbf{x}_{i-j+1}^h + h\, \beta_0\, \mathbf{q}(t_{i+1}) \\ \mathbf{r}(t_{i+1}) \end{bmatrix} ,$

where

(5.7b) $\qquad \mathbf{J} = \begin{bmatrix} \mathbf{I} - \dfrac{\beta_0}{\alpha_0} h\mathbf{A} & -\dfrac{\beta_0}{\alpha_0} h\mathbf{B} \\ \mathbf{C} & \mathbf{D} \end{bmatrix} .$

The latter matrix is nonsingular, except for a small number of step size choices h.

Property 5.8. *If we have sufficiently accurate starting values, then the order-k BDF method (5.5) for the index-1 semi-explicit DAE (5.4) converges with order k.*

Proof: We only sketch the proof for $k=1$ and the linear case (5.6). The inverse of the matrix $\mathbf{J}$ in (5.7) is given by

$$\mathbf{J}^{-1} = \begin{bmatrix} \mathbf{I} + h(\mathbf{A} - \mathbf{B}\,\mathbf{D}^{-1}\,\mathbf{C}) & h\,\mathbf{B}\,\mathbf{D}^{-1} \\ -\mathbf{D}^{-1}\,\mathbf{C}\left(\mathbf{I} + h(\mathbf{A} - \mathbf{B}\,\mathbf{D}^{-1}\,\mathbf{C})\right) & (\mathbf{I} - h\,\mathbf{D}^{-1}\,\mathbf{C}\,\mathbf{B})\,\mathbf{D}^{-1} \end{bmatrix} +$$

$$+ \; O(h^2) \; .$$

So the homogeneous part of the recursion (5.7a) can be considered as an Euler discretisation of $\dot{\mathbf{x}} = (\mathbf{A} - \mathbf{B}\,\mathbf{D}^{-1}\,\mathbf{C})\,\mathbf{x}$, which is the homogeneous part of the underlying ODE. This implies both first order consistency and stability on finite intervals, i.e. convergence. For the nonlinear case one should consider (5.6) as a linearisation. See also §VI.6.   □

Remark 5.9. From the proof of Property 5.8 it follows that round-off errors in both $\{\mathbf{x}_i\}$ and $\{\mathbf{y}_i\}$ are proportional to the machine error MACHEPS. This means that semi-explicit index-1 DAE behave as well-conditioned as ODE.

A linear semi-explicit index-2 DAE looks like (5.6) with $\mathbf{D} = 0$ and $\mathbf{CB}$ nonsingular. Note that $\mathbf{J}$ is still nonsingular (see Exercise 12). Introducing the projection matrix

(5.10) $\mathbf{P} := \mathbf{B}(\mathbf{CB})^{-1}\,\mathbf{C}$,

gives

(5.11) $$\mathbf{J}^{-1} = \begin{bmatrix} (\mathbf{I}-\mathbf{P})\left(\mathbf{I}+\hat{h}\mathbf{A}(\mathbf{I}-\mathbf{P})\right) & [\mathbf{I}+\hat{h}(\mathbf{I}-\mathbf{P})\mathbf{A}]\,\mathbf{B}(\mathbf{CB})^{-1} \\ -\hat{h}^{-1}(\mathbf{CB})^{-1}\,\mathbf{C}\left(\mathbf{I}+\hat{h}\mathbf{A}(\mathbf{P}-\mathbf{I})\right) & \hat{h}^{-1}(\mathbf{CB})^{-1}\,[\mathbf{I}-\hat{h}\mathbf{CAB}(\mathbf{CB})^{-1}] \end{bmatrix} ,$$

where $\hat{h} := h\,\beta_0/\alpha_0$.

From (5.11) it can now be seen that rounding errors will be $O(\text{MACHEPS})$ for $\{\mathbf{x}_i\}$, but $O(\text{MACHEPS}/h)$ for $\{\mathbf{y}_i\}$. This shows the ill-conditioning of a higher index (semi-explicit) DAE.

Example IX.9.
Consider the following DAE:

$$\begin{cases} \dot{x} = -y + 2\sin t + \delta(x) \\ 0 = x - \cos t + \varepsilon(x) \end{cases}$$

$$x(0) = 1 \; , \qquad y(0) = 0 \; .$$

Here $\delta(x)$ and $\varepsilon(x)$ are randomly chosen numbers in the range $[-0.5\ 10^{-4}, 0.5\ 10^{-4}]$ to simulate a MACHEPS. If $\delta = \varepsilon \equiv 0$ one immediately sees that $x = \cos t$, $y = 2\sin t$. On the interval $[0, 10^{-3}]$ we

Table IX.1

h	e_1	e_2	res
10^{-3}	$0.47\ 10^{-4}$	$0.47\ 10^{-1}$	$0.47\ 10^{-4}$
10^{-4}	$0.45\ 10^{-4}$	$0.15\ 10^{-1}$	$0.45\ 10^{-4}$
10^{-5}	$0.49\ 10^{-4}$	$0.95\ 10^{1}$	$0.49\ 10^{-4}$
10^{-6}	$0.47\ 10^{-4}$	$0.20\ 10^{1}$	$0.47\ 10^{-4}$
10^{-7}	$0.47\ 10^{-4}$	$0.19\ 10^{2}$	$0.47\ 10^{-4}$
10^{-8}	$0.46\ 10^{-4}$	$0.67\ 10^{3}$	$0.46\ 10^{-4}$

obtain from discretisation with Euler backward the data in Table IX.1, where $e_1 = \max\limits_{i \leq T/h} |x(ih) - x_i^h|$, $e_2 = \max\limits_{i \leq T/h} |y(ih) - y_i^h|$, res $= \max\limits_{i \leq T/h} |x_i^h - \cos ih|$ (the residual left in this constraint).

6. Regularisation and stabilisation

Higher index problems (in particular for index 3 and up) may cause problems when numerical methods are applied. There is still active research in this area, where certain implicit Runge-Kutta methods, in particular, seem to be promising (cf. [8]). Yet the relative simplicity of BDF methods (e.g. low complexity if the step size choice is sophisticated enough, and the variable order mechanism) does not rule them out, even for higher index problems. This holds in particular if they are combined with some of the techniques we describe below.

A straightforward first idea is to lower the index by differentiation. In particular for semi-explicit systems, such as Hessenberg systems, this is easy to carry out.

Example IX.10.
Consider the simple pendulum (see Fig. IX.2). This problem will be extensively studied in Chapter XI. Here we shall describe the motion using Cartesian coordinates. Let the gravitational force (in the y-direction) be given by $-g$. Let the tension in the string be proportional to the position of the mass m. Finally let the string have length L. Then we have

$$\begin{cases} \ddot{x} = -\lambda x\ , \\ \ddot{y} = -\lambda y - g\ , \\ 0 = x^2 + y^2 - L^2\ . \end{cases}$$

Here $\lambda(t)$ is the so-called *Lagrange multiplier*. The pendulum problem

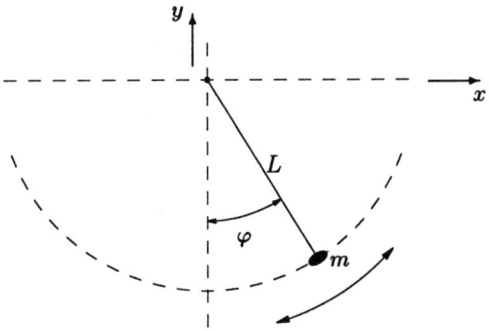

Figure IX.2 Mathematical pendulum of length L and mass m.

can be rewritten in standard form by defining $x_1 := x$, $x_2 := \dot{x}$, $x_3 := y$ and $x_4 := \dot{y}$:

$$(6.1a) \quad \begin{cases} \dot{x}_1 = x_2 \ , \\[1mm] \dot{x}_2 = -\lambda x_1 \ , \\[1mm] \dot{x}_3 = x_4 \ , \\[1mm] \dot{x}_4 = -\lambda x_3 - g \ , \end{cases}$$

$$(6.1b) \qquad 0 = x_1^2 + x_3^2 - L^2 \ .$$

If we differentiate the constraint once, we find

$$0 = x_1 x_2 + x_3 x_4 \ .$$

If we differentiate the latter constraint in turn, we obtain

$$(6.1b)' \qquad 0 = x_2^2 - \lambda x_1^2 + x_4^2 - \lambda x_3^2 - x_3 g = x_2^2 + x_4^2 - \lambda L^2 - x_3 g \ .$$

One easily verifies that differentiating once more would yield an ODE with $\dot{\lambda}$ expressed in x_1, x_2, x_3 and x_4. Hence the ODE (6.1a) together with (6.1b)$'$ constitutes an index-1 DAE.

The next example demonstrates some numerical problems with this so called *index reduction* (i.e. reducing the index by differentiation).

Example IX.11.
Of course the solution x, describing the motion of the pendulum, is given by a sinus function. If we assume $x(0) = \sin\varphi$ we should find $x(t) = L\sin(t+\varphi)$. In Fig. IX.3 we have given the value of $x^2 + y^2 - L^2$, for $L=1$ and $ABSTOL=RELTOL=10^{-4}$ after reducing the pendulum problem to an index-1 DAE. As can be seen the actual constraint is not satified very well. This is also demonstrated by Fig. IX.4, where we have displayed the numerical solution for $20 \leq t \leq 30$ and for $50 \leq t \leq 60$.

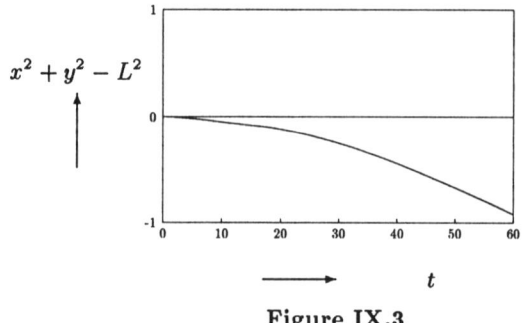

Figure IX.3

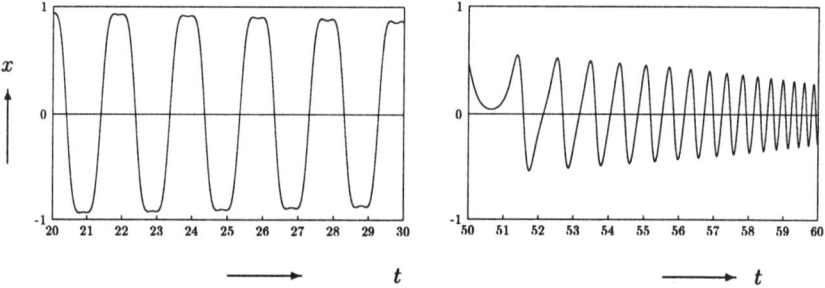

Figure IX.4 x-component of the undamped pendulum. The deviation from the actual, sinusoidal solution increases with time.

The phenomenon we encounter in Example IX.11 is typical for index reduction. The defect in the original constraint is called the *drift* and is related to the fact that the differentiated constraint is not precisely the same as the original one. This is in particular due to the fact that approximate solutions are used, where exact solutions and their derivatives should be employed.

There are a number of *stabilisation* techniques, each of which aims at alleviating this drift problem. One such method is Baumgarte regularisation[11]. If we denote the constraint equation as (cf. (4.3c))

(6.2) $0 = \mathbf{f}_3(\mathbf{x}_2)$,

then we require a linear combination of this and the two (hidden) constraints to be satisfied, i.e.

(6.3) $0 = \ddot{\mathbf{f}}_3 + 2\alpha\,\dot{\mathbf{f}}_3 + \beta^2\,\mathbf{f}_3$, $\alpha, \beta \in \mathbb{R}$.

One can easily verify that (4.2a), (4.2b), (6.3) have index 1. Now α and β are chosen such that (6.3) represents a damped harmonic oscillator (cf. Example IV.2). If $\alpha = \beta > 0$, then one has so-called critical damping. Although damping is stronger (and thus drift is more suppressed) if α is larger, the latter choice introduces (additional) stiffness in the system. As a general tool it is not easy to use, but it may work for some mechanical systems.

Example IX.12.
If we use Baumgarte regularisation for the pendulum problem with $ABSTOL = RELTOL = 10^{-4}$ we notice the increasing stiffness for the choice $\alpha = \beta$, if this parameter becomes larger. To indicate this we have given the number of function evaluations on a log-log scale in Fig. IX.5. The straight line denotes the complexity of the method (6.5), to be discussed below; the latter turns out to be more efficient. For $\alpha > 10^4$ the Baumgarte method simply breaks down.

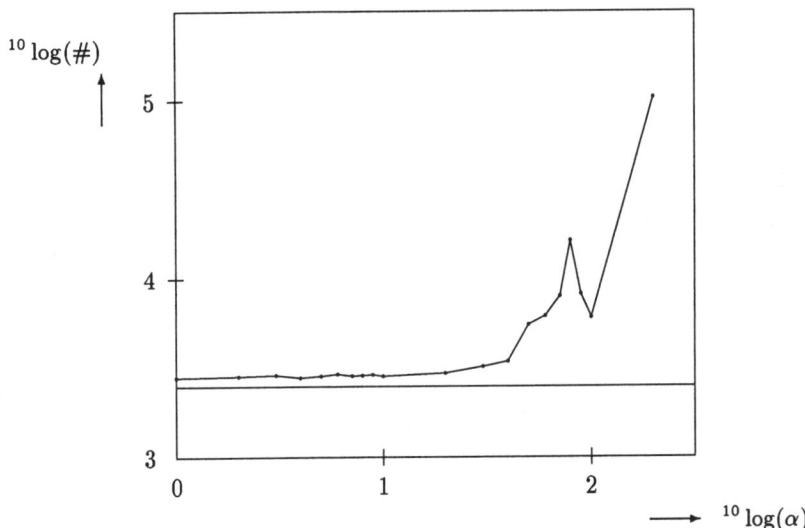

Figure IX.5 Number of Baumgarte iterations as a function of the parameter α for $t = 50$.

Another technique stabilizes the problem by adding suitable "parameter" terms. Consider the following general semi-explicit DAE:

(6.4a) $\dot{\mathbf{x}} = \mathbf{f}(t, \mathbf{x})$

(6.4b) $\mathbf{0} = \mathbf{g}(t, \mathbf{x})$.

Now assume that the constraints are linearly independent, i.e. $\partial \mathbf{g}/\partial \mathbf{x}$ has full rank (this is quite often the case for mechanical systems). Then consider the associated semi-explicit DAE

(6.5a) $\dot{\mathbf{x}} = \mathbf{f}(t, \mathbf{x}) + \left[\dfrac{\partial \mathbf{g}}{\partial \mathbf{x}} \right]^T \boldsymbol{\mu}$

(6.5b) $\mathbf{0} = \mathbf{g}(t, \mathbf{x})$.

Theorem 6.6. *The DAE (6.5) has index 2, and for a solution $(\mathbf{x}^T, \boldsymbol{\mu}^T)^T$ of (6.5), $\mathbf{x}$ satisfies (6.4) and $\boldsymbol{\mu} \equiv \mathbf{0}$.*

Proof: By requiring that $\mathbf{x}$ satisfies the constraint (6.5b), i.e. (6.4b), and by prescribing a suitable initial value it follows that $\dot{\mathbf{x}} = \mathbf{f}(t, \mathbf{x})$. Premultiplying then (6.5a) by $\partial \mathbf{g}/\partial \mathbf{x}$ gives $\mathbf{0} = [\partial \mathbf{g}/\partial \mathbf{x}][\partial \mathbf{g}/\partial \mathbf{x}]^T \boldsymbol{\mu} = \mathbf{0}$. The full rank requirement then implies $\boldsymbol{\mu} \equiv \mathbf{0}$. Differentiation of (6.5b) gives

$$ \mathbf{0} = \frac{\partial \mathbf{g}}{\partial t} + \frac{\partial \mathbf{g}}{\partial \mathbf{x}} \dot{\mathbf{x}} = \frac{\partial \mathbf{g}}{\partial t} + \frac{\partial \mathbf{g}}{\partial \mathbf{x}} \mathbf{f} + \frac{\partial \mathbf{g}}{\partial \mathbf{x}} \left[\frac{\partial \mathbf{g}}{\partial \mathbf{x}} \right]^T \boldsymbol{\mu} \ . $$

The matrix in front of the 'algebraic variable' $\boldsymbol{\mu}$ is nonsingular, so clearly the index must be 2.   □

Example IX.13.
If Fig. IX.6 we have drawn the solution x from numerically obtained values, using $ABSTOL = RELTOL = 10^{-4}$ and employing the parameter technique (6.5). In contrast to Baumgarte for larger values of α, this technique does not require an excessive number of steps, cf. Fig. IX.5.

Note that one may as well use least squares techniques to regularise in a more classical sense. These methods are then related to some form of projection; they are more complicated than using (6.5) of course.

Finally we remark that some implicit Runge-Kutta methods are able to perform some form of projection implicitly. We refer to [8].

Exercises Chapter IX

1.   Consider the DAE

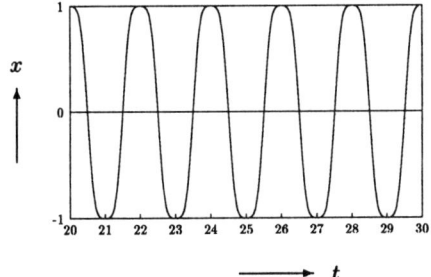

Figure IX.6 x-component of the pendulum problem obtained with Baumgarte regularisation.

$$\begin{bmatrix} 0 & 0 \\ 1 & 0 \end{bmatrix} \dot{\mathbf{x}} + \begin{bmatrix} 1 & 0 \\ 0 & 1 \end{bmatrix} \mathbf{x} = \mathbf{f} \ .$$

Determine the index and characterise the solution space.

2. Consider the DAE

$$\begin{bmatrix} 0 & 0 \\ 1 & 0 \end{bmatrix} \dot{\mathbf{x}} + \begin{bmatrix} 0 & 0 \\ 1 & 0 \end{bmatrix} \mathbf{x} = \mathbf{f} \ .$$

Determine the index. How many initial conditions are required?

3. Let φ be smooth and such that $\begin{cases} \varphi(t) > 0 \ , & t > 0 \ , \\ \varphi(t) = 0 \ , & t \le 0 \ . \end{cases}$

Similarly let ψ be smooth and such that $\begin{cases} \psi(t) = 0 \ , & t \ge 0 \ , \\ \psi(t) > 0 \ , & t < 0 \ . \end{cases}$

Determine the index and characterise the solution space of

$$\begin{bmatrix} 0 & \varphi \\ \psi & 0 \end{bmatrix} \dot{\mathbf{x}} + \mathbf{x} = \mathbf{f} \ .$$

4. Compute the solution of the problem in Exercise 1 with Euler backward for $\mathbf{f} := (1,1)^T$. Take an arbitrary initial value. What do you see?

5. a) Use Euler forward for the dynamical part only and an implicit approach for the algebraic part in the DAE

$$\begin{bmatrix} 0 & 0 \\ 1 & 0 \end{bmatrix} \dot{\mathbf{x}} + \begin{bmatrix} 0 & 1 \\ 1 & 0 \end{bmatrix} \mathbf{x} = \begin{pmatrix} 1 \\ 1 \end{pmatrix} .$$

b) What should be a reasonable step size for the problem and method above if $\mathbf{f} := (\sin t, \cos t)^T$ and we have an absolute tolerance TOL $= 10^{-2}$?

6. Consider the semi-explicit nonlinear system

$$\begin{cases} \mathbf{f}(\mathbf{x}_1, \dot{\mathbf{x}}_1, \mathbf{x}_2) = \mathbf{0} \ , \\ \mathbf{g}(\mathbf{x}_1, \mathbf{x}_2) = \mathbf{0} \ . \end{cases}$$

Let $(\partial \mathbf{f}/\partial \dot{\mathbf{x}}_1)^{-1}$ exist and be bounded in a neighbourhood of the solution. Further, let $\partial \mathbf{g}/\partial \mathbf{x}_2$ have full rank. Show that this is an index-2 system.

7. a) Consider the semi-explicit index-2 problem (4.2) with $\mathbf{x}_1(t) \in \mathbb{R}^p$ and $\mathbf{x}_2(t) \in \mathbb{R}^q$ for all t. Show that $p \geq q$.

b) For a semi-explicit index-3 problem with $\mathbf{u}_3(t) \in \mathbb{R}^r$, show, moreover, that we need $p \geq q \geq r$.

8. Prove that the midpoint rule, although implicit, is not suitable for DAE.

9. Consider the problem (cf. Example IX.8)

$$\dot{x} = \alpha(t) y \ , \qquad \alpha(t) \neq 0 \ ,$$
$$\dot{y} = \alpha(t) z \ ,$$
$$0 = x - q(t) \ .$$

Show that $\alpha(t)$ can be absorbed in the time step and thus that applying a numerical method to this DAE with constant step size is equivalent to applying the method to a problem with variable step size and $\alpha \equiv 1$.

10. Write out the general form of a linear Hessenberg DAE.

11. When is the DAE

$$\begin{cases} \dot{\mathbf{x}} = \mathbf{f}(\mathbf{x}, \mathbf{y}) \ , \\ \dot{\mathbf{y}} = \mathbf{g}(\mathbf{x}, \mathbf{y}, \mathbf{z}) \ , \\ \mathbf{0} = \mathbf{h}(\mathbf{x}) \ , \end{cases}$$

index-3 ?

12. Prove that $\mathbf{J}$ in (5.7b) is nonsingular in the index-2 case $(\alpha_0, \beta_0 \neq 0)$.

13. The θ-method is a weighted mean of Euler forward and Euler backward, with weights θ and $1 - \theta$ respectively; in particular, if $\theta = \frac{1}{2}$ we obtain the trapezoidal rule. Apply this method to the DAE (cf. Exercise 1)

$$\begin{bmatrix} 1 & 0 \\ 0 & 0 \end{bmatrix} \dot{\mathbf{x}} + \begin{bmatrix} 1 & 0 \\ 0 & 1 \end{bmatrix} \mathbf{x} = \begin{bmatrix} 1 \\ 1 \end{bmatrix} .$$

For which values of θ do we have stability?

14. Consider a linear index-2 DAE

$$\begin{cases} \dot{\mathbf{x}} = \mathbf{A}\,\mathbf{x} + \mathbf{B}\,\mathbf{y} + \mathbf{f}(t) \ , \\ 0 = \mathbf{C}\,\mathbf{x} + \mathbf{g}(t) \ . \end{cases}$$

Derive the hidden constraint for the case that $\mathbf{A}$, $\mathbf{B}$ and $\mathbf{C}$ are constant and show that this, together with the constraint above, constitutes a rank-$2q$ linear system ($\mathbf{y}(t) \in \mathbb{R}^q$ for all t).

X

Boundary Value Problems

This chapter is different from the previous ones in that it deals with ODE for which boundary conditions, rather than initial conditions, are given. In §1 we survey the class of problems that will be discussed. Then in §2 we treat Fredholm's alternative for existence of solutions and also introduce Green's functions. The latter play an important role in §3 where we investigate the conditioning of boundary value problems (BVP); it is shown how an analogue of stability for IVP is given by the notion of dichotomy. A simple (but often naive) approach for solving BVP is guessing the missing part of the initial condition, solving the corresponding IVP and iteratively refining the solution through the boundary condition; this so-called shooting technique is considered in §4 for the linear case and in §5 for the nonlinear case. Single shooting may both lack sufficient numerical stability and suffer from convergence problems (in the nonlinear case for Newton's method). Therefore we consider an important improvement, multiple shooting, in §6.

1. Introduction

Although the emphasis of this book is on initial value problems, it is useful to pay some attention to boundary value problems (BVP) as well. There are quite a lot of practical ODE problems that arise as BVP. As it will turn out there are some essential distinctions between BVP and IVP, the most important one being the fact that the former are not related to something like an *evolution* of a phenomenon; consequently space rather than time should be the most appropriate independent variable. Nevertheless, in order to maintain a universal notation throughout this book we will also use for this variable the symbol t here. Consider the ODE

$$(1.1) \qquad \frac{d\mathbf{x}}{dt} = \mathbf{f}(t, \mathbf{x}) , \qquad \mathbf{x}(t) \in \mathbb{R}^n ,$$

subject to *boundary conditions* (BC). We shall exclusively deal with *two-point* BC, i.e. information about the solution is given at the boundaries of the interval on which the problem is defined, (a, b) say:

(1.2) $g\Big(x(a), x(b)\Big) = 0$.

Since the BVP (1.1), (1.2) is non-evolutionary in character (neither conditions at $t = a$ nor those at $t = b$ completely determine the solution by integrating forward and backward respectively), this solution has a *global nature* (i.e. $x(t)$ depends on $x(\tau)$ for $\tau > t$ and $\tau < t$). If the BC are linear we may write

(1.3) $g\Big(x(a), x(b)\Big) = B_a\, x(a) + B_b\, x(b) - \beta = 0$,

where $B_a, B_b \in I\!\!R^{n^2}$, $\beta \in I\!\!R^n$. Often the BC are even *separated*, e.g.

(1.4) $B_a = \begin{bmatrix} 0 \\ B_{a2} \end{bmatrix} \updownarrow k$, $B_b = \begin{bmatrix} B_{b1} \\ 0 \end{bmatrix} \updownarrow n - k$.

Example X.1.
Consider the BVP

$$\begin{cases} \dot{x} = \begin{bmatrix} 2 & 1 \\ 1 & 2 \end{bmatrix} x\, , \\[2mm] \begin{bmatrix} 0 & 0 \\ 0 & 1 \end{bmatrix} x(0) + \begin{bmatrix} 1 & 0 \\ 0 & 0 \end{bmatrix} x(1) = \begin{bmatrix} e \\ -1 \end{bmatrix} . \end{cases}$$

The BC are clearly linear and separated. One simply verifies that the solution is given by $x(t) = e^t \begin{bmatrix} 1 \\ -1 \end{bmatrix}$, $0 \le t \le 1$.

Often we encounter BVP consisting of (systems of) higher order *scalar* ODE satisfying some BC. In particular second order scalar ODE occur, say

(1.5) $\ddot{x} = G(t, x, \dot{x})$.

The BC for (1.5) are usually of the form $x(a)$, $x(b)$ given or $\dfrac{\partial x}{\partial t}(a)$, $\dfrac{\partial x}{\partial t}(b)$ given, or a combination of these. In the first case we call them Dirichlet conditions, in the second case Neumann conditions. It goes without saying that (1.5) and appropriate scalar BC can be written into the form (1.1) and (1.4) respectively.

Example X.2.
Consider the BVP

$$\begin{cases} \ddot{x} = \lambda x + r(t)\, , \\ x(0) = x(1) = 0 \, . \end{cases}$$

In standard form this reads

$$\begin{cases} \dot{\mathbf{x}} = \begin{bmatrix} 0 & 1 \\ \lambda & 0 \end{bmatrix} \mathbf{x} + \begin{bmatrix} 0 \\ r(t) \end{bmatrix} , & \text{where } \mathbf{x} := \begin{bmatrix} x \\ \dot{x} \end{bmatrix} , \\ \begin{bmatrix} 0 & 0 \\ 1 & 0 \end{bmatrix} \mathbf{x}(0) + \begin{bmatrix} 1 & 0 \\ 0 & 0 \end{bmatrix} \mathbf{x}(1) = \begin{bmatrix} 0 \\ 0 \end{bmatrix} . \end{cases}$$

2. Existence of solutions

In Chapter II we considered the existence of solutions of IVP. This question is much more difficult for BVP, due to the delicate way the BC 'control' the solution. We shall mainly look at linear problems, which demonstrate the latter problem sufficiently well. Consider the ODE

(2.1) $\dot{\mathbf{x}} = \mathbf{A}\,\mathbf{x} + \mathbf{r}(t)$.

Let $\mathbf{F}$ be a fundamental solution, i.e. $\dot{\mathbf{F}} = \mathbf{A}\,\mathbf{F}$, $\mathbf{F}(0)$ nonsingular, and let $\mathbf{p}$ be some particular solution (cf. §IV.2) then we can write the general solution of (2.1) as (*superposition*, cf. (2.2.10))

(2.2) $\mathbf{x}(t) = \mathbf{F}(t)\,\mathbf{c} + \mathbf{p}(t)$,

for some $\mathbf{c} \in I\!R^n$. From the linear BC (1.3) we then find

(2.3) $[\mathbf{B}_a\,\mathbf{F}(a) + \mathbf{B}_b\,\mathbf{F}(b)]\,\mathbf{c} = \beta - \mathbf{B}_a\,\mathbf{p}(a) - \mathbf{B}_b(\mathbf{p}(b))$.

This gives rise to the following:

Property 2.4. *The BVP (2.1), (1.3) has a unique solution iff*

(2.4a) $\mathbf{Q} := \mathbf{B}_a\,\mathbf{F}(a) + \mathbf{B}_b\,\mathbf{F}(b)$

is nonsingular.

The property above is called the *Fredholm alternative*: a solution of a linear problem has either a unique solution or infinitely many solutions depending on the (non)singularity of $\mathbf{Q}$ in (2.4a).

A BVP is called *well-posed* if the solution exists and is unique. For linear problems we find from (2.2)

(2.5) $\mathbf{c} = \mathbf{Q}^{-1}\,[\beta - \mathbf{B}_a\,\mathbf{p}(a) - \mathbf{B}_b\,\mathbf{p}(b)]$.

Example X.3.
Consider the BVP

$$\begin{cases} \ddot{x} + x = 0 , \\ x(0) = 0 , \qquad x(b) = \beta , \end{cases}$$

which reads in standard form

$$\begin{cases} \dot{x} = \begin{bmatrix} 0 & 1 \\ -1 & 0 \end{bmatrix} x , \\ \begin{bmatrix} 0 & 0 \\ 1 & 0 \end{bmatrix} x(0) + \begin{bmatrix} 1 & 0 \\ 0 & 0 \end{bmatrix} x(1) = \begin{pmatrix} \beta \\ 0 \end{pmatrix} . \end{cases}$$

A fundamental solution is apparently given by

$$F(t) := \begin{bmatrix} \cos t & \sin t \\ -\sin t & \cos t \end{bmatrix} ,$$

whence

$$Q = \begin{bmatrix} 1 & 0 \\ \cos \beta & \sin b \end{bmatrix} .$$

From this we conclude that the solution uniquely exists provided $b \neq k\pi$, $k \in I\!N$.

In §IV.2 we introduced a fundamental solution Y with the property that $x(t) = Y(t) x_0$ for homogeneous problems, i.e. $Y(0) = I$. Quite similarly it makes sense here to normalise F such that for a well-posed problem

$$(2.6) \qquad B_a F(a) + B_b F(b) = I .$$

This is trivially obtained through scaling with the matrix Q in (2.4a). Let us choose p such that

$$(2.7) \qquad B_a p(a) + B_b p(b) = 0 .$$

In order to indicate p more precisely we introduce the analogue of the transition matrix $Y(t, s)$ in §IV.2: *Green's matrix* $G(t, s)$ is defined as

$$(2.8) \qquad G(t, s) := \begin{cases} F(t) B_a F(a) F^{-1}(s) , & t > s \\ -F(t) B_b F(b) F^{-1}(s) , & t < s . \end{cases}$$

One may show that (cf. (IV.3.5))

$$(2.9) \qquad \mathbf{p}(t) = \int_a^b \mathbf{G}(t,s)\,\mathbf{r}(s)\,ds \ .$$

Concluding, we have found that the superposition in the linear case results in

$$(2.10) \qquad \mathbf{x}(t) = \mathbf{F}(t)\,\boldsymbol{\beta} + \int_a^b \mathbf{G}(t,s)\,\mathbf{r}(s)\,ds \ .$$

For nonlinear problems, the existence question is often related to linearisations around the (potential) solution. If such a solution uniquely exists in a neighbourhood of $\mathbf{x}$ then the solution is called *isolated*. Sometimes it is useful to relate this to the following IVP question. Consider the family of IVP

$$\begin{cases} \dot{\mathbf{y}}_\mathbf{s} = \mathbf{f}(t, \mathbf{y}_\mathbf{s}) \ , \\ \mathbf{y}_\mathbf{s}(a) = \mathbf{s} \ , \end{cases}$$

where the boundary condition $\mathbf{s}$ at $t = a$ is read as a parameter. Varying $\mathbf{s}$ the complete BC

$$(2.11) \qquad \mathbf{g}\Big(\mathbf{s}, \mathbf{y}_\mathbf{s}(b)\Big) = 0$$

will (hopefully) be fulfilled for some s^*, say. If s^* is a single solution of (2.11) the corresponding solution $x = y_{s^*}$ is called *isolated*.

3. Well-conditioning and dichotomy

The importance of (2.10) is that it shows us how the solution depends on the parameters. Defining

$$(3.1\text{a}) \qquad \kappa_1 := \max_{t \in [a,b]} \|\mathbf{F}(t)\| \ ,$$

$$(3.1\text{b}) \qquad \kappa_2 := \max_{t,s \in [a,b]} \|\mathbf{G}(t,s)\| \ ,$$

we obtain from (2.10)

$$(3.2) \qquad \|\mathbf{x}(t)\| \le \kappa_1 \|\boldsymbol{\beta}\| + \kappa_2 \int_a^b \|\mathbf{r}(s)\|\,ds \ , \qquad t \in [a,b] \ .$$

We see that the latter expression is an analogue of what we saw in Definition V.1.4 for total stability. Of course if (a, b) is a finite interval well-posedness would give us finite κ_1, κ_2. However, in a context where either b may grow unboundedly, or the ODE depends on a (small) parameter, as in singularly perturbed problems, we would like to have κ_1, κ_2 uniformly bounded. From a practical point of view it makes sense to require κ, defined by

$$(3.3) \qquad \kappa := \max\left(\kappa_1, \kappa_2\right),$$

to be *moderate* for having a so-called *well-conditioned* problem, i.e. one where small perturbations in the data do not give large effects (in an absolute sense). We shall henceforth call κ the *stability constant*.

In particular the bound on Green's matrix is of interest for us. Consider, for simplicity, the separated BC (1.4). Then it follows from the normalisation of $\mathbf{F}$ that

$$(3.4) \qquad \mathbf{B}_a\,\mathbf{F}(a) = \begin{bmatrix} 0 & 0 \\ 0 & \mathbf{I}_n \end{bmatrix} =: \mathbf{P}\;; \qquad \mathbf{B}_b\,\mathbf{F}(b) = \begin{bmatrix} \mathbf{I}_{n-k} & 0 \\ 0 & 0 \end{bmatrix} = \mathbf{I} - \mathbf{P}\,.$$

The matrix $\mathbf{P}$ is clearly a projection. Apparently the expressions in (2.8) simplify somewhat in this case and we obtain

$$(3.5) \qquad \begin{cases} \|\mathbf{F}(t)\,\mathbf{P}\,\mathbf{F}^{-1}(s)\| \le \kappa_2 & , \quad t > s \\ \|\mathbf{F}(t)(\mathbf{I} - \mathbf{P})\,\mathbf{F}^{-1}(s)\| \le \kappa_2 &, \quad t < s\,. \end{cases}$$

More generally, if for some fundamental solution $\mathbf{F}$ there exists a projection $\mathbf{P}$ and a constant κ_2, such that (3.5) holds, then this fundamental solution is called *dichotomic* with threshold κ_2. Though slightly more complicated one can show that also the more general BC (1.3) induce dichotomy, with nearly the same threshold. The terminology is explained in the following:

Property 3.6. *Let $\mathbf{F}$ be dichotomic, as in (3.5). Define the solution spaces*

$$(3.6a) \qquad S_1 := \{\mathbf{F}(t)\,\mathbf{P}\,\mathbf{c}\,|\,\mathbf{c} \in I\!\!R^n\}$$

$$(3.6b) \qquad S_2 := \{\mathbf{F}(t)\,(\mathbf{I} - \mathbf{P})\,\mathbf{c}\,|\,\mathbf{c} \in I\!\!R^n\}\,.$$

Let $\varphi \in S_1$ and $\psi \in S_2$, then

(i) $\dfrac{\|\varphi(t)\|}{\|\varphi(s)\|} \le \kappa_2, t > s$ *and* $\dfrac{\|\psi(t)\|}{\|\psi(s)\|} \le \kappa_2, t < s.$

(ii) *Let ϑ be the minimum angle between $\varphi(t)$ and $\psi(t)$ for all $t \in [a, b]$, then*

$$\cot \vartheta \le \kappa_2\,.$$

Proof: We only prove (i). For the rest see [4]. For some $\mathbf{c}$ we have

$$\frac{\|\varphi(t)\|}{\|\varphi(s)\|} = \frac{\|\mathbf{F}(t)\,\mathbf{P}\,\mathbf{c}\|}{\|\mathbf{F}(s)\,\mathbf{P}\,\mathbf{c}\|} = \frac{\|\mathbf{F}(t)\,\mathbf{P}\,\mathbf{F}^{-1}(s)\,\mathbf{F}(s)\,\mathbf{P}\,\mathbf{c}\|}{\|\mathbf{F}(s)\,\mathbf{P}\,\mathbf{c}\|} \le$$

$$\le \|\mathbf{F}(t)\,\mathbf{P}\,\mathbf{F}^{-1}(s)\| \le \kappa_2\,. \qquad\qquad \square$$

So $\mathcal{S}_1$ represents a solution subspace of modes which do not grow more than the threshold value κ_2 for increasing argument, and likewise $\mathcal{S}_2$ for decreasing argument. Of course, $\mathcal{S}_1$ may contain fast decaying modes. Moreover, these subspaces are geometrically well-separated since they cannot make an angle smaller than $\cot^{-1} \kappa_2$. Roughly speaking we thus have found that this kind of BVP in a way consists of a 'stable' *initial value problem* (i.e. having modes not growing much) and a 'stable' *terminal value problem.*

Example X.4.

(i) Consider the ODE

$$\dot{\mathbf{x}} = \begin{bmatrix} \lambda \cos 2\omega t & -\lambda \sin 2\omega t + \omega \\ -\lambda \sin 2\omega t - \omega & -\lambda \cos 2\omega t \end{bmatrix} \mathbf{x}, \qquad \lambda > 0,$$

having a (non-normalised) fundamental solution

$$\mathbf{F}(t) := \begin{bmatrix} \cos \omega t & \sin \omega t \\ -\sin \omega t & \cos \omega t \end{bmatrix} \begin{bmatrix} e^{\lambda t} & 0 \\ 0 & e^{-\lambda t} \end{bmatrix}.$$

After normalisation we use $\hat{\mathbf{F}} := \mathbf{F} \mathbf{Q}^{-1}$, where $\mathbf{Q}$ is given by (2.4a). An obvious choice for a projection matrix as meant in (3.5) is given by $\mathbf{P} := \begin{bmatrix} 0 & 0 \\ 0 & 1 \end{bmatrix}$. Since the first matrix in $\mathbf{F}$ is orthogonal we even find $\kappa_2 = 1$ since we use the Euclidian $\| \cdot \|_2$; note that this κ_2 is uniform in λ and ω! In order to appreciate this, we note that the *eigenvalues* of the system matrix of the ODE above are given by $\pm \sqrt{\lambda^2 - \omega^2}$. In other words, if $|\omega| > \lambda$ these eigenvalues are purely imaginary and are not even indicative for the growth behaviour of the two very different types of fundamental modes (i.e. $\sim e^{\lambda t}$, $\sim e^{-\lambda t}$).
If the properly normalised fundamental mode had been used, the projection matrix would have read:

$$\hat{\mathbf{P}} := \mathbf{Q} \mathbf{P} \mathbf{Q}^{-1}.$$

(ii) Consider the ODE

$$\dot{\mathbf{x}} = \begin{bmatrix} -2t/\varepsilon & 0 \\ 0 & 1 \end{bmatrix} \mathbf{x}, \qquad -1 < t < 1.$$

A (non-normalised) fundamental solution is given by

$$\mathbf{F}(t) := \begin{bmatrix} e^{-t^2/\varepsilon} & 0 \\ 0 & e^t \end{bmatrix}.$$

One can simply verify that this growth behaviour prohibits the existence of a projection and a threshold κ_2 such that there is dichotomy uniformly in ε, say $\varepsilon \leq \varepsilon_0$. On the other hand one may also verify directly that Green's matrix cannot be bounded uniformly in ε!

4. Single shooting

The availability of sophisticated software for solving IVP numerically makes it seemingly obvious to try and solve BVP by associated IVP. Using the superposition principle this indeed looks particularly attractive for linear problems.

Recall the fundamental solution $\mathbf{Y}$ with $\mathbf{Y}(a) = \mathbf{I}$ and the particular solution $\mathbf{p}$ with $\mathbf{p}(a) = \mathbf{0}$. Then the solution $\mathbf{x}$ of the BVP

$$(4.1) \qquad \begin{cases} \dot{\mathbf{x}} = \mathbf{A}\,\mathbf{x} + \mathbf{r}(t)\ , \\ \mathbf{B}_a\,\mathbf{x}(a) + \mathbf{B}_b\,\mathbf{x}(b) = \boldsymbol{\beta}\ , \end{cases}$$

is found as

$$(4.2a) \qquad \mathbf{x}(t) = \mathbf{Y}(t)\,\mathbf{s} + \mathbf{p}(t)\ ,$$

where, in turn, $\mathbf{s}$ is found from solving

$$(4.2b) \qquad [\mathbf{B}_a + \mathbf{B}_b\,\mathbf{Y}(b)]\,\mathbf{s} = \boldsymbol{\beta} - \mathbf{B}_b\,\mathbf{p}(b)\ .$$

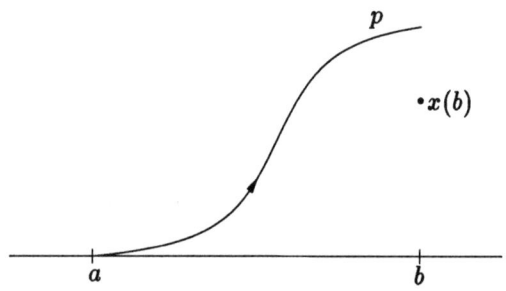

Figure X.1   Single shooting when $\mathbf{y}(a) = \mathbf{0}$.

Note that (4.2b) has a unique solution if the BVP is well-posed. This method is called *(single) shooting*. In Fig. X.1 we have drawn a picture for a system like (1.5) (G linear in x, $\dot{x}$) and Dirichlet BC $x(a) = 0$, $x(b) \neq 0$, so $\mathbf{p}(a) = (p(a), p'(a))^T = \mathbf{0}$ and $p(b) \neq x(b)$ in general! Note that if $x(a) \neq 0$, we had better chosen $p(a) = x(a)$ to simplify the computation somewhat.

Example X.5.
Consider the BVP

$$\begin{cases} \dot{\mathbf{x}} = \begin{bmatrix} 0 & 1 \\ \lambda^2 & 0 \end{bmatrix} \mathbf{x} + (1 - \lambda^2) \begin{bmatrix} 0 \\ e^t \end{bmatrix}, & 0 < t < 1, \\[2mm] \begin{bmatrix} 0 & 0 \\ 1 & 0 \end{bmatrix} \mathbf{x}(0) + \begin{bmatrix} 1 & 0 \\ 0 & 0 \end{bmatrix} \mathbf{x}(1) = \begin{bmatrix} 1 \\ e \end{bmatrix}. \end{cases}$$

One easily verifies that

$$\begin{cases} \mathbf{Y}(t) = \begin{bmatrix} \cosh \lambda t & \lambda^{-1} \sinh t \\ \lambda \sinh t & \cosh \lambda t \end{bmatrix}, \\[4mm] \mathbf{p}(t) = \begin{bmatrix} e^t - \cosh \lambda t - \lambda^{-1} \sinh \lambda t \\ e^t - \lambda \sinh \lambda t - \cosh \lambda t \end{bmatrix}. \end{cases}$$

The linear equation for s then reads:

$$\begin{bmatrix} 1 & 0 \\ \cosh \lambda & \lambda^{-1} \sinh \lambda \end{bmatrix} \mathbf{s} = \begin{bmatrix} 1 \\ \cosh \lambda + \lambda^{-1} \sinh \lambda \end{bmatrix},$$

from which we find $\mathbf{x}(t) = (e^t, e^t)^T$.

The previous example was worked out analytically only to demonstrate the idea. In practice we use of course a numerical integration method, such as a one-step method (cf. Chapter III). Such a method can be written as

$$(4.3) \qquad \mathbf{x}_{i+1} = \mathbf{x}_i + h_i \, \Phi(t_i, \mathbf{x}_i, h_i), \qquad i = 1, ..., N-1.$$

Let us denote the thus found fundamental solution by $\{\mathbf{Y}_i\}_{i=1}^N$ and the particular solution as $\{\mathbf{p}_i\}_{i=1}^N$. We recall that we may view the numerical solution as an exact solution of a perturbed ODE. So denoting this (continuously embedded) solution as $\mathbf{y}$ we find

$$(4.4) \qquad \begin{cases} \dot{\mathbf{y}} = \mathbf{A}\,\mathbf{y} + \mathbf{r}(t) + \boldsymbol{\varepsilon}(t), \\ \mathbf{B}_a\,\mathbf{y}(a) + \mathbf{B}_b\,\mathbf{y}(b) = \boldsymbol{\beta}. \end{cases}$$

Here $\boldsymbol{\varepsilon}(t)$ is some local error vector, bounded in norm by $C\,h^p$, $C \in \mathbb{R}_+$ if we use a p-th order method. We may now define a global error vector, $\mathbf{e}$ say, satisfying the same BVP as $\mathbf{y}$ but for $\boldsymbol{\beta} = 0$. Hence:

Theorem 4.5. *Let κ_2 be the stability constant defined in (3.1b), then the global error $\mathbf{e}_i := \mathbf{x}(t_i) - \mathbf{x}_i$ satisfies*

$$\|\mathbf{e}_i\| \le \kappa_2(b-a)\,C\,h^p.$$

Proof: Apparently we have

$$e(t) = -\int_a^b \mathbf{G}(t,s)\,\boldsymbol{\epsilon}(s)\,ds\,,$$

so

$$\|e(t)\| \le \kappa_2 \int_a^b \|\boldsymbol{\epsilon}(s)\|\,ds\,. \qquad\qquad \square$$

This result is satisfactory. However, one should realise that we have assumed *exact arithmetic*: if we include rounding errors (4.4) is not justified, as we shall show now. Assume that calculations with computer-represented quantities are contaminated with rounding errors with a relative error bound MACHEPS; so any number $a \ne 0$ is represented as $\bar{a}$, where

(4.6) $\left| \dfrac{a - \bar{a}}{a} \right| \le \text{MACHEPS}\,.$

The actually (numerically) computed fundamental solution $\{\bar{\mathbf{Y}}_i\}$ can then be assumed to satisfy

(4.7a) $\bar{\mathbf{Y}}_i := \mathbf{Y}_i(\mathbf{I} + \mathbf{E}_i)\,, \qquad \|\mathbf{E}_i\| \le C \cdot \text{MACHEPS}\,,$

where C is a moderate constant.

The presence of (possibly unstable) modes, generated by rounding errors during the forward integration, means that $\{\mathbf{p}_i\}$ is likewise contaminated with rounding errors, say

(4.7b) $\bar{\mathbf{p}}_i := \mathbf{p}_i + \mathbf{Y}_i\,\tilde{\mathbf{E}}_i\,\mathbf{d}_i\,, \qquad \|\tilde{\mathbf{E}}_i\| \le \tilde{C} \cdot \text{MACHEPS}\,, \qquad \|\mathbf{d}_i\| = 1\,.$

Neglecting further errors, the actually computed vector $\bar{\mathbf{s}}$, say, satisfies

(4.8) $[\mathbf{B}_a + \mathbf{B}_b\,\mathbf{Y}_n(\mathbf{I} + \mathbf{E}_N)]\,\bar{\mathbf{s}} = \boldsymbol{\beta} - \mathbf{B}_b\,\mathbf{p}_N - \mathbf{B}_b\,\mathbf{Y}_n\,\tilde{\mathbf{E}}_N\,\mathbf{d}_N\,.$

Identifying $\mathbf{Y}_N$ and $\mathbf{p}_N$ with $\mathbf{Y}(b)$ and $\mathbf{p}(b)$, respectively, yields (cf. (4.2b) and (4.8))

$$\bar{\mathbf{s}} - \mathbf{s} = [\mathbf{B}_a + \mathbf{B}_b\,\mathbf{Y}_N]^{-1}\,\mathbf{B}_b\,\mathbf{Y}_n\,[\mathbf{E}_N\,\bar{\mathbf{s}} + \tilde{\mathbf{E}}_N\,\mathbf{d}_N]\,.$$

As a consequence the *computed* $\mathbf{x}$ is contaminated with errors equal to

$$\mathbf{Y}_i(\bar{\mathbf{s}} - \mathbf{s}) = -\mathbf{Y}_i\,[\mathbf{B}_a + \mathbf{B}_b\,\mathbf{Y}_N]^{-1}\,\mathbf{B}_b\,\mathbf{Y}_N\,[\mathbf{E}_N\,\bar{\mathbf{s}} + \tilde{\mathbf{E}}_N\,\mathbf{d}_N]\,,$$

whence

(4.9) $\|\mathbf{Y}_i(\bar{\mathbf{s}} - \mathbf{s})\| \le \kappa_1\,\|\mathbf{B}_b\|\,\|\mathbf{Y}_N\,[\mathbf{E}_N\,\bar{\mathbf{s}} + \tilde{\mathbf{E}}_N\,\mathbf{d}_N]\|\,.$

Note that $\mathbf{Y}(t)\,[\mathbf{B}_a\,\mathbf{I}+\mathbf{B}_b\,\mathbf{Y}(b)] = \mathbf{F}(t)$ is the normalised fundamental solution. Our major concern in (4.9) is the factor $\|\mathbf{Y}_N\,(\mathbf{E}_N\,\bar{\mathbf{s}}+\tilde{\mathbf{E}}_N\,\mathbf{d}_N)\|$. Since neither $\mathbf{E}_N$ nor $\tilde{\mathbf{E}}_N$ has a special structure (in fact they are built up by rather random-like rounding errors), we may say that factor this is fairly likely to be of the order $\|\mathbf{Y}_N\|$. MACHEPS. However, $\|\mathbf{Y}_N\|$ may be very large, being the terminal value of a fundamental solution with *biased* normalisation at the initial point (e.g. of the order $e^{\lambda(b-a)}$ in Example X.5). We therefore conclude that this single shooting is *potentially unstable*! Only given a required tolerance TOL, say where $\|\mathbf{Y}_N\|$. MACHEPS $\leq$ TOL, are the results acceptable.

Example X.6.
Consider the BVP

$$\begin{cases} \dot{\mathbf{x}} = \begin{bmatrix} 0 & \lambda \\ \lambda & 0 \end{bmatrix} \mathbf{x} + \begin{bmatrix} \lambda \cos^2 \pi t + \dfrac{2}{\lambda}\pi^2 \cos 2\pi t \end{bmatrix} , \qquad \lambda > 0 \\[2mm] \begin{bmatrix} 0 & 0 \\ 1 & 0 \end{bmatrix} \mathbf{x}(0) + \begin{bmatrix} 1 & 0 \\ 0 & 0 \end{bmatrix} x(1) = \begin{bmatrix} 0 \\ 0 \end{bmatrix} . \end{cases}$$

One easily verifies that the conditioning constant is moderate (and independent of λ) and

$$\mathbf{x}(t) = \begin{bmatrix} (e^{\lambda(t-1)} + e^{-\lambda t})\,(1+e^{-\lambda})^{-1} - \cos^2 \pi t \\[2mm] (e^{\lambda(t-1)} - e^{-\lambda t})\,(1+e^{-\lambda})^{-1} + \dfrac{\pi}{\lambda} \sin 2\pi t \end{bmatrix} .$$

The fundamental solution $\mathbf{Y}(t)$ (with $\mathbf{Y}(0) = \mathbf{I}$) is given by

$$\mathbf{Y}(t) = \begin{bmatrix} \cosh \lambda t & \sinh \lambda t \\ \sinh \lambda t & \cosh \lambda t \end{bmatrix} .$$

Hence for λ large we have $\mathbf{s} \approx \begin{bmatrix} 0 \\ -1 \end{bmatrix}$.

Note that $\mathbf{Y}(1) = e^\lambda \begin{bmatrix} 1 & 1 \\ 1 & 1 \end{bmatrix}$, so that single shooting will certainly be unacceptable for λ large. In Table X.1 we have displayed global errors for $\lambda = 20$ and $\lambda = 5$, using Fehlberg with ABSTOL $= 10^{-10}$.

5. Single shooting for nonlinear problems

Now consider the nonlinear BVP

$$(5.1) \qquad \begin{cases} \dot{\mathbf{x}} = \mathbf{f}(t, \mathbf{x}) , \\[2mm] \mathbf{g}\big(\mathbf{x}(a), \mathbf{x}(b)\big) = \mathbf{0} . \end{cases}$$

Table X.1   Errors in single shooting.

t	error		t	error
0.1	.44E-9		0.1	.19E-7
0.2	.30E-8		0.2	.28E-5
0.3	.22E-7		0.3	.41E-3
0.4	.26E-6		0.4	.61E-1
0.5	.12E-5		0.5	.90E+1
0.6	.87E-5		0.6	.13E+4
0.7	.66E-4		0.7	.20E+6
0.8	.48E-3		0.8	.29E+8
0.9	.35E-2		0.9	.44E+10
1.0	.26E-1		1.0	.65E+12

$$\lambda = 20 \qquad\qquad\qquad \lambda = 50$$

Let $\mathbf{s}$ be an estimate for $\mathbf{x}(a)$ and denote by $\mathbf{y_s}$ the solution of the IVP

$$(5.2) \qquad \begin{cases} \dot{\mathbf{y}}_{\mathbf{s}} = \mathbf{f}(t, \mathbf{y_s}) \,, \\ \mathbf{y_s}(a) = \mathbf{s} \,. \end{cases}$$

One can measure 'how good' the approximation $\mathbf{y_s}$ is by the defect

$$(5.3) \qquad \mathbf{d}(\mathbf{s}) := \mathbf{g}\Big(\mathbf{s}, \mathbf{y_s}(b)\Big) \,.$$

If $\mathbf{x}$ is an isolated solution then, by definition, $\mathbf{d}(\mathbf{s})$ has a simple zero $\mathbf{s}^* = \mathbf{x}(a)$ (cf. (2.11)). We can try to find $\mathbf{s}^*$ approximately by Newton's method. Let $\mathbf{s}^0$ be given. Determine a sequence $\{\mathbf{s}^m\}$ such that

$$(5.4a) \qquad \mathbf{s}^{m+1} = \mathbf{s}^m + \boldsymbol{\xi}^m \,,$$

where

$$(5.4b) \qquad \mathbf{J}(\mathbf{s}^m)\,\boldsymbol{\xi}^m = -\mathbf{d}(\mathbf{s}^m) \,; \qquad \mathbf{J}(\mathbf{s}) := \frac{\partial \mathbf{d}}{\partial \mathbf{s}} \,.$$

Denoting by $\mathbf{u}$ and $\mathbf{v}$ the first and second variable, of $\mathbf{g}$, respectively, we find

$$(5.5) \qquad \mathbf{J}(\mathbf{s}) = \frac{\partial}{\partial \mathbf{s}} \mathbf{g}\Big(\mathbf{s}, \mathbf{y_s}(b)\Big) =$$

$$= \left[\frac{\partial}{\partial \mathbf{u}} \mathbf{g}(\mathbf{u}, \mathbf{v}) + \frac{\partial}{\partial \mathbf{v}} \mathbf{g}(\mathbf{u}, \mathbf{v}) \frac{\partial}{\partial \mathbf{s}} \mathbf{y_s}(b) \right]_{\substack{\mathbf{u} = \mathbf{s} \\ \mathbf{v} = \mathbf{y_s}(b)}} \,.$$

This Jacobian matrix can be found as follows. From (5.2) we deduce

(5.6)
$$\begin{cases} \dfrac{d}{dt}\dfrac{\partial \mathbf{y_s}(t)}{\partial \mathbf{s}} = \dfrac{\partial}{\partial \mathbf{y}}\mathbf{f}(t,\mathbf{y_s}) \cdot \dfrac{\partial \mathbf{y_s}(t)}{\partial \mathbf{s}} \;, \\[2mm] \dfrac{\partial}{\partial \mathbf{s}}\mathbf{y_s}(a) = \mathbf{I} \;. \end{cases}$$

Write

(5.7a) $\dfrac{\partial}{\partial \mathbf{u}}\mathbf{g}\Big(\mathbf{s},\mathbf{y_s}(b)\Big) =: \mathbf{B}_a \;;\qquad \dfrac{\partial}{\partial \mathbf{v}}\mathbf{g}\Big(\mathbf{s},\mathbf{y_s}(b)\Big) =: \mathbf{B}_b \;;$

(5.7b) $\dfrac{\partial \mathbf{y_s}}{\partial \mathbf{s}} = \mathbf{Y_s} \;;\qquad \dfrac{\partial}{\partial \mathbf{y_s}}\mathbf{f}(t,\mathbf{y_s}) = \mathbf{A_s}(t) \;.$

Then after first computing the fundamental solution of the IVP

(5.8a)
$$\begin{cases} \dot{\mathbf{Y}}_\mathbf{s} = \mathbf{A_s}\,\mathbf{Y_s} \;, \\ \mathbf{Y_s}(a) = \mathbf{I} \;, \end{cases}$$

$\mathbf{J}(\mathbf{s})$ can be determined as

(5.8b) $\mathbf{J}(\mathbf{s}) = \mathbf{B}_a + \mathbf{B}_b\,\mathbf{Y_s}(b) \;.$

Example X.7.
Consider the BVP

$$\begin{cases} \dot{\mathbf{x}} = \begin{bmatrix} \dot{x} \\ -e^x \end{bmatrix} \;, \qquad \text{where } \mathbf{x} := (x,\dot{x})^T \;, \\[4mm] \begin{bmatrix} 0 & 0 \\ 1 & 0 \end{bmatrix}\mathbf{x}(0) + \begin{bmatrix} 1 & 0 \\ 0 & 0 \end{bmatrix}\mathbf{x}(1) = \mathbf{0} \;. \end{cases}$$

In this case the matrices $\mathbf{B}_a$ and $\mathbf{B}_b$ are trivially known. For $\mathbf{A_s}$ we obtain

$$\mathbf{A_s}(t) = \begin{bmatrix} 0 & 1 \\ -e^{y_s} & 0 \end{bmatrix}, \qquad \mathbf{y}_s = \begin{bmatrix} y_s \\ \dot{y}_s \end{bmatrix}, \qquad \mathbf{s} = \begin{bmatrix} 0 \\ s \end{bmatrix}.$$

Since the first component of $\mathbf{d}(\mathbf{s})$ is always zero, we can restrict ourselves to the second component of $\mathbf{s}$, viz. s; the exact value of s is 0.54935 to 5 decimals accuracy (and for the physically stable solution, cf. [4]). Table X.2 presents a comparison of the work done using a fourth order RK method with fixed step size h.

The Jacobian matrix in (5.4b) may be computed numerically as follows: apart from computing the solution $\mathbf{y}$ (we omit the index $\mathbf{s}$) we also compute n solutions $\mathbf{z}_1,...,\mathbf{z}_n$ satisfying

Table X.2　Step size, # iterations and error at $t = \frac{1}{2}$.

s	h	#	error
0	.2	6	.43E-5
0	.1	5	.28E-6
0	.05	5	.17E-8
0	.025	5	.11E-8
5	.2	8	.43E-5
5	.1	7	.28E-6
5	.05	7	.17E-7
5	.025	7	.11E-8

$$(5.9) \quad \begin{cases} \dot{\mathbf{z}}_j = \mathbf{f}(t, \mathbf{z}_j) , \\ \mathbf{z}_j(a) = \mathbf{s} + \delta\, \mathbf{e}_j \end{cases} \quad j = 1, ..., n .$$

Here $\mathbf{e}_j$ is the j-th unit vector $(0, ..., 0, 1, 0, ..., 0)^T$ and δ a small number $\approx \sqrt{\text{MACHEPS}}$. Now define

$$(5.10a) \quad \hat{\mathbf{y}}_j(t) := \Big(\mathbf{z}_j(t) - \mathbf{y}(t)\Big)/\delta ,$$

then $\hat{\mathbf{Y}}(t)$ defined by

$$(5.10b) \quad \hat{\mathbf{Y}}(t) := [\hat{\mathbf{y}}_1(t) \,|\, ... \,|\, \hat{\mathbf{y}}_n(t)]$$

is a fairly good approximation of $\mathbf{Y}(t)$, the exact fundamental solution of the problem.

In order to investigate convergence of the Newton process we use the following famous theorem:

Theorem 5.11 (Newton-Kantorovich). *Let* $\mathbf{d}(\mathbf{s})$ *be continuously differentiable in a neighbourhood* $\mathcal{B}(\mathbf{s}^0, \rho)$ *of* $\mathbf{s}^0$ *(i.e. a ball with centre* $\mathbf{s}^0$ *and radius* ρ*) and let the constants* C_1, C_2 *and* C_3 *be such that*

$$\|\mathbf{J}^{-1}(\mathbf{s}^0)\, \mathbf{d}(\mathbf{s}^0)\| \le C_1 ,$$

$$\|\mathbf{J}^{-1}(\mathbf{s}^0)\| \le C_2 ,$$

$$\|\mathbf{J}(\mathbf{s}_1) - \mathbf{J}(\mathbf{s}_2)\| \le C_3 \|\mathbf{s}_1 - \mathbf{s}_2\| , \qquad for \ \mathbf{s}_1, \mathbf{s}_2 \in \mathcal{B}(\mathbf{s}^0, \rho) .$$

Define $\sigma := C_1 \cdot C_2 \cdot C_3$.
If $\sigma \le \frac{1}{2}$ *and* $\rho \ge \rho_- := \dfrac{1}{C_2 C_3} (1 - \sqrt{1 - 2\sigma})$ *then the following holds:*

(i)   *There is a unique root* $s^* \in \mathcal{B}(s^0, \rho_-)$.

(ii)   *The sequence of Newton iterates* $\{s^m\}$ *converges to* s^*.

If $\sigma < \frac{1}{2}$ *then moreover*

(iii) s^* *is the unique root* $\in \mathcal{B}(s^0, \min(\rho, \rho_+))$, *where* $\rho_+ := \dfrac{1}{C_2 C_3}(1 + \sqrt{1 - 2\sigma})$.

(iv) $\|s^m - s^*\| \leq (2\sigma)^{2^m} \dfrac{1}{C_2 C_3}$. $\square$

From this theorem we conclude that $\sigma \leq 1/2$ if only s^- is close enough to the root s^*. Now let the Lipschitz constant of $\mathbf{f}$ be L; then we find

$$(5.12) \quad \|\mathbf{y}_{s_1}(b) - \mathbf{y}_{s_2}(b)\| \leq \|s_1 - s_2\| + \int_a^b L \|\mathbf{y}_{s_1}(t) - \mathbf{y}_{s_2}(t)\| \, dt \leq$$

$$\leq e^{L(b-a)} \|s_1 - s_2\| .$$

It follows that $\mathbf{Y}_s$ may have a similar sensitivity to the initial value as $\mathbf{y}$, so we expect the constant C_3 in Theorem 5.11 to be something like $e^{L(b-a)}$. Though only a sufficient condition it is indicative for the potential (lack of) convergence for the Newton algorithm if this quantity is large. It is interesting to note that the same type of quantity should also remain fairly small in order to have stability.

We conclude with a remark on Newton's method. Quite often the method does not converge. One of the ways to improve this is to use the so-called *damped Newton method*. Instead of (5.4a) one takes

$$(5.13) \quad s^{m+1} = s^m + \lambda \xi^m , \qquad 0 < \lambda \leq 1 .$$

One may choose λ in various ways. A simple one is halving the parameter when no decrease of an appropriate criterion function is monitored (one may e.g. use $\|\mathbf{f}(s)\|$ or $\|\mathbf{J}^{-1}(s)\,\mathbf{d}(s)\|$ for the latter). Once some convergence becomes visible λ should be updated and of course eventually become 1 to have the (de facto) quadratic local convergence of the full Newton process. Finally we note that one may save on computing time by keeping Jacobian matrices fixed during some iterations. For more details see [4, 60].

6. Multiple shooting

Both the instability and the convergence problem can be alleviated by 'shooting' on smaller intervals. Indeed, the stability constant reduces exponentially if the integration interval (a, b) is reduced, and likewise the estimate for the constant C_3 in Theorem 5.10.

The idea is conceptually very simple: divide the interval (a, b) into N subintervals and choose an initial value s_j at each of the intervals (t_j, t_{j+1}), $j = 1, ..., N$ $(t_1 = a, t_{N+1} = b)$. See Fig. X.2. So we now have the set of IVP for $j = 1, ..., N$:

$$(6.1) \qquad \begin{cases} \dot{\mathbf{y}}_{\mathbf{s}_j} = \mathbf{f}(t, \mathbf{y}_{\mathbf{s}_j}) \,, & t_j < t < t_{j+1} \,, \\ \mathbf{y}_{\mathbf{s}_j}(t_j) = \mathbf{s}_j \,. \end{cases}$$

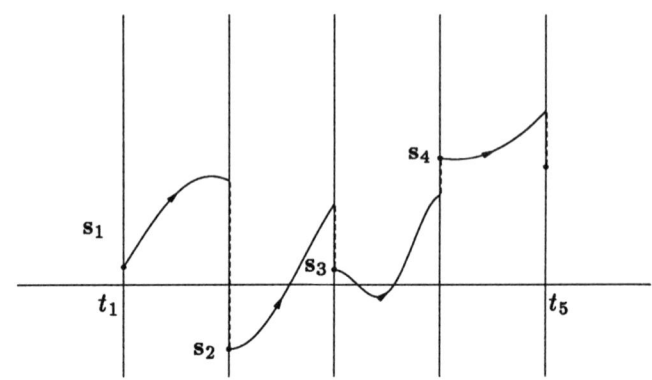

Figure X.2 Multiple shooting.

It is interesting to note that this so-called *multiple shooting* can be done in *parallel*. Since we require continuity of the solution at intermediate points, we have $(N - 1)$ matching conditions

$$(6.2) \qquad \mathbf{y}_{\mathbf{s}_{j-1}}(t_j) = \mathbf{y}_{\mathbf{s}_j}(t_j) = \mathbf{s}_j \,, \qquad j = 2, ..., N \,.$$

Of course we also have the BC which now gives

$$(6.3) \qquad \mathbf{g}\Big(\mathbf{s}_i, \mathbf{y}_{\mathbf{s}_N}(b)\Big) = 0 \,.$$

The resulting system can be written as

$$(6.4a) \qquad \mathcal{F}(\mathbf{S}) = 0 \,,$$

where

$$(6.4b) \qquad \mathbf{S} := \begin{bmatrix} \mathbf{s}_1 \\ \vdots \\ \mathbf{s}_{N-1} \\ \mathbf{s}_N \end{bmatrix}, \qquad \mathcal{F}(\mathbf{S}) := \begin{bmatrix} \mathbf{s}_2 - \mathbf{y}_{\mathbf{s}_1}(t_2) \\ \vdots \\ \mathbf{s}_N - \mathbf{y}_{\mathbf{s}_{N-1}}(t_N) \\ \mathbf{g}\big(\mathbf{s}_1, \mathbf{y}_{\mathbf{s}_N}(b)\big) \end{bmatrix}.$$

Again we use Newton to obtain the solution $\left(\mathbf{x}^T(a), \mathbf{x}^T(t_2), ..., \mathbf{x}^T(t_N)\right)^T$. The Jacobian matrix reads

$$(6.5) \quad \mathcal{J}(\mathcal{S}) := \begin{bmatrix} -\mathbf{Y}_{\mathbf{s}_1}(t_2) & \mathbf{I} & & & \\ & -\mathbf{Y}_{\mathbf{s}_2}(t_3) & \mathbf{I} & & \\ & & \ddots & \ddots & \\ & & & -\mathbf{Y}_{\mathbf{s}_{N-1}}(t_N) & \mathbf{I} \\ \mathbf{B}_a & & & & \mathbf{B}_b \mathbf{Y}_{\mathbf{s}_N}(b) \end{bmatrix},$$

where $\mathbf{B}_a$ and $\mathbf{B}_b$ are defined as in (5.7a) and $\mathbf{Y}_{\mathbf{s}_j}$ is the fundamental solution of the linearised problem on (t_j, t_{j+1}).

In the linear case Newton converges in one iteration. In fact by choosing $\mathcal{S}^0 := 0$ we find the linear version of multiple shooting, viz.

$$(6.6) \quad \mathcal{J}(0)\,\mathcal{S}^1 = -\mathcal{F}(0),$$

from which we find $\mathbf{x}(t_i) = \mathbf{Y}_i(t_i)\,\mathbf{s}_i^1$. Note that in this case the fundamental solution on (t_i, t_{i+1}) is simply denoted by $\mathbf{Y}_i$ as there is no dependence on $\mathbf{s}_i$.

Example X.8.
For a satisfactory implementation of multiple shooting one needs not only to optimise the Newton process (e.g. using damping etc.), but also to avoid unnecessary computations as much as possible. One way to achieve this is by keeping the Jacobian matrices fixed during some iterations. Another way is not to ask too high an accuracy of these initial approximations (which are not very good anyway!). This idea may lead to a strategy where the initial tolerance is low; once Newton has converged within this tolerance, a new (sharper) tolerance is employed etc.
The example here involves the BVP

$$\begin{cases} \dot{\mathbf{x}} = \begin{bmatrix} \dot{x} \\ \lambda \sin \lambda x \end{bmatrix}, \quad \lambda > 0, \quad \mathbf{x} := \begin{bmatrix} x \\ \dot{x} \end{bmatrix}, \\ \begin{bmatrix} 0 & 0 \\ 1 & 0 \end{bmatrix} \mathbf{x}(0) + \begin{bmatrix} 1 & 0 \\ 0 & 0 \end{bmatrix} \mathbf{x}(1) = \begin{bmatrix} 0 \\ 1 \end{bmatrix}. \end{cases}$$

The first component, x, remains almost 0, starting from $t = 0$, until it very rapidly approaches 1 in the neighbourhood of $t = 1$.
As initial solution approximation we use

$$\mathbf{y}(t) = \begin{bmatrix} y_1(t) \\ y_2(t) \end{bmatrix},$$

where

$$y_1(t) = \frac{\sinh \lambda t}{\sin \lambda}, \quad y_2(t) = [2\cosh \lambda\, y_1(t) - 2]^{\frac{1}{2}}.$$

As shooting points we take $t = 0$, 0.8, 0.85, 0.9, 0.95 and 1.0. The results in Table X.3 show the norm of the (weighted) residual $\boldsymbol{\xi} := \mathcal{J}^{-1}(\mathcal{S})\,\mathcal{F}(\mathcal{S})$ and the number of function calls for a given tolerance of 10^{-8} (first column), initially 10^{-4} and in the second stage of the process 10^{-8} (second column), and finally for three values of the tolerance, viz. 10^{-2}, 10^{-4} and 10^{-8}. Note that the numbers are always totals. One clearly sees the advantage of the aforementioned strategy (which had been more pronounced in the third column if the tolerance would have been sharper). The results were obtained by the FORTRAN code MUSN [4].

Table X.3   Residual, # function calls.

$\|\xi\|_2$	#	$\|\xi\|_2$	#	$\|\xi\|_2$	#
$\text{TOL}_1 = 10^{-8}$		$\text{TOL}_1 = 10^{-4}$		$\text{TOL}_1 = 10^{-2}$	
38.06	4587	38.09	1218	35.65	781
36.03	7017	36.03	1938	33.58	1321
1.91	9447	1.92	2658	1.94	1861
$3.0\ 10^{-2}$	11877	$3.0\ 10^{-2}$	2658	1.94	1861
$1.3\ 10^{-5}$	14307	$3.0\ 10^{-5}$	3378	$3.1\ 10^{-2}$	2401
$1.7\ 10^{-12}$	16737				
		$\text{TOL}_2 = 10^{-8}$		$\text{TOL}_2 = 10^{-4}$	
		$6.0\ 10^{-5}$	6106	$1.3\ 10^{-5}$	3575
		$1.6\ 10^{-10}$	7638		
				$\text{TOL}_3 = 10^{-8}$	
				$2.6\ 10^{-4}$	5583
				$1.7\ 10^{-10}$	7113

Example X.9.
We consider the BVP from Example X.8 once more, again with the code MUSN. For $s^0 = 0$ and two shooting points $t = 0$ and $t = 1$ only (i.e. single shooting), we obtain the results of Table X.4. For $s^0 = 5$ we get an error message (even damping does not work). With 3 extra shooting points, viz. at $t = 0.25$, 0.5 and 0.75, we obtain the results given in Table X.5.

The step from single shooting to multiple shooting can be extended to the extreme by choosing every grid point as a 'shooting point'. In that case the 'course grid' of the shooting points coincides with the 'fine grid' of actual

Table X.4 $s^0 = 0$; single shooting.

TOL	$\|\xi\|$	# iterations
10^{-1}	.27E-1	1
10^{-2}	.14E-4	2
10^{-4}	.14E-4	2

Table X.5 $s^0 = 0$; multiple shooting.

TOL	$\|\xi\|$	# iterations
10^{-1}	.46E-3	2
10^{-2}	.46E-3	2
10^{-4}	.24E-6	3

discretisation points. Having done this there is no need any more for explicit discretisations, and methods like implicit Runge-Kutta are equally efficient. This opens a new territory of stability control as implicit RK schemes can be designed to handle stiff IVP. Less trivial, but more interesting is that even suitable implicit RK schemes can be designed to handle singularly perturbed BVP to some extent. Note that dichotomy does not exclude very fast modes of increasing or decreasing type. All such RK schemes can actually be thought of as (condensed) forms of so-called *collocation methods*. We refer the interested reader to [4].

Exercises Chapter X

1. Consider the BVP

$$\begin{cases} \ddot{x} = 100x \ , \\ x(0) = x(b) = 1 \ . \end{cases}$$

a) Write this BVP in standard (matrix vector) form.

b) Let $\mathbf{Y}(t)$ be the fundamental solution of the BVP with $\mathbf{Y}(0) = \mathbf{I}$. Indicate why $\mathbf{Y}(b)$ will be ill-conditioned for values of b away from zero (the condition number of a matrix $\mathbf{M}$ is defined as $\|\mathbf{M}\| \|\mathbf{M}^{-1}\|$).

c) Show that the matrix $\mathbf{Q} := \mathbf{B}_a + \mathbf{B}_b \mathbf{Y}(b)$ is ill-conditioned (cf. (2.4a)).

2. Show that the conditioning constant κ_2 (see (3.1b)) is independent of the choice of fundamental solution.

3. Consider the BVP

$$\begin{cases} \dot{\mathbf{x}} = \begin{pmatrix} 1 & 2 \\ 2 & 1 \end{pmatrix} \mathbf{x}\,, \\ \mathbf{x}(0) + \mathbf{x}(b) = \beta\,. \end{cases}$$

Show that the stability constant κ in (3.3) is uniformly bounded in b.

4. Let $\mathbf{B}_a\,\mathbf{B}_a^T + \mathbf{B}_b\,\mathbf{B}_b^T = \mathbf{I}$.

   a) Show that $\|\mathbf{F}(t)\|_2^2 = \|\mathbf{G}(t,a)\,\mathbf{G}(t,a)^T + \mathbf{G}(t,b)\,\mathbf{G}(t,b)^T\|_2$ (cf. (2.8)).
   b) Show that $\kappa - 1 \le \sqrt{2}\,\kappa_2$ (cf. (3.3)).

5. Consider the BVP

$$\begin{cases} \dddot{x} = 2\ddot{x} + \dot{x} - 2x\,, \quad 0 < t < b\,, \\ x(0) = 1\,, \quad x(b) - \dot{x}(b) = 0\,, \quad \ddot{x}(b) = 1\,. \end{cases}$$

Take b 'large'.

   a) Write this BVP as a system in standard form and determine the fundamental solution $\mathbf{Y}$ with $\mathbf{Y}(0) = \mathbf{I}$.
   b) Show that the stability constant is moderate.
      Hint: Use a clever fundamental solution, e.g.

$$\mathbf{F}(t) := \begin{bmatrix} e^{-t} & e^{t-b} & e^{2(t-b)} \\ -e^{-t} & e^{t-b} & 2e^{2(t-b)} \\ e^{-t} & e^{t-b} & 4e^{2(t-b)} \end{bmatrix},$$

and find an estimate for $\mathbf{F}(t)\Big(\mathbf{B}_a\,\mathbf{F}(a) + \mathbf{B}_b\,\mathbf{F}(b)\Big)^{-1}$.
   c) Determine the coefficients of the matrix $\mathbf{B}_a + \mathbf{B}_b\,\mathbf{Y}(b)$ and show that (4.2b) is an ill-conditioned system.

6. Let $\mathcal{J}$ be the multiple shooting matrix as in (6.5) for a linear problem, i.e.

$$\begin{cases} \mathbf{s}_{i+1}^1 = \mathbf{Y}_i\,\mathbf{s}_i^1 + \mathbf{s}_1^0 \qquad (\mathbf{Y}_i := \mathbf{Y}_i(t_{i+1}))\,, \\ \mathbf{B}_a\,\mathbf{s}_1^1 + \mathbf{B}_b\,\mathbf{Y}_{N+1}\,\mathbf{s}_N^1 = \beta\,. \end{cases}$$

Let $\{\mathbf{F}_i\}_{i=1}^N$ be a fundamental solution of the Δ-equation above, and define $\mathbf{F}_{N+1} := \mathbf{Y}_{N+1}\,\mathbf{F}_N$, where $\mathbf{B}_a\,\mathbf{F}_1 + \mathbf{B}_b\,\mathbf{F}_{N+1} = \mathbf{I}$. Define a *discrete Green's function* $\{\mathbf{G}_{ij}\}$ by

$$
\mathbf{G}_{ij} := \left\{ \begin{array}{ll} \mathbf{F}_i \, \mathbf{B}_a \, \mathbf{F}_1 \, \mathbf{F}_{j+1}^{-1} & , \; j \leq i \\ -\mathbf{F}_i \, \mathbf{B}_b \, \mathbf{F}_{N+1} \, \mathbf{F}_{j+1}^{-1} & , \; j > i \end{array} \right. , \quad 1 \leq i, j \leq N - 1 .
$$

Show that

$$
\mathbf{G}_{i+1,j} = \mathbf{Y}_i \, \mathbf{G}_{ij} + \delta_{ij} \, \mathbf{I}, \;\; 1 \leq i \leq N - 1
$$

$$
\mathbf{B}_a \, \mathbf{G}_{1j} + \mathbf{B}_b \, \mathbf{G}_{Nj} = 0
$$

(δ_{ij} is the Kronecker index).

7. Let $\mathcal{J}$ be the multiple shooting matrix as in (6.5). Show with $\mathbf{G}_{ij}$ as in Exercise 6 that

$$
\mathcal{J}^{-1} = \left[\begin{array}{cccc} \mathbf{G}_{11} & \cdots & \mathbf{G}_{1\,N-1} & \mathbf{F}_1 \\ \vdots & & & \vdots \\ \mathbf{G}_{N1} & \cdots & \mathbf{G}_{N\,N-1} & \mathbf{F}_N \end{array} \right] .
$$

8. Consider the BVP

$$
\left\{ \begin{array}{l} \dot{\mathbf{x}} = \left[\begin{array}{ccc} -10 & 0 & 0 \\ 0 & 20 & 0 \\ 20 & 0 & 0 \end{array} \right] \mathbf{x} + \left[\begin{array}{c} 10 \\ 20 \\ -30 \end{array} \right], \; 0 < t < b, \; \mathbf{x} := \left[\begin{array}{c} x_1 \\ x_2 \\ x_3 \end{array} \right], \\ x_1(0) + x_3(b) = 2, \;\; x_2(0) + x_2(b) = 2, \;\; x_3(0) = 1 . \end{array} \right.
$$

a) Show that $\mathbf{x}(t) \equiv (1, 1, 1)^T$.

b) Show that this BVP has a moderate stability constant.

c) This so-called *partially separated* BVP can be solved by computing a partial fundamental solution only. Rather than computing the fundamental solution $\mathbf{Y}$ with $\mathbf{Y}(0) = \mathbf{I}$, we need to compute $\mathbf{Y}' \in \mathbb{R}^{3 \times 2}$ only, determined from $[0\;0\;1]\,\mathbf{Y}'(0) = 0$. We can now use superposition employing a particular solution $\mathbf{p}$ with $[0\;0\;1]\,\mathbf{p}(0) = 1$. Can you work out this procedure further for the example above?

XI

Concepts from Classical Mechanics

In this chapter we give a summary of basic concepts of classical mechanics. The starting point is Newton's second law, which is included in §1. If kinematic constraints apply, the Lagrange formalism, dealt with in §2, is the appropriate framework to derive the equations of motion. An alternative formalism, presented in §3, stems from Hamilton and has certain advantages if the system is autonomous and the constraints are not time dependent. Then the energy of the system is conserved and orbits in phase space can be found from contour lines of the Hamiltonian. In §4 the application of this to the phase plane is shown. In §5 continuous media are briefly dealt with. They give rise to partial differential equations (PDE) instead of ODE. By an example, in which the so-called method of lines is used, we show that the numerical treatment of PDE may lead to ODE.

1. Introduction

The study of differential equations is largely inspired by observations and descriptions of mechanical systems. Experiments on earth and observation of the solar system led to the concepts of position, velocity, acceleration, force, and the mathematical language describing their relations. A lot of knowledge about ODE has been gained within the framework of classical mechanics, but these insights are more generally applicable too. In this chapter we present some central issues of classical mechanics which have also proved their value elsewhere. The emphasis will be on the ideas behind the formalisms, rather than on the details of the derivations, which can be found in many other works. See, e.g., [6, 32, 51]. We mainly deal with systems consisting of rigid bodies, because they are the ones giving rise to ODE. The mechanics of continua, such as vibrating strings, beams, and plates, leads to partial differential equations (PDE). Within the present context the latter are relevant only if these equations can be reduced to ODE.

Let us summarise the fundamental starting points of classical mechanics.

- Physical quantities such as *space, time, mass*, and *force* are introduced by indicating how they are measured.

- In practice even the smallest rigid bodies have certain spatial dimensions. The position of such a body is characterised by the position of its centre of mass and its orientation with respect to some coordinate system attached to its centre of mass. So, in principle 6 coordinates are involved in total. It is customary to introduce the concept of *point mass*. This is an artifact, where all the mass is thought to be concentrated, so that its orientation is irrelevant. Hence we only need 3 coordinates to describe its position. We shall refer to such an idealized object by *particle*.

- A position in space is always measured with respect to some coordinate system. It is assumed that there exists at least one coordinate system in which a completely isolated body, i.e. a body on which no force is exerted, has constant velocity. *Newton's second law* then implies that the body has constant velocity in all coordinate systems which have constant speed with respect to that particular frame. Those coordinate systems are called *inertial frames*. The assumption of the existence of inertial frames is sometimes referred to as *Newton's first law*. For practical purposes one often considers a frame attached to the earth's surface as an inertial frame for localised motions on earth. From a pure point of view this is a poor concept, because the earth rotates around both its axis and the sun. Furthermore, the sun is not at rest, but rotates within our galaxy, while the latter in its turn follows an orbit through the universe.

- In all inertial frames *Newton's second law* holds: the force $\mathbf{F}$ exerted on a body equals the time derivative of its momentum $\mathbf{p}$:

$$(1.1) \qquad \mathbf{F} = \dot{\mathbf{p}},$$

with the *momentum* $\mathbf{p}$ defined as the product of mass m and velocity $\mathbf{v} := \dot{\mathbf{x}}$:

$$(1.2) \qquad \mathbf{p} = m\mathbf{v} \ .$$

If the mass m is constant, Newton's second law takes on the famous form

$$(1.3) \qquad \mathbf{F} = m\mathbf{a},$$

where $\mathbf{a} := \dot{\mathbf{v}} = \ddot{\mathbf{x}}$ is the *acceleration* of the body. Equations (1.1) and (1.3) are also called *equations of motion*. The fact that they are the same in all inertial frames is related to the fact that they involve only second derivatives of position $\mathbf{x}$ with respect to time t. This *relativity* or *equivalence principle* does not hold if accelerated frames, such as rotating coordinates, are involved. Then extra acceleration terms appear in the equations of motion. These are usually included in the left-hand side of (1.1) or (1.3), and then interpreted as so-called *pseudo-forces*. An illustration of this can be found in Examples XI.3 and XI.4.

- For completeness' sake we mention *Newton's third law*: if particle 1 exerts a force $\mathbf{F}_{12}$ on particle 2, and particle 2 a force $\mathbf{F}_{21}$ on particle 1, and if these forces are parallel, then these forces have equal strength: $\mathbf{F}_{12} = -\mathbf{F}_{21}$. A popular summary of this law is: *action is minus reaction*. This third law is much less general than the second one, because it applies to a restricted class of forces only. This class does not include, e.g., electromagnetic forces.

The validity of (1.1) (or (1.3)) implies that all motions in classical mechanics are governed by ODE of maximum order 2. The dependence of the force on particle positions and velocities determines the character of these ODE. In the following example some forces are introduced which are frequently met in practice.

Example XI.1.

(i)   *Gravitation.* The gravitational force $\mathbf{F}_{12}$ of a particle of mass m_1 at position $\mathbf{x}_1$ on a particle with mass m_2 at position $\mathbf{x}_2$ is given by

$$\mathbf{F}_{12} = G \frac{m_1 m_2}{\|\mathbf{x}_1 - \mathbf{x}_2\|^3} (\mathbf{x}_1 - \mathbf{x}_2) \,,$$

with G the gravitational constant. This force satisfies Newton's third law. It generally gives rise to nonlinear equations of motion. However, in some cases a linear approximation may be reliable. For example, if particle 1 is the earth and particle 2 a falling body, this force has magnitude

$$F = m_2 \, g \,,$$

and is directed downwards. Here, g is the gravitational acceleration. This reduction is possible, because in this situation the relative displacement is very small.

We remark that the masses m_1 and m_2 appearing in the gravitational law are identical to the masses in the equations of motion (1.1) and (1.3). This equivalence between 'gravitational mass' and 'inertial mass' is far from being straightforward; it is empirical evidence.

(ii) *Harmonic force.* A harmonic force satisfies *Hooke's law*: the force exerted on a particle connected to a fixed point by means of a spring is linearly proportional to its distance from that fixed point.

If the spring is along the x-axis and the position of the particle at rest is taken as the origin, the force on the particle is

$$F = -k\,x \; ,$$

with the spring constant $k > 0$ determining the strength of the force. Harmonic forces give rise to oscillations around the rest position which are sinusoidal in time. Many oscillating systems can reliably be described as a harmonic motion as long as the amplitude remains small. Harmonic systems are easy to cope with, because the equations of motion are linear.

(iii) *Friction.* The frictional force exerted on a particle if it moves along or through a medium tends to reduce its velocity. The force is directed in the opposite direction of the velocity **v** and usually modelled as

$$\mathbf{F} = -c(v)\,\mathbf{v} \; ,$$

with $v = \|\mathbf{v}\|$ and $c(v) > 0$ a function depending on the geometry and material under consideration. One often makes the assumption that $c(v)$ is independent of v and thus constant.

(iv) *Lorentz force.* A particle with charge e in a magnetic field **B** experiences the electromagnetic force

$$\mathbf{F} = e\,\mathbf{v} \times \mathbf{B} \; ,$$

where $\times$ stands for the outer product of two vectors, which is orthogonal to these vectors. The Lorentz force is thus orthogonal to the directions of both the velocity and the magnetic field. The presence of the outer product is the reason that, apart from static situations, electromagnetic forces do not satisfy Newton's third law.

2. Lagrangian formalism

The kinetic energy T of a system consisting of N particles is the sum of the kinetic energies of the individual particles:

$$(2.1) \qquad T = \sum_{i=1}^{N} \frac{1}{2}\, m_i \, \|\dot{\mathbf{x}}_i\|^2 \; .$$

This energy measures the amount of work the system can perform in principle, thanks to the motions of its parts. In most practical systems these are hindered by *kinematic constraints*, e.g. because the body is forced to move along a surface or to rotate around an axis. In general, constraints are algebraic in the positions and velocities of the particles. A system of N particles in three dimensions without constraints can be specified in terms of $3N$ (scalar) coordinates and $3N$ (scalar) velocities. The $6N$-dimensional space spanned by the positions and the velocities is usually referred to as the *configuration space*. Such a system is said to have $3N$ *degrees of freedom*. The presence of constraints decreases the number of degrees of freedom. In the following we restrict ourselves to *holonomic constraints*, which are of the form

$$(2.2) \qquad f_k(t, \mathbf{x}_1, ..., \mathbf{x}_N) = 0 , \qquad k = 1, ..., K .$$

In non-holonomic constraints velocities and inequalities are also involved, which hamper the derivation of a general formalism. The effect of constraints can be represented by introducing *reaction forces* $\mathbf{R}_i$, $i = 1, ..., N$. By definition they force the system to satisfy the constraints. Newton's second law for particle i reads:

$$(2.3) \qquad \dot{\mathbf{p}}_i = \frac{d}{dt}\left(\frac{\partial T}{\partial \dot{\mathbf{x}}_i}\right) = \mathbf{R}_i + \mathbf{F}_i ,$$

where the total force consists of the *reaction force* $\mathbf{R}_i$ and the so-called *applied force* $\mathbf{F}_i$, which is the sum of all other forces. We can get rid of the reaction force in (2.3) by applying the *principle of d'Alembert*, which can be proved for rigid bodies. It states that the total work of the reaction forces vanishes under a *virtual displacement* of the system. A displacement $(\delta \mathbf{x}_1, ..., \delta \mathbf{x}_N)^T$ of the particles is called virtual if, for $k = 1, ..., K$,

$$(2.4) \qquad f_k(t, \mathbf{x}_1, ..., \mathbf{x}_N) = f_k(t, \mathbf{x}_1 + \delta \mathbf{x}_1, ..., \mathbf{x}_N + \delta \mathbf{x}_N) = 0 .$$

Note that the time is kept fixed in (2.4). Expanding the second term in (2.4) up to first order we find that a virtual displacement satisfies

$$(2.5) \qquad \sum_{i=1}^{N} \mathbf{c}_{ki} \cdot \delta \mathbf{x}_i := \sum_{i=1}^{N} \frac{\partial f_k}{\partial \mathbf{x}_i} \cdot \delta \mathbf{x}_i = 0 , \qquad k = 1, ..., K,$$

where the dot denotes the inner product. Taking the inner product of both sides of (2.3) with $\delta \mathbf{x}_i$ and summing over all particles, we obtain

$$(2.6) \qquad \sum_{i=1}^{N} \left(\frac{d}{dt}\left(\frac{\partial T}{\partial \dot{\mathbf{x}}_i}\right) - \mathbf{F}_i\right) \cdot \delta \mathbf{x}_i = \sum_{i=1}^{N} \mathbf{R}_i \cdot \delta \mathbf{x}_i = 0$$

where the $\delta \mathbf{x}_i$ are assumed to satisfy (2.5) and the last equality expresses the principle of d'Alembert.

We shall deal with (2.6) in two ways. The first approach uses the so-called *Lagrange multipliers*, the second one leads to the *Lagrange equations*. In both approaches it is important to realize that only $N - K$ of the displacements $\delta \mathbf{x}_i$ in (2.6) can be varied independently. To take into account restriction (2.5) we add a linear combination of the constraints (2.5) to (2.6). This yields

$$(2.7) \qquad \sum_{i=1}^{N} \left\{\frac{d}{dt}\left(\frac{\partial T}{\partial \dot{\mathbf{x}}_i}\right) - \mathbf{F}_i + \sum_{k=1}^{K} \lambda_k \mathbf{c}_{ki}\right\} \cdot \delta \mathbf{x}_i = 0 .$$

The time dependent coefficients $\lambda_k(t)$ are the *Lagrange multipliers*. They are chosen such that $N - K$ terms in the summation over $i = 1, ..., N$ in (2.7) vanish identically. Suppose that the vanishing terms are the last K ones, thus

$$(2.8a) \qquad \frac{d}{dt}\left(\frac{\partial T}{\partial \dot{\mathbf{x}}_i}\right) - \mathbf{F}_i + \sum_{k=1}^{K} \lambda_k c_{ki} = 0 , \qquad i = N - k + 1, ..., N .$$

Because the first $N - K$ displacements $\delta\mathbf{x}_1, ..., \delta\mathbf{x}_{N-K}$ can now be varied independently, we find

$$(2.8b) \qquad \frac{d}{dt}\left(\frac{\partial T}{\partial \dot{\mathbf{x}}_i}\right) - \mathbf{F}_i + \sum_{k=1}^{K} \lambda_k c_{ki} = 0 , \qquad i = 1, ..., N - K .$$

Note that the linear combinations $\sum_k \lambda_k c_{ki}$ can be interpreted as the forces which make the system satisfy the constraints (2.2). So, in this approach the reaction forces seem to be eliminated via the principle of d'Alembert, but they are still present, however, and form a part of the answer.

Equations (2.2), (2.8a), and (2.8b) form a system of $3N + K$ equations for $3N + K$ unknowns, namely the $3N$ components of the positions $\mathbf{x}_i(t)$, together with the K Lagrange multipliers $\lambda_k(t)$. Equations (2.8) are ODE, which are coupled to the algebraic equations (2.2). So we meet with a set of differential-algebraic equations (DAE). The theory of DAE is extensively dealt with in Chapter IX. Here we illustrate the multiplier approach with the following simple example.

Example XI.2.

Consider a particle moving along a frictionless, inclined plane. For convenience we assume the particle to move in a vertical plane for which we take the (x, z) plane. Its position $\mathbf{x}$ is thus characterized by the Cartesian coordinates $\mathbf{x} := (x, z)^T$. The plane makes an angle φ with the horizontal and is given by

$$z - x \tan\varphi = 0 .$$

This equation forms a kinematic constraint for the particle. Clearly this constraint is holonomic. To evaluate equations (2.8) we need the kinematic energy T, which in this case is given by

$$T = \tfrac{1}{2} m(\dot{x}^2 + \dot{z}^2) .$$

The coefficients $\mathbf{c}$ in (2.5) are obtained from differentiating the constraint. This yields $\mathbf{c} = (-\tan\varphi, 1)^T$. Evaluation of (2.8) for the present system leads to

$$\begin{cases} m\ddot{x} - \lambda\tan\varphi &= 0 , \\ m\ddot{z} + \lambda &= mg , \end{cases}$$

where mg is the gravitational force in the z-direction. Differentiating the constraint twice we find

$$\ddot{z} = \ddot{x} \tan \varphi .$$

Substitution of this into the equations of motion makes it possible to solve for λ:

$$\lambda = mg \cos^2 \varphi .$$

From this we find that

$$\left\{ \begin{array}{rcl} x(t) & = & \frac{1}{2} gt^2 \sin \varphi \cos \varphi , \\ z(t) & = & -\frac{1}{2} gt^2 \sin^2 \varphi , \end{array} \right.$$

for a particle starting in the origin with vanishing initial velocity.

Instead of introducing Lagrange multipliers relations (2.2) can be used to eliminate K of the $3N$ coordinates directly. The resulting $M :=$ $3N-K$ coordinates $q_1, ..., q_M$ are independent and referred to as *generalised coordinates* and the time derivatives $\dot{q}_1, ..., \dot{q}_M$ as *generalised velocities*, together spanning a $2M$-dimensional generalised configuration space. If no constraints apply, the kinetic energy T depends on the velocities of the particles only. It should be realised that this changes when constraints come in. This is illustrated in the following example.

Example XI.3.
Consider a bead sliding on a moving hoop as sketched in Fig. XI.1. The hoop has radius R and remains in a vertical plane while moving. To describe the system conveniently, we take the vertical plane of the hoop as the $y = 0$ plane of Cartesian coordinates. The centre of the hoop has then coordinates $(x_0(t), 0, z_0(t))^T$. Without constraints the system would have three degrees of freedom, However, the coordinates $((x(t), y(t), z(t))^T$ of the bead have to satisfy, for all t, the kinematic constraints

$$\left\{ \begin{array}{l} y(t) = 0 \\ \Big(x(t) - x_0(t)\Big)^2 + \Big(y(t) - y_0(t)\Big)^2 = R^2 . \end{array} \right.$$

These constraints are holonomic and can be used to eliminate two degrees of freedom, so $M = 1$. An appropriate choice for the generalised coordinate is the angle φ indicated in Fig. XI.1. The coordinates $x(t)$ and $z(t)$ of the bead can be expressed in φ:

$$\left\{ \begin{array}{l} x(t) = x_0(t) + R \sin \varphi(t) \\ z(t) = z_0(t) - R \cos \varphi(t) . \end{array} \right.$$

Differentiation of these expressions and substitution into (2.1) yields the kinetic energy as a function of t, φ, $\dot{\varphi}$:

$$T = T(t, \varphi, \dot{\varphi}) =$$

$$= \tfrac{1}{2} m \left(\dot{x}_0^2 + \dot{z}_0^2 + 2R\dot{\varphi}(\dot{x}_0 \cos \varphi + \dot{z}_0 \sin \varphi) + R^2 \dot{\varphi}^2 \right) .$$

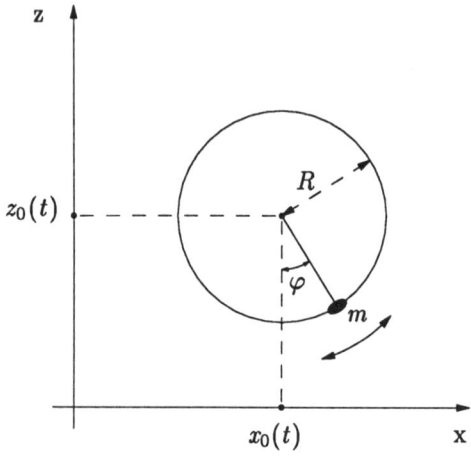

Figure XI.1   The bead-on-a-hoop system discussed in Examples XI.3, 4.

When the N coordinates $\mathbf{x}_i$ are reduced to the M generalized coordinates q_j, equation (2.6) transforms into the *Lagrange equations*, which are given by

$$(2.9) \qquad \frac{d}{dt}\left(\frac{\partial T}{\partial \dot{q}_j} \right) - \frac{\partial T}{\partial q_j} = \sum_{i=1}^{N} \mathbf{F}_i \cdot \frac{\partial \mathbf{x}_i}{\partial q_j}, \qquad j = 1, ..., M .$$

It is easily checked that equations (2.9) reduce to the well-known form (1.3) if no constraints are present. The q_j are then precisely the components of the positions $\mathbf{x}_i$. The right-hand side of (2.9) expresses the idea that only those components of the applied forces $\mathbf{F}_i$ are relevant that are consistent with the constraints. We illustrate this as follows.

Example XI.4.
Consider again the bead-on-a-hoop system introduced in Example XI.3 and depicted in Fig. XI.1. We assume that gravitation acts on the bead and that friction may be ignored. In the coordinates chosen the

gravitational force is given by $\mathbf{F} = (0, 0, mg)^T$, with m the bead mass. The right-hand side of (2.9) is then given by

$$-mg \frac{\partial z}{\partial \varphi} \ .$$

To work out the left-hand side of (2.9) we calculate the derivatives

$$\left\{ \begin{array}{l} \dfrac{\partial T}{\partial \varphi} = -m\,R\,\dot{\varphi}\,(\dot{x}_0 \sin \varphi - \dot{z}_0 \cos \varphi) \ , \\[3mm] \dfrac{\partial T}{\partial \dot{\varphi}} = m\,R\,(\dot{x}_0 \cos \varphi + \dot{z}_0 \sin \varphi) + m\,R^2\,\dot{\varphi} \ . \end{array} \right.$$

Adding these results we obtain the following equation of motion for the bead:

$$R\,\ddot{\varphi} + \ddot{z}_0 \sin \varphi + \ddot{x}_0 \cos \varphi + g \sin \varphi = 0 \ .$$

If the hoop has constant velocity (or is at rest) this reduces to the famous pendulum equation

$$\ddot{\varphi} = -\frac{g}{R} \sin \varphi \ .$$

This reduction illustrates the relativity principle that the equation of motion is the same in all inertial systems. A constant acceleration of the hoop in the z direction, say $\ddot{z}_0 = \bar{g}$ and $\ddot{x}_0 = 0$, leads to the same equation with g replaced by $g+\bar{g}$. This can be interpreted as a correction of the gravitational force and is an illustration of the occurrence of a pseudo-force due to accelerated kinematic constraints.

The Lagrange equations (2.9) greatly reduce, if the applied forces $\mathbf{F}_i$ can be derived from a potential $V = V(t_1, \mathbf{x}_1, ..., \mathbf{x}_N)$:

$$(2.10) \qquad \mathbf{F}_i = -\frac{\partial V}{\partial \mathbf{x}_i} \ , \qquad i = 1, ..., N \ .$$

A system with forces satisfying relations (2.10) is called *conservative*, if the potential V does not explicitly depend on time t.

The *Lagrangian* is defined as the difference between kinetic energy T and potential energy V:

$$(2.11) \qquad L := T - V \ .$$

L generally depends on time, positions and velocities. In the case of K constraints we have

$$(2.12) \qquad L = L(t, q_1, ..., q_M, \dot{q}_1, ..., \dot{q}_M) \ .$$

Substitution of (2.10) and (2.11) into (2.9) yields the *Euler-Lagrange equations*

$$(2.13) \qquad \frac{d}{dt}\left(\frac{\partial L}{\partial \dot{q}_j}\right) = \frac{\partial L}{\partial q_j} , \qquad j = 1, ..., M .$$

Equations (2.13) remain valid if the potential V is velocity dependent and the applied forces $\mathbf{F}_i$ follow from the relation

$$(2.14) \qquad \mathbf{F}_i = -\frac{\partial V}{\partial \mathbf{x}_i} + \frac{d}{dt}\left(\frac{\partial V}{\partial \dot{\mathbf{x}}_i}\right) .$$

This kind of potential is met if electromagnetic forces are involved.

The (Euler-)Lagrange equations (2.9) and (2.13) are preferable to the original formulation (1.3), even if no constraints are present, because only the scalar function L is involved instead of N vector functions $\mathbf{F}_i$.

3. Hamiltonian formalism

The Euler-Lagrange equations (2.13) can be put into an alternative form. An important motivation is that the Lagrangian L has no physical interpretation. Moreover, a disadvantage of the Lagrangian formulation is that the resulting equations of motion are of second order, while first order equations are often more convenient to handle. To illustrate the first argument we consider the (total) time derivative of L:

$$(3.1) \qquad \frac{dL}{dt} = \frac{\partial L}{\partial t} + \sum_{j=1}^{M}\left(\frac{\partial L}{\partial q_j}\dot{q}_j + \frac{\partial L}{\partial \dot{q}_j}\ddot{q}_j\right) .$$

We see that L is generally not conserved if time evolves, not even if L does not explicitly depend on t. Using (2.13) we may rewrite (3.1) as

$$(3.2) \qquad \frac{dL}{dt} = \frac{\partial L}{\partial t} + \frac{d}{dt}\left[\sum_j \frac{\partial L}{\partial \dot{q}_j}\dot{q}_j\right] .$$

This suggests introducing the *Hamiltonian*, which is defined by

$$(3.3a) \qquad H := \sum_j \frac{\partial L}{\partial \dot{q}_j}\dot{q}_j - L = \sum_j p_j\dot{q}_j - L ,$$

with the momentum p_j defined by

$$(3.3b) \qquad p_j := \frac{\partial L}{\partial \dot{q}_j} .$$

If no constraints apply, the latter definition reduces to the one given in (1.2).

For the Hamiltonian H we have from (3.2)

(3.4) $\quad \dfrac{dH}{dt} = \dfrac{\partial H}{\partial t} = \dfrac{\partial L}{\partial t} \; ,$

so that H is conserved in time if either H or L does not depend explicitly on t. In the Lagrangian formalism the independent variables are t, q_j, and $\dot{q}_j$, $j = 1, ..., M$. In the Hamiltonian formalism the $\dot{q}_j$ are replaced by the momenta p_j. The corresponding transformation is only possible under certain conditions (see Exercise 1). If these conditions are fulfilled, the $\dot{q}_j$ can be expressed in terms of t, q_j, and p_j, $j = 1, ..., M$, and we have $H = H(t, q_1, ..., q_M, p_1, ..., p_M)$.

The *Hamiltonian equations*, which are equivalent to the Euler-Lagrange equations (2.13), read for $j = 1, ..., M$:

(3.5)
$$\begin{cases} \dot{q}_j = \dfrac{\partial H}{\partial p_j} \; , \\[2ex] \dot{p}_j = -\dfrac{\partial H}{\partial q_j} \; . \end{cases}$$

The Hamiltonian H represents the total energy of the system if

(i)   The potential is velocity independent,

(ii) The constraints are time independent,

because then we have

$$H = \sum_j \frac{\partial L}{\partial \dot{q}_j} \dot{q}_j - L = \sum_j \frac{\partial T}{\partial \dot{q}_j} \dot{q}_j - L = 2T - L = T + V \; .$$

A function $f(t, q_1, ..., q_M, p_1, ..., p_M)$ represents a *conserved quantity* if during the motion

$$\frac{df}{dt} = 0 \; .$$

From the Hamiltonian equations (3.5) we directly conclude that p_j is conserved if H does not depend on q_j, and that q_j is conserved if H does not depend on p_j. The total time derivative of f may generally be written as

(3.6) $\quad \dfrac{df}{dt} = \dfrac{\partial f}{\partial t} + \{f, H\} \; ,$

where the *Poisson brackets* are defined as

$$\{f_1, f_2\} := \sum_{j=1}^{M} \left[\frac{\partial f_1}{\partial q_j} \frac{\partial f_2}{\partial p_j} - \frac{\partial f_1}{\partial p_j} \frac{\partial f_2}{\partial q_j} \right] \; .$$

The choice $f = H$ in (3.6) yields the first equality in (3.4), because $\{f, f\} = 0$ for any f. The Hamiltonian equations (3.5) in terms of Poisson brackets read:

(3.7)
$$\begin{cases} \dot{q}_j = \{q_j, H\}, \\ \dot{p}_j = \{p_j, H\}. \end{cases}$$

The *configuration space* in the Lagrangian formalism is replaced by the *phase space* in the Hamiltonian formalism. It consists of the $2M$-dimensional vectors $(q_1, ..., q_M, p_1, ..., p_M)^T$. In the notation used above the coordinates and momenta are still treated separately. The symmetry between the q_j and p_j can be emphasised by introducing vectors $\mathbf{z}$:

$$\mathbf{z} := (q_1, ..., q_M, p_1, ..., p_M)^T .$$

Equations (3.7) can then be summarised as

(3.8) $\dot{\mathbf{z}} = \{\mathbf{z}, H\}$.

Another, frequently used way to present the Hamiltonian equations uses the *symplectic matrix*

(3.9a) $\mathbf{S} := \begin{pmatrix} 0 & \mathbf{I} \\ -\mathbf{I} & 0 \end{pmatrix}$,

where $\mathbf{I}$ is the $M \times M$ unit matrix. An alternative form of (3.7) and (3.8) is then

(3.9b) $\dot{\mathbf{z}} = \mathbf{S}\, \boldsymbol{\nabla} H$,

where the gradient $\boldsymbol{\nabla} H$ is defined by $\boldsymbol{\nabla} H := (\partial H/\partial z_1, ..., \partial H/\partial z_{2M})^T$.

If H does not depend explicitly on t, the system is *autonomous* and the Hamiltonian is a conserved quantity. Its value, called the *energy* E of the system, follows from the initial values $\mathbf{z}(t_0)$. The orbits in the phase space of an autonomous system satisfy the equation

(3.10) $H\big(\mathbf{z}(t)\big) = H\big(\mathbf{z}(t_0)\big) = E$,

and thus coincide with contour lines corresponding to energy E. These contour lines form together a surface of dimension $\leq 2M - 1$ in the $2M$-dimensional phase space. A contour line, which does not contain a stationary point, is called *regular*. Otherwise it is said to be *singular*.

The Hamiltonian equations (3.8) (or (3.9b)) generate a flow in the phase space. Starting at an arbitrary point $\mathbf{z}_0$ we may follow its time evolution given by the solution of (3.9b). In the next theorem it is proved that Hamiltonian systems are volume preserving. This has the consequence that these systems cannot be asymptotically stable (cf. §V.1).

Theorem 3.11 (Liouville). *The phase flow of a Hamiltonian system is volume preserving.*

Proof: Consider at arbitrary time $t = t_0$ an arbitrary region $A(t_0)$ in phase space with coordinates $\mathbf{z}(t_0)$. Its volume $V(t_0)$ is given by

$$V(t_0) = \int_{A(t_0)} d\mathbf{z}(t_0) .$$

After the system has evolved to time t, the region to $A(t)$, and the coordinates to $\mathbf{z}(t)$, the volume $V(t)$ is given by

$$V(t) = \int_{A(t)} d\mathbf{z}(t_0) = \int_{A(t_0)} d\mathbf{z}(t) .$$

The phase flow $\mathbf{z}(t) := \mathbf{\Psi}(t\,;\,t_0, \mathbf{z}(t_0))$ (cf. (I.2.4)) is here read as a coordinate transformation. Applying the formula for changing variables in a multiple integral we may write

$$V(t) = \int_{A(t_0)} \det\left(\frac{\partial \mathbf{z}(t)}{\partial \mathbf{z}(t_0)}\right) d\mathbf{z}(t_0) ,$$

where $\partial \mathbf{z}(t)/\partial \mathbf{z}(t_0)$ is the Jacobian of the transformation. Because

$$\mathbf{z}(t) = \mathbf{z}(t_0) + (t - t_0)\,\dot{\mathbf{z}}(0) + o((t - t_0)^2) ,$$

we have from (3.9b)

$$\frac{\partial \mathbf{z}(t)}{\partial \mathbf{z}(t_0)} = \mathbf{I} + (t - t_0)\,\nabla^T \mathbf{S}\,\nabla H + o((t - t_0)^2) ,$$

with the derivatives evaluated at $t = t_0$. As shown in Theorem IV.2.14 we may write

$$\det\left(\mathbf{I} + (t-t_0)\,\nabla^T \mathbf{S}\,\nabla H + o((t-t_0)^2)\right) = 1 + (t-t_0)\,\mathrm{Tr}(\nabla^T \mathbf{S}\,\nabla H) + o((t-t_0)^2).$$

In view of the special form of $\mathbf{S}$ in (3.9a), we have for all H

$$\mathrm{Tr}(\nabla^T \mathbf{S}\,\nabla H) = 0 .$$

From this it follows that

$$V(t) = V(t_0) + o((t - t_0)^2) ,$$

and thus $\dot{V}(t_0) \equiv 0$. Because t_0 and A_0 are arbitrary, the phase flow is volume preserving. $\qquad\qquad\square$

4. Phase plane analysis

Here we consider Hamiltonian systems with one degree of freedom, so $M = 1$. The phase space is then a plane with vectors $(q, p)^T$. The system is assumed to be autonomous, so that the energy E is conserved. For convenience the mass of the system is taken equal to unity. The Hamiltonian is assumed to have the form

(4.1) $\qquad H(q, p) = \frac{1}{2}p^2 + V(q)$.

For these systems a detailed classification of the possible orbits can be given. The stationary points of (4.1) follow from $\dot{V}(q) = 0$ and $p = 0$, where $\dot{V}$ denotes differentiation of V with respect to q. Let $(q_0, 0)^T$ be a stationary point. To investigate the behaviour of the system in its neighbourhood, we choose q_0 as the origin of the q-axis and approximate $V(q)$ by:

(4.2) $\qquad V(q) \doteq V(0) + \frac{1}{2}\ddot{V}(0)\, q^2$.

Orbits in the vicinity of a stationary point correspond to energies $E \approx V(0)$. Combining equations (4.1) and (4.2) we find that these orbits satisfy

(4.3) $\qquad E = \frac{1}{2}p^2 + V(0) + \frac{1}{2}\ddot{V}(0)\, q^2$.

For convenience we write this in the form

(4.4) $\qquad p^2 + \ddot{V}(0)\, q^2 = 2\Big(E - V(0)\Big)$.

The possible orbits can be classified according to the signs of the parameters $\ddot{V}(0)$ and $E - V(0)$:

- $\ddot{V}(0) > 0$.

  - $E < V(0)$.   No orbits possible.
  - $E = V(0)$.   The stationary point itself.
  - $E > V(0)$.   Ellipse.

- $\ddot{V}(0) < 0$.

  - $E < V(0)$.   Hyperbola with the p-axis as principal axis.
  - $E = V(0)$.   Degenerate hyperbola: two straight lines through the stationary point.
  - $E > V(0)$.   Hyperbola with the q-axis as principal axis.

In view of this, stationary points with $\ddot{V}(0) > 0$ are said to be *elliptic*. For $\ddot{V}(0) < 0$ these points are called *hyperbolic*. In the last case we have a saddle point. The two branches of the degenerate hyperbola found for $E = V(0)$

coincide with the eigenvectors of the linearised equations of motion (see also Exercise 3). Orbits approaching the stationary point for $t \to \infty$ or $t \to -\infty$ will be tangent to these eigenvectors.

Let us consider the possible orbits corresponding to given energy E, which is not necessarily close to the energy of a stationary point. Because $p = \dot{q}$ for the system under consideration, we have

(4.5) $\qquad \frac{1}{2}\dot{q}^2 + V(q) = E$.

Because (4.5) is invariant for reflection with respect to the q-axis, the orbits are symmetric around this axis. The part of the orbit with $\dot{q} \geq 0$ satisfies

(4.6) $\qquad \dot{q} = \sqrt{2\Big(E - V(q)\Big)}$.

Separation of the variables q and t then yields

(4.7) $\qquad \displaystyle\int_{q(t_0)}^{q(t)} \frac{1}{\sqrt{E - V(q)}}\, dq = \sqrt{2}\,t$.

When the orbit approaches a point with $V(q) = E$, the integral converges only if $\dot{V}(q) \neq 0$ in such points. *Bounded regular orbits* are periodic. From (4.7) we find that the period T is given by

(4.8) $\qquad T(E) = \sqrt{2} \displaystyle\int_{q_0}^{q_1} \frac{1}{\sqrt{E - V(q)}}\, dq$,

where q_0 and q_1 are the points where the orbit crosses the q-axis and the velocity thus vanishes. *Bounded singular orbits* can be classified as follows:

– The orbit converges for $t \to \pm\infty$ to one and the same hyperbolic stationary point. Such an orbit is called *homoclinic*.

– The orbit converges for $t \to \pm\infty$ to different hyperbolic stationary points. Such an orbit is called *heteroclinic*.

We note that a singular orbit can never reach a stationary point in finite time. This would contradict uniqueness, because a stationary point is also an orbit. This is consistent with the observation that the integral in (4.7) diverges in those cases (see also Exercise 5). The ideas above are illustrated in Fig. XI.2 for an arbitrary potential $V(q)$. The energy E_1 gives rise to regular orbits, which are unbounded. Energy E_2 is just equal to two relative maxima of $V(q)$ and may correspond to a homoclinic, a heteroclinic, and an unbounded singular orbit. Energy E_3 corresponds to three regular orbits, two of which are ellipsoidal and one is unbounded.

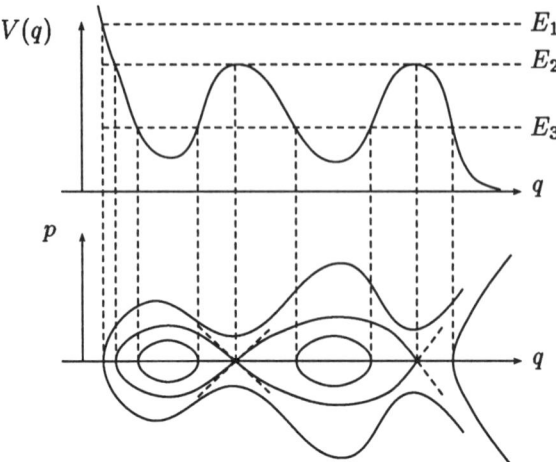

Figure XI.2 Example of a possible potential V (above) and the corresponding orbits in phase space (below).

Example XI.5.

We consider a mathematical pendulum as shown in Fig. XI.3. We note that the bead-on-a-hoop system in Examples XI.3 and XI.4 reduces to such a pendulum if the hoop is at rest.

The mass m of the pendulum moves in a uniform gravitational field. The zero level of the corresponding potential $V(\varphi)$ can be chosen arbitrarily. Our present choice is $V(\varphi) = 0$ if $\varphi = \pi/2$. If frictional forces are ignored, the Lagrangian L of the system is

$$L := T - V = \tfrac{1}{2} m R^2 \dot\varphi^2 + mg R \cos \varphi .$$

The momentum p is given by

$$p := \frac{\partial L}{\partial \dot\varphi} = m R^2 \dot\varphi .$$

From (3.3a) we find the Hamiltonian

$$H(\varphi, p) = \frac{p^2}{2m R^2} - mg R \cos \varphi .$$

The system is autonomous. H represents the energy E of the system and this energy is conserved. The Hamiltonian equations (3.5) read in the present case

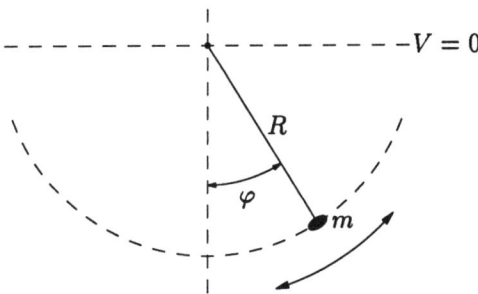

Figure XI.3 Mathematical pendulum of length R and mass m.

$$\begin{cases} \dot{\varphi} & = & \dfrac{1}{m\,R^2}\,p\,, \\[2mm] \dot{p} & = & -mg\,R\,\sin\varphi\,. \end{cases}$$

These two first order equations are equivalent to the second order pendulum equation in Example XI.4. Apart from some constants the Hamiltonian H has the form as assumed in (4.1) with the potential given by

$$V(\varphi) = -mg\,R\,\cos\varphi\,.$$

Stationary points are thus $\varphi = 0$, corresponding to the pendulum in the downward vertical position, and $\varphi = \pm\pi$, where the pendulum is in the upward vertical position. Because $\ddot{V}(0) > 0$ and $\ddot{V}(\pm\pi) < 0$, the former position is elliptic and thus stable, whereas the latter is hyperbolic and unstable. For $-mg\,R < E < mg\,R$ the pendulum oscillates around its lowest position. For $E > mg\,R$ the pendulum has so much energy that it rotates.

Singular orbits are found for $E = \pm mg\,R$. For $E = -mg\,R$ the system is at rest. For $E = mg\,R$ two situations may occur. Firstly, if the initial values are $p_0 = 0$ and $\varphi_0 = \pm\pi$, the pendulum starts and remains at rest in the highest position. Secondly, if $p_0 \neq 0$, the pendulum will eventually converge to the upward vertical position if $t \to \pm\infty$. The corresponding orbit is thus *homoclinic*.

The oscillation time T of the solutions for $-mg\,R < E < mg\,R$ can be found from (4.8). The maximum amplitude φ_0 follows from

$$H(\varphi_0, 0) = -mg\,R\,\cos\,\varphi_0 = E \ .$$

Application of (4.8) yields

$$T = \sqrt{\frac{2R}{g}} \int\limits_{-\varphi_0}^{\varphi_0} \frac{1}{\sqrt{\cos\,\varphi - \cos\,\varphi_0}} \, d\varphi = 4K(\varphi_0) \sqrt{\frac{R}{g}} \ ,$$

where K is the elliptic integral defined by

$$K(\varphi_0) := \int\limits_0^1 \frac{1}{\sqrt{1 - s^2}} \, \frac{1}{\sqrt{1 - s^2 \sin^2 \frac{1}{2} \varphi_0}} \, ds \ .$$

Here the substitution $\sin \frac{1}{2} \varphi := s \sin \frac{1}{2} \varphi_0$ is used. For $\varphi_0 \ll 1$ we have $K(\varphi_0) \approx \dfrac{\pi}{2}$. This leads to the the well-known approximation

$$T \approx 2\pi \sqrt{\frac{R}{g}} \ .$$

5. Continuous media

In systems consisting of discrete particles one is interested in the positions and velocities of the particles as functions of time. This information follows in principle from Newton's second law. In continuous media the question is to calculate the motion of a continuum – e.g. the motion of a vibrating string – as a function of time and position. So, both time and space coordinates act as independent variables. This leads to partial differential equations (PDE). We deal with PDE only as far as they can be reduced to ODE. This is often the case if we can use the method of *separation of variables*. In general the PDE can then be written as an initial boundary value problem, with an initial value in time and boundary values in space. We illustrate this below.

Example XI.6.
Consider a string with length l, tension s, and uniform mass density ρ, when at rest. When the string vibrates, the amplitude $u(t, x)$, with x measured along the string at rest, satisfies the equation of motion

$$u_{tt} = c^2 \, u_{xx},$$

with $c^2 = s/\rho$. This so-called *wave equation* is valid under the following assumptions:

(i) Longitudinal waves, i.e. mass density variations along the string, propagate much faster along the string than transverse waves described by $u(t, x)$. Under this condition the tension is uniform along the string. Experiments have confirmed this.

(ii) The amplitude remains so small with respect to l that the variations in length of the string can be neglected.

The wave equation allows for separation of the variables t and x. To show this we assume $u(t, x) = v(t) \, w(x)$. Substitution of this and division by u leads to the equation

$$\frac{1}{c^2} \frac{v_{tt}}{v} = \frac{w_{xx}}{w} \, .$$

Because the right- and left-hand sides depend on different variables, we conclude that they are both equal to the same constant, λ say. The time dependent part v satisfies the IVP

$$v_{tt} = \lambda \, c^2 \, v \, ,$$

with initial values $v(0) = 1$, $v_t(0) = 0$. Because the solutions must remain bounded, we conclude that $\lambda < 0$, say $\lambda = -\mu^2$. The admissible values for μ follow from the BVP

$$\begin{cases} w_{xx} = -\mu^2 \, w \, , \\[2mm] w(0) = w(l) = 0. \end{cases}$$

The solutions are $w = \sin \mu_n x$ with $\mu_n = n\pi/l$, $n = 1, 2, \ldots$. It is easily checked that the general solution can be written as the linear combination

$$u(t, x) = \sum_{n=1}^{\infty} c_n \, \sin(\mu_n x) \, \cos(\mu_n c \, t) \, ,$$

with the coefficients c_n given by the Fourier expansion of the initial profile $u(0, x)$. The theory of *Sturm-Liouville equations*, which deals with this type of problem, is beyond the scope of this book. See, e.g., [63].

A useful strategy to derive the equation of motion of a continuous, flexible system is to discretise it. Then the system is described as a set of rigid bodies and the flexibility properties of the system are expressed in terms of forces between the bodies. The discretisation should be such that the resulting set of equations converges to the original equations if the number of bodies is increased, under the condition that the total mass is preserved. This is why continuous media are sometimes referred to as systems with *infinite degrees of freedom*. The discretisation approach described here is just one of the possibilities to deal with PDE numerically.

Example XI.7.
We apply the discretisation method described above to the wave equation
introduced in Example XI.6. The interval $[0, l]$ is subdivided using $N+2$
equispaced points $x_i = il/(N+1)$, $i = 0, ..., N+1$. The boundary values
are $x_0 = 0$, $x_{N+1} = l$ with vanishing velocities at these end points. The
uniform mass distribution of the string is discretised by assigning a mass
$m_i = \rho l/(N + 2)$ to each point x_i. Under the assumptions mentioned
in Example XI.6 it is reasonable to assume that these masses can move
only in the direction orthogonal to the string at rest. The amplitude of
the i-th mass point is denoted by $u_i(t) = u(t, x_i)$. The first derivative
u_x is replaced by

$$u_x(t, x_i) \doteq \frac{u_{i+1}(t) - u_i(t)}{\Delta l}, \qquad i = 0, ..., N$$

with mesh width $\Delta l = l/(N+1)$. Hence we use for the second derivative
the discretisation

$$u_{xx}(t, x_i) \doteq \frac{u_{i+1}(t) - 2u_i(t) + u_{i-1}(t)}{(\Delta l)^2}.$$

We then obtain the system of equations

$$m_i \ddot{u}_i = \gamma(u_{i+1} - 2u_i + u_{i-1}), \qquad i = 1, ..., N,$$

with $\gamma = sl/(\Delta l^2(N + 2))$. The force $F_{i+1,i}$ exerted from mass m_{i+1} on
mass m_i is given by

$$F_{i+1,i} = \gamma(u_{i+1} - u_i).$$

From this we see that the equation of motion can be written as

$$m_i \ddot{u}_i = F_{i+1,i} + F_{i-1,i} =: F_i.$$

The forces tend to give neighbouring masses the same amplitude. In
the Lagrangian approach the u_i and $\dot{u}_i$, $i = 1, ..., N$, act as generalised
coordinates and velocities. The kinetic energy T is given by

$$T = \sum_{i=1}^{N} \frac{1}{2} m_i \dot{u}_i^2.$$

The total force F_i can be derived from a potential. If we define

$$V(u_0, ..., u_{N+1}) := \gamma \sum_{i=0}^{N} (u_{i+1} - u_i)^2,$$

then

$$F_i = -\frac{\partial V}{\partial u_i} , \qquad i = 1, ..., N .$$

The Hamiltonian H is given by

$$H = \sum_{i=0}^{N} \left[\frac{p_i^2}{2m_i} + \gamma(u_{i+1} - u_i)^2 \right] ,$$

where the generalized momenta are given by $p_i := m_i \dot{u}_i$. Note that $p_0 = p_{N+1} = 0$. The system is autonomous and the Hamiltonian H is conserved. The possible orbits of the discretised system in phase space lie on contour lines of H. Because $H \to +\infty$ if any of the u_i or p_i goes to $\pm\infty$, we conclude that all orbits corresponding to a finite energy are bounded. The origin of the phase space is a stationary point. It corresponds to the rest position of the string. Because H vanishes at the origin and is positive elsewhere, we conclude that the rest position is a stable, elliptic stationary point. Orbits starting in its vicinity will oscillate around it, in agreement with the observed behaviour of a vibrating string.

Exercises Chapter XI

1. The transition from the Lagrangian to the Hamiltonian formalism gives rise to a transformation from the variables $(t, q_1, ..., q_M, \dot{q}_1, ..., \dot{q}_M)$ to the variables $(t, q_1, ..., q_M, p_1, ..., p_M)$. This transformation is possible only if $\det(\mathbf{J}) \neq 0$, with $\mathbf{J}$ the corresponding $(2M+1) \times (2M+1)$ Jacobian matrix. Check that

$$\det(\mathbf{J}) = \det(\mathbf{J}^*) ,$$

with the elements of the reduced $M \times M$ matrix $\mathbf{J}^*$ given by

$$\mathbf{J}_{ij}^* = \frac{\partial^2 L}{\partial \dot{q}_i \partial \dot{q}_j} .$$

2. Both the Euler-Lagrange equations (2.13) and the Hamiltonian equations (3.5) yield the equations of motion of systems, for which the forces can be deduced from a potential. Derive equations (3.5) from (2.13) and vice versa using definitions (3.3a,b) for the Hamiltonian and the momentum.

3. Derive the Hamiltonian equations for a system with one degree of freedom and with H given by (4.1). Let q_0 be a stationary point of the solution. Linearise the resulting equations of motion around this stationary point,

and find the eigenvalues and eigenvectors of the Jacobian matrix in terms of $\ddot{V}(q_0)$. Show that for a hyperbolic point the orbits with energy $E = V(q_0)$ are tangent to the eigenvectors of the linearised problem while approaching the stationary point.

4. Let q_0 be an elliptic stationary point of a system with one degree of freedom with a Hamiltonian given by (4.1). Show that for the oscillation time $T(E)$ of a periodic orbit around q_0 we have

$$T(E) \to \frac{2\pi}{\sqrt{\ddot{V}(q_0)}}, \qquad \text{if } E \downarrow V(q_0) .$$

Hint: Use expression (4.8) and expand $V(q)$ around q_0 in a Taylor series. If $E \downarrow V(q_0)$ the integrand becomes singular, but the integral can be calculated using complex function theory. The integration contour has to be deformed to the upper half part of a circle around q_0 with radius $E - V(q_0)$.

5. Use (4.7) to show that an orbit converging to a hyperbolic stationary point q_0 of a Hamiltonian system of the form (4.1) cannot reach it in finite time. Hint: Expand in (4.7) the potential $V(q)$ in a Taylor series around the stationary point q_0 and use the facts that $\dot{V}(q_0) = 0$ and $\ddot{V}(q_0) < 0$ at that point. As was done in Exercise 4, the limit $E \downarrow V(q_0)$ has to be considered.

6. Consider a particle with mass m, which moves under the influence of a conservative force with potential $V(\mathbf{x})$. Assume that V only depends on the distance to the origin, so $V(\mathbf{x}) = V(x)$ with $x = \|\mathbf{x}\|$. The corresponding forces are then radially directed. One can think of the sun-earth system with the sun assumed to be in a fixed position.

   a) Show that the particle moves in a plane. Hint: First show that the *angular momentum*

   $$\mathbf{L} := m\dot{\mathbf{x}} \times \mathbf{x}$$

   is a conserved quantity. Here $\times$ denotes the outer product.
   Hint: Use the facts that the outer product of two vectors $\mathbf{a}$ and $\mathbf{b}$ is orthogonal to both $\mathbf{a}$ and $\mathbf{b}$, and that $d/dt(\mathbf{a} \times \mathbf{b}) = \dot{\mathbf{a}} \times \mathbf{b} + \mathbf{a} \times \dot{\mathbf{b}}$.

   b) The plane of motion is determined by the initial speed of the particle. Introduce polar coordinates in this plane. Derive the Euler-Lagrange and Hamiltonian equations for this system. Find which momentum is conserved.

   c) Take $V = -k/x$ with k constant. This is the gravitational potential as in the sun-earth system. Determine the stationary points and their stability properties. Analyse the possible regular and singular orbits. Determine globally the topology of the contour lines of the Hamiltonian and find for which energy the orbits are bounded and periodic.

7. Derive the equations of motion of the bead-on-a-hoop system in Examples XI.3 and XI.4, introducing a potential for the gravitational force as done in Example XI.5, and using the Hamiltonian formalism in §3. Are the Hamiltonian and the momentum conserved quantities?

XII

Mathematical Modelling

In this chapter we first consider the basic concepts of mathematical modelling in §1. In reducing a real-life problem to a manageable mathematical problem it is important to know which parameters are relevant. The resulting dimension analysis and Buckingham's theorem are treated in §2. The subsequent sections each contain an interesting application. We consider vehicle control through steering of rear wheels in §3. Next a mechanical problem, viz. inverse resonance, is dealt with in §4. Water waves and in particular solitary solutions of the Korteweg-de Vries equation are considered in §5. The next two sections are devoted to epidemics models: diffusive effects are neglected in §6 and taken into account in §7. In §8 we consider nerve impulse propagation and solitary wave solutions of the Fitzhugh-Nagumo equation. The next case study is the torsion in a crank shaft, which is carried out in §9. Finally, chaos is shown to play an important rôle in studying the dripping of a faucet, cf. §10.

1. Introduction

When mathematics is applied to real-life problems, a 'translation' is needed to put the subject into a mathematically tractable form. This process is usually referred to as mathematical modelling, and a possible definition reads:
mathematical modelling is the description of an experimentally verifiable phenomenon by means of the mathematical language.
See also [12, 18]. The phenomenon to be described will be called *the system*, and the mathematics used, together with its interpretation in the context of the system, will be called *the mathematical model*. In general mathematical models contain two classes of quantities:

(i) Variables
Within this class we distinguish between dependent and independent variables. For example, one may be interested in the temperature (dependent variable) of a system as a function of time and position (independent variables). Or, one

may look for the position (dependent variable) of an object as a function of time (independent variable). If the dependent variables are differentiable functions of the independent ones, the model might comprise a set of differential equations. The number of independent variables then determines whether one is concerned with ordinary differential equations (one independent variable) or with partial differential equations (more than one independent variable).

(ii) **Parameters**
In this class we distinguish between parameters which are in fact constant and parameters which can be adjusted by the experimentalist. The acceleration of gravity and the decay time of radioactive material are examples of the first kind. The temperature of chemical reactions is an example of the second kind: the experimentalist can often control the reaction by means of the environmental temperature, because in many systems the reaction rates strongly depend on this parameter.

A mathematical model is said to be *solved* if the dependent variables are in principle known as functions of the independent variables and the parameters. The *solution* may be obtained either analytically or numerically.

Mathematical modelling generally consists of the following steps:

1: *Orientation*
2: *Formulation of relations between variables and parameters*
3: *Non-dimensionalisation*
4: *Reduction*
5: *Asymptotic analysis*
6: *Numerical analysis*
7: *Verification*
8: *Reiteration of Steps 2–7*
9: *Implementation.*

We discuss these stages separately:

Step 1: The modelling process always starts with an orientation stage, in which the modeller gets acquainted with the system under consideration through observations and by information from experts and the literature. Large scale experiments are not relevant in this stage. In fact, most experiments are premature if they are not based on some model. Testing of models, developed later on, frequently leads to the conclusion that some essential parameter may not have been measured.

Step 2: The next step is to formulate relations between variables and parameters. For certain classes of systems these relations have been established already long ago. We may mention the laws of Newton in classical mechanics, the Maxwell equations in electromagnetism, and the Navier-Stokes equations in fluid mechanics. In many other cases rules of thumb are in use, which do not have the same status as the well-accepted fundamental laws. These rules may certainly be reliable and useful, yet always in a restricted context. It may also happen that the modeller has to start from scratch. An important condition is that of *consistency*: the proposed relations should not contain contradictions. For differential equations one has, e.g., to specify not too few but also not too many boundary and/or initial conditions in order to ensure that the model has a unique solution.

Step 3: The relevant quantities often have physical dimensions like cm, sec, kg etc. The specific measuring system should, of course, not be relevant to the actual behaviour of these quantities. The consequences of this principle are extensively dealt with in §2.

Step 4: Most models tend to be too complicated initially. Therefore, it is often necessary to look for reduced models, which are still reliable under certain restrictions. Such simplified models may be useful in many ways. For example, a simple model can definitely show that it is possible to bring a rocket into an orbit around the moon, whereas the complexity of the calculations, needed to evaluate the full model, may be extremely high. The technique of reducing a model requires much experience, because one easily leaves out seemingly unimportant, yet essential parts. In mechanical models, e.g., friction is often neglected, but in modelling the clutch or the brakes of a car it might be the heart of the matter.

Step 5: The models describing real-life problems are usually too complicated to be solved analytically. In the asymptotic analysis of a model one tries to find analytical solutions, which are approximately valid for certain ranges of the parameter values only. The results may serve as a first check on the reliability of the model.

Step 6: In this step a quantitative solution of the problem is calculated by numerical methods. Since quite often a host of methods are available, the analysis of Step 5 is essential for determining which method has to be used, and – in the case of iterative methods – how to obtain a starting value.

Step 7: If the (reduced) model has been (approximately) solved, it remains to explore the solution as a function of the independent variables and the adjustable parameters. Not all possibilities are of practical use. In most cases only certain tendencies or special cases are relevant. These features have to be checked against the data. At this stage the modeller may propose to perform specific experiments. A model that is too complex to be verified experimentally is useless. Also, a model containing too many parameters is not interesting from a practical point of view.

Step 8: After having compared the measured data and the calculated solution, one has often gained enough insight to improve the model and to reiterate Steps 2–7. In view of this iteration process, it is convenient to follow a *modular* approach from the very beginning, i.e. one should rather start with the most simple, but still relevant, model, gradually adding more and more aspects. It seldom happens that a model describes all dependent variables equally well. Of course, the specification of the 'value' of a model is a subtle matter, which may be a matter of taste, too. In practice, it is for the user and not the designer of a model to determine when the iteration process is to be stopped.

Step 9: If a model suffices for a specific application, the implementation phase starts. In general the results are used by non-mathematicians, who need instruction about the power and the poverty of the model. The appropriate presentation of the results is an important part of each project, and the attention and time it requires should not be underestimated.

We close this section with some general remarks:

(i) One and the same system may be described satisfactorily by different mathematical models. These models may complement each other, as is the case, e.g., with the particle and the wave models for light and elementary particles. Identification of a system with one particular model is a serious misunderstanding of the character of 'modelling'. Another misinterpretation, often met in popular scientific publications, is the statement that a phenomenon is 'understood' if a satisfactory model exists. Mathematical models relate to 'how' and not to 'why' things happen.

(ii) Not all systems can suitably be modelled by mathematics. Notable examples appear in physics and chemistry. This is increasingly less true for the systems studied in biology, economics, and political science, in this order. Many, if not most, real-life systems are governed by laws which can hardly be put into mathematical form.

(iii) One and the same model may apply to totally different systems. This has actually proved to be one of the powers of the mathematical approach. In particular this may become clear after non-dimensionalisation.

(iv) Mathematical modelling is merely descriptive and does not necessarily have direct, practical implications in society. However, nearly all successful models will, sooner or later, be used in technical applications for the design and control of practical systems.

2. Dimensional analysis

The variables and parameters in a mathematical model have physical dimensions in general. Most dimensions are obvious, like time, length, mass, temperature etc., while others can be deduced from the rule that all terms in a particular equation must have the same dimensions. This rule stems from the condition that no equation may depend on the units used. The dimensionalities of constants of proportionality directly follow from these considerations. For instance, if a frictional force is introduced with a strength proportional to the velocity of the object under consideration, then the constant of proportionality will have the dimensions of the quotient of force and velocity.

The technique of non-dimensionalising is an extremely powerful tool in mathematical modelling. See, e.g., [48]. Its importance is only fully appreciated through examples, which account for the largest part of this chapter. We summarise some striking advantages:

(i) The number of variables and parameters decreases.

(ii) Dimensional analysis may yield insight into the general scaling properties of the system. In the first example below this point is illustrated by deriving Pythagoras' theorem from dimensional analysis.

(iii) Mathematical models for describing seemingly different systems may appear to be identical if they are put into dimensionless form.

(iv) The reduction of a model (Step 4 in the modelling scheme of §1) is often accomplished by neglecting those terms in the model equations that are much smaller than the other terms. It should be realised, however, that comparing magnitudes only makes sense if the model is in dimensionless form.

(v) Quite often it is attractive to perform experiments on systems that have been scaled down in size. Only non-dimensionalisation of the model can really show whether the results of such experiments are still meaningful with respect to the original system. We illustrate this point in Example XII.3 below.

Non-dimensionalising a model first implies that a (nonlinear) transformation is applied to the set of variables and parameters, which yields dimensionless combinations of them. Next, the model is made dimensionless by rewriting all equations in terms of these dimensionless quantities. By no means is it clear in advance that this will always be possible, but the existence of such a transformation is proved by the theorem given below. This theorem also shows how such transformations can be found. Note that this theorem deals with scalar variables and parameters only. The components of vector-valued variables and parameters should thus be treated separately.

Theorem 2.1 (Buckingham). *Consider a system with (scalar) variables* $x_1, \ldots, x_k$, *and (scalar) parameters* $p_1, \ldots, p_l$. *The physical dimensions occurring in the system are denoted by* $d_1, \ldots, d_m$. *Then each relation*

$$f(x_1, \ldots, x_k, p_1, \ldots, p_l) = 0$$

can be rewritten in the equivalent, dimensionless form

$$\bar{f}(q_1, \ldots, q_n) = 0 ,$$

with $q_1, \ldots, q_n$ *dimensionless products and quotients of the* $x_i, i = 1, .., k$, *and* $p_j, j = 1, .., l$. *The number* n *is given by*

$$n = k + l - m .$$

Proof: Write $M := k + l$. Let us introduce the set V with elements v of the form

$(*)$ $v = x_1^{r_1} \ldots x_k^{r_k} \, p_1^{r_{k+1}} \ldots p_l^{r_M} ,$

with $r_i \in \mathbb{R}$, $i = 1, \ldots, M$. There is an obvious one-to-one correspondence between the elements of V and the vectors $(r_1, \ldots, r_M)^T \in \mathbb{R}^M$. The corresponding map is denoted by $T_1 : \mathbb{R}^M \to V$. If the x_i and p_j on the right-hand side of $(*)$ are replaced by their associated dimensions, a set W is obtained with elements w of the form

$$w = d_1^{s_1} \ldots d_m^{s_m} ,$$

with $s_i \in \mathbb{R}$, $i = 1, \ldots, m$. This replacement procedure induces a map $T_2 : V \to W$. This mapping is surjective, because each dimension occurs in the system. There is an obvious one-to-one correspondence between the elements of W and the vectors $(s_1, \ldots, s_m)^T \in \mathbb{R}^m$. This map is denoted by $T_3 : W \to \mathbb{R}^m$. The composite map

$$T = T_3 T_2 T_1 , \qquad T : \mathbb{R}^M \to \mathbb{R}^m,$$

is linear and surjective. Its null space $N_0 \subset \mathbb{R}^M$ has dimension $n = M - m$ $= k + l - m$. The elements of N_0 precisely correspond to the dimensionless elements of W. We choose a basis $q_1, \ldots, q_n$ in N_0 and extend it to a basis in $\mathbb{R}^M$ by adding linearly independent elements $q_{n+1}, \ldots, q_M$. Each x and p can be written as a unique, linear combinations of these basis elements. This implies that every relation $f(x_1, \ldots, x_k, p_1, \ldots, p_l) = 0$ can be rewritten in the form $\bar{f}(q_1, \ldots, q_M) = 0$ in a unique way. However, the function $\bar{f}$ has to be independent of the units used to measure the dimensions. This can only be the case if $\bar{f}$ does not contain any of the basis elements $q_{n+1}, \ldots, q_M$, because their values may attain any value if the units are varied. Thus $\bar{f} = \bar{f}(q_1, \ldots, q_n)$. □

The following points should be realised if this theorem is applied in practice:

(i) The choice of the q_j is not unique in most systems. Different choices may lead to quite different dimensionless models and the analysis of one model (cf. Steps 5 and 6 of the modelling process) may be much more convenient than some other.

(ii) A relatively simple version of non-dimensionalisation is often already quite effective. A typical example is the transformation that makes each variable dimensionless by dividing it by a convenient combination of parameters, e.g. where all variables with the dimension 'length' are divided by a characteristic length of the system, all variables with the dimension 'time' are divided by a characteristic time of the system, etc. Non-dimensionalisation is then nothing but a scaling of the variables by the parameters, and the number of q_j is equal to the number of variables. We illustrate this in Example XII.4.

(iii) Dimensional analysis yields more insight if it is possible to find a set of q_j such that at least one q_j contains more than one variable. The corresponding transformation is sometimes called a *similarity transformation*. Such a transformation is applied in Example XII.4.

Example XII.1.
The theorem of Pythagoras can be derived by means of arguments from dimensional analysis, cf. [10]. We consider a rectangular triangle ABC as drawn in Fig. XII.1. Such a triangle is uniquely determined by two of its sides, say a and c. This system has no variables and two parameters, both with the dimension 'length'. So, there is only one dimensionless quantity, and an obvious choice is to take

$$q := a/c \ .$$

The area O of the triangle is a function of a and c and we may write

$$O = f(a,c) = c^2(f(a,c)/c^2) \ .$$

Because both O and c^2 have the dimensions of area, we conclude that the quotient f/c^2 is dimensionless and can only depend on q. So, we write

$$\bar{f}(q) := f(a,c)/c^2 \quad \Rightarrow \quad O = c^2 \, \bar{f}(q) \ .$$

This argument holds for all triangles congruent to triangle ABC. For all these triangles the same function $\bar{f}$ and the same value of q must describe the surface area. So, for these triangles the factor $\bar{f}(q)$ is a constant, say $\bar{f}_0$. As seen from the figure, triangles DBA and DAC are congruent to the original triangle ABC. We may thus write

$$O_{ABC} = c^2\,\bar{f}_0 = O_{DBA} + O_{DAC} = a^2\,\bar{f}_0 + b^2\,\bar{f}_0\;,$$

from which the well-known relation $a^2 + b^2 = c^2$ follows directly.

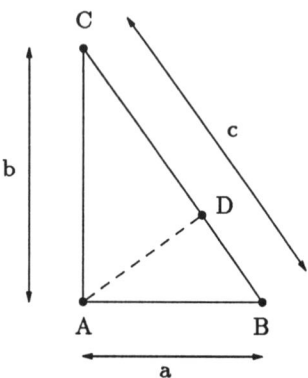

Figure XII.1

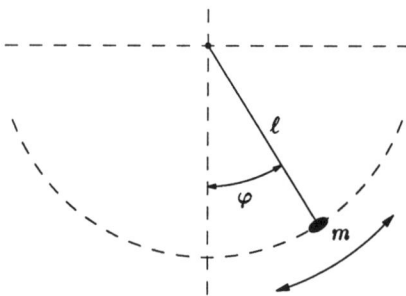

Figure XII.2

Example XII.2.
Here, we study the consequences of dimensional analysis if applied to a mathematical pendulum as presented in Fig. XII.2. We restrict the motion of the pendulum to a vertical plane. In practice the swinging

behaviour of the pendulum will damp out because of friction. We assume
the strength of the friction force to be proportional to the velocity
squared.

What are the variables, parameters, and dimensions of this system?
The independent variable is the time t; its dimension is denoted by $[t]$.
The dependent variable is the dimensionless angle φ, indicated in the
figure. The parameters are the length l of the pendulum with dimension
$[l]$, its mass m with dimension $[m]$, the acceleration of gravity g with
dimension $[l]/[t]^2$, the dimensionless initial angle φ_0 ($\neq 0$), and the
friction coefficient α. The dimension of α is given by the quotient of
force and velocity squared, thus $[m]/[l]$. In the system, modelled this
way, 7 variables and parameters and 3 dimensions are involved. We
have thus 4 dimensionless quantities q_1, q_2, q_3, and q_4. Because φ and
φ_0 are already dimensionless, obvious choices are $q_1 := \varphi$, and $q_2 := \varphi_0$.
To find q_3 and q_4 we form the products

$$q = t^{r_1} \, l^{r_2} \, m^{r_3} \, \alpha^{r_4} \, g^{r_5} \ .$$

The condition $[q] = 0$ leads to three linear equations for $r_1, \ldots, r_5$:

$$\begin{cases} r_1 - 2r_5 = 0 \\ r_2 - r_4 + r_5 = 0 \\ r_3 + r_4 = 0 \ . \end{cases}$$

We can choose two r_i independently. The natural choices $(r_1, r_2) = (1, 0)$
and $(r_1, r_2) = (0, 1)$ yield $q_3 := t\sqrt{\alpha g/m}$ and $q_4 := l\alpha/m$ respectively.
Each property of this system can thus be expressed by an equation of
the form

$$f(q_1, \ldots, q_4) = f(\varphi_1, \varphi_0, \, t\sqrt{\alpha g/m}, l\alpha/m) = 0 \ .$$

Our first conclusion is that all pendulums of given length and given ratio
α/m behave identically, if started at the same φ_0.

Assume we may rewrite this relation as

$$q_3 = \bar{f}(q_1, q_2, q_4) \ ,$$

i.e.

$$(*) \qquad t = \sqrt{\frac{m}{\alpha g}} \ \bar{f}(\varphi_1, \varphi_0, l\alpha/m) \ .$$

We note that this inversion will not be possible for all t. From the physics
of the system we know that φ is a single-valued function of t. However,
this will hold for t as a function of φ only as long as the pendulum has
not changed its direction of motion. Let us direct our attention to the
time $t_{1/2}$, at which the amplitude of the pendulum has been halved for
the first time. Thus, $\varphi(t_{1/2}) := \frac{1}{2}\varphi_0$. From $(*)$ we may write

$$t_{1/2} = \sqrt{m/\alpha g}\ \bar{f}(\varphi_0, l\alpha/m)\ .$$

Our second conclusion from dimensional analysis is that if we vary the values of $l\alpha$ and m, but keep the value of the quotient $l\alpha/m$ fixed, $t_{1/2}$ scales with $\sqrt{m/\alpha}$ or, equivalently, with $\sqrt{l}$.

Let us reduce the model by neglecting the friction. This implies that $r_4 = 0$, and thus also $r_3 = 0$. From this a third conclusion follows: the behaviour of a frictionless pendulum is independent of its mass. For the reduced model we find unambiguously $q_3 = t\,\sqrt{g/l}$, and we may write

$$t = \sqrt{l/g}\ \bar{f}(\varphi, \varphi_0)\ .$$

Dimensional analysis itself does not tell us that φ is a periodic function of time in the frictionless system. If we take this for granted, our fourth conclusion is that the period τ satisfies

$$\tau = \sqrt{l/g}\ \bar{f}(\varphi_0)$$

and thus scales with $\sqrt{l/g}$. It requires the explicit solution of the equation of motion to find that $\bar{f}(\varphi_0)$ is given by an elliptic integral, which reduces to 2π if $|\varphi_0| \ll 1$. For this, see Example XI.5.

Example XII.3.
Let us model a ship sailing at constant speed. Obvious parameters are its length l with dimension $[l]$ and its velocity v with dimension $[l]/[t]$. While the ship moves through the water, energy is transferred from the ship to the water as a result of viscous friction. This energy is used partly to induce surface waves and partly to generate turbulent motion of the water. The viscosity α has dimension $[\alpha] = [m]/([l][t])$. Because of these effects the acceleration of gravity g, with $[g] = [l]/[t]^2$, and the density of water ρ, with $[\rho] = [m]/[l]^3$, will also play a rôle. If we assume that the ship is streamlined such that its height and width are not of importance in the present analysis, the system has 5 variables and parameters. Because 3 dimensions are involved, the number of dimensionless quantities is 2. As above, we form the products

$$q = v^{r_1}\,\rho^{r_2}\,l^{r_3}\,\alpha^{r_4}\,g^{r_5}\ .$$

The condition $[q] = 0$ yields three equations:

$$\begin{cases} r_1 - 3r_2 + r_3 + r_5 = 0 \\ r_1 + r_4 + 2r_5 = 0 \\ r_2 + r_4 = 0\ . \end{cases}$$

The choice $(r_1, r_2) = (1, 0)$ yields $q_1 := v/\sqrt{lg}$. This is called the *Froude number* after William Froude, a famous ship builder. The choice $(r_1, r_2) = (0, 1)$ yields $\bar{q}_2 := \rho l \sqrt{lg}/\alpha$. For historical seasons one prefers to introduce $q_2 := q_1 \bar{q}_2 = v\rho l/\alpha$. The latter dimensionless quantity is called the *Reynolds number*, after Osborne Reynolds, a researcher in fluid mechanics. Because real-life experiments are hard for these systems, it is very attractive to perform experiments on (physical) models in which all sizes are scaled down by a certain factor. The conclusions from these experiments are valid for the original system only if both systems are described by the same dimensionless (mathematical) model. So, q_1 and q_2 have to remain constant upon scaling. In practice the values of g, ρ, and α can hardly be adjusted. To keep q_1 constant, $v/\sqrt{l}$ may not change, and to keep q_2 constant, vl must be preserved. These requirements cannot be satisfied at the same time. So one may be led to conclude that (physical) scaling does not make sense here. However, under certain conditions scaling may still be useful. If the generation of surface waves is unimportant compared to the other mechanisms of energy dissipation, we may ignore the Froude number. In that case one only has to keep the Reynolds number constant, which implies that the velocity of the scaled model must be larger than the velocity of the real system. On the other hand, if the Froude number is much larger than the Reynolds number, the latter may be ignored. Then the speed of the scaled model must be smaller than that of the original ship.

Example XII.4.
We consider heat diffusion in a rod of length l. The rod is thermally isolated everywhere, except at one of the end points. The system acts as a one-dimensional conductor. Initially the rod has uniform temperature τ_0. From a certain moment onwards one end of the rod is brought into contact with a heat reservoir, which keeps that end of the rod at constant temperature τ_1. We are interested in the temperature $\tau(t, x)$ in the rod as a function of time t and position x. This well-known system is described by the heat diffusion equation

$$\frac{\partial \tau}{\partial t} = k \frac{\partial^2 \tau}{\partial x^2} ,$$

with boundary condition

$$\tau(t, 0) = \tau_1 , \qquad t \geq 0 ,$$

and initial condition

$$\tau(0, x) = \tau_0 , \qquad 0 < x \leq l .$$

The independent variables are time t with dimension $[t]$, and position x with dimension $[x]$. The dependent variable is the temperature τ

with dimension $[\tau]$. As origin of the temperature scale we take τ_0. The parameters are the length of the rod l with dimension $[l]$, the temperatures τ_0 and τ_1 with dimension $[\tau]$, and the thermal conductivity k with dimension $[l]^2/[t]$. An obvious way to obtain dimensionless quantities is to scale the variables using characteristic quantities for length, time, and temperature, respectively:

$$\begin{cases} q_1 := x/l \ , \\ q_2 := t/(l^2/k) \ , \\ q_3 := (\tau - \tau_0)/(\tau_1 - \tau_0) \ . \end{cases}$$

For the mathematical model it makes sense to assume the rod to be infinitely long. In practice, this assumption is reasonable only if the time period l^2/k is much larger than the period during which we wish to observe the system.

For $l \to \infty$ dimensional analysis gives rise to a far reaching conclusion. We form the products

$$q = t^{r_1} x^{r_2} \tau^{r_3} k^{r_4} (\tau_1 - \tau_0)^{r_5} \ .$$

From the condition $[q] = 0$ we obtain

$$\begin{cases} r_1 - r_4 = 0 \ , \\ r_2 + 2r_4 = 0 \ , \\ r_3 + r_5 = 0 \ . \end{cases}$$

From these equations we may find

$$\begin{cases} q_1 := x^2/kt \ , \\ q_2 := (\tau - \tau_0)/(\tau_1 - \tau_0) \ . \end{cases}$$

Because the physics of the system is such that the temperature $\tau(t, x)$ will be a unique function of the pair (t, x), we may write

$$q_2 = f(q_1) \ ,$$

or

$$\tau = \tau_0 + (\tau_1 - \tau_0)\, f(x^2/kt) \ .$$

The important conclusion is that in an infinitely long rod the solution $\tau(t, x)$ of the heat equation depends on the quotient x^2/kt only. This implies that this partial differential equation transforms into an ODE if written in dimensionless form. It appears that the function $f(q_1)$ above satisfies the equation

$$4q_1 \frac{\partial^2 f}{\partial q_1^{\,2}} + (q_1 + 2)\,\frac{\partial f}{\partial q_1} = 0\,, \qquad 0 < q_1 < \infty\,,$$

with the boundary conditions

$$f(0) = 0\,, \qquad f(\infty) = 1\,.$$

The solution is the so-called *complementary error function* (cf. [1]):

$$f(q_1) = \text{erfc}\left(\tfrac{1}{2}\sqrt{q_1}\right) = \frac{1}{\sqrt{\pi}} \int\limits_{\tfrac{1}{2}\sqrt{q_1}}^{\infty} e^{-s^2}\,ds\,.$$

3. Rear-wheel steering

Why is it so difficult to drive a long vehicle backward into a gateway? Why is parking a hard job if the parking space is limited? Is rear-wheel steering favourable when driving around a corner? These questions have one theme in common: what is the relation between the rear-wheel and the front-wheel trajectories of a vehicle? To answer this question, we shall work out a model system. Let us first deal with modelling Steps 1 (*Orientation*) and 2 (*Formulation of relations between variables and parameters*).

For long vehicles it seems reasonable to assume that the number of front and rear wheels is of only minor importance, and hence we restrict our model to have only one front and one rear wheel. Both are free to turn. The model vehicle is shown in Fig. XII.3. The front and rear wheels touch the ground at the points A and B respectively, and the distance between these points is l. To simulate rear-wheel steering, we assume that the rear wheel turns over an angle $\alpha\varphi$, where $-1 < \alpha < +1$, if the front wheel turns over an angle φ to the right or the left. For $\alpha > 0$ front and rear wheels both steer to the same side, whereas for $\alpha < 0$ they turn in opposite directions. For $\alpha = 0$ the vehicle strongly resembles a bicycle.

Given the trajectory of the front wheel, the derivation of the equation of motion of the rear wheel requires the application of some differential geometry in the plane. The equation of motion will be obtained from the observation that the velocity of the rear wheel must always be tangential to the trajectory of the rear wheel. We denote the path of the front wheel by $\mathbf{x}(s)$, where s is the arc length. At each point $\mathbf{x}(s)$ we choose a local, orthonormal coordinate system $(\mathbf{t}, \mathbf{n})$, where $\mathbf{t}$ is the tangent and $\mathbf{n}$ the normal vector. Note the difference between time t and tangent $\mathbf{t}$. To fix the orientation we take the direction of $\mathbf{n}$ pointing to the left-hand side of the driver. This implies that the front wheel path has positive curvature if the driver turns to the left. The tangent $\mathbf{t}$ and normal $\mathbf{n}$ are given by

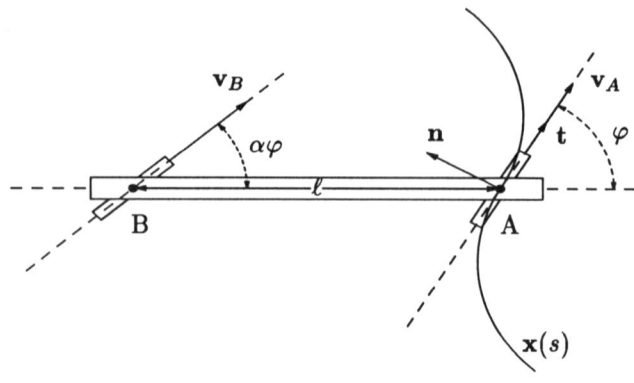

Figure XII.3 A two-wheel vehicle.

$$t = \frac{d}{ds}\, \mathbf{x} \,, \qquad \|\mathbf{t}\| = 1 \,,$$

$$\mathbf{n} = \frac{1}{k}\frac{d}{ds}\, \mathbf{t} \,, \qquad \|\mathbf{n}\| = 1 \,,$$

where k is the (s-dependent) curvature of the path $\mathbf{x}(s)$. We shall provide quantities referring to the wheel at points A and B in Fig. XII.3 with an index A and B respectively. The velocity $\mathbf{v}_A$ of the front wheel satisfies the equation

$$\mathbf{v}_A = \frac{d\mathbf{x}_A}{dt} \equiv \dot{\mathbf{x}}_A = \dot{s}\,\frac{d\mathbf{x}_A}{ds} = v_A\, \mathbf{t} \,,$$

where $v_A := \|\mathbf{v}_A\|$ and $ds/dt \equiv \dot{s} \equiv v_A$. In the theory that follows we shall need the time derivatives $\dot{\mathbf{t}}$ and $\dot{\mathbf{n}}$. Because $\mathbf{t}$ and $\mathbf{n}$ are unit vectors, they satisfy the (2-dimensional) *Serret-Frenet formulae* $d\mathbf{t}/ds = k\mathbf{n}$ and $d\mathbf{n}/ds = -k\mathbf{t}$. We then obtain the expressions

(3.1)
$$\begin{cases} \dot{\mathbf{t}} = \dot{s}\,\dfrac{d\mathbf{t}}{ds} = v_A\, k\, \mathbf{n} \,, \\[2mm] \dot{\mathbf{n}} = \dot{s}\,\dfrac{d\mathbf{n}}{ds} = -v_A\, k\, \mathbf{t} \,. \end{cases}$$

The position $\mathbf{x}_B$ of the rear wheel with respect to the position $\mathbf{x}_A$ of the front wheel is given by

(3.2a) $\mathbf{x}_B = -l(\cos\varphi\, \mathbf{t} + \sin\varphi\, \mathbf{n}) \,,$

in which φ is taken to be positive in the first quadrant of $(\mathbf{t}, \mathbf{n})$. By differentiating with respect to time t we find an equation for the velocity $\mathbf{v}_B$:

(3.2b) $\mathbf{v}_B = \left(v_A + l \sin \varphi(\dot{\varphi} + k\, v_A)\right) \mathbf{t} - l \cos \varphi(\dot{\varphi} + k\, v_A) \mathbf{n}$.

From Fig. XII.3 we see that $\mathbf{t}_B$ is given by

$$\mathbf{t}_B = \cos(1 - \alpha)\, \varphi \mathbf{t} + \sin(1 - \alpha)\, \varphi \mathbf{n} .$$

The equation of motion is obtained from the condition that this velocity is tangential to the path of the rear wheel, i.e. $\mathbf{v}_B$ should be parallel to $\mathbf{t}_B$. We thus arrive at the autonomous equation of motion

(3.3) $\dot{\varphi} = -\dfrac{v_A}{l} \left(\sin \varphi - \tan \alpha \varphi \cos \varphi + lk\right)$.

Let us turn now to modelling Step 3 (*Non-dimensionalisation*). The dimensional quantities in this equation are time t, length l, velocity v_A and curvature k. From the dimensional analysis presented in §2 we may derive the following dimensionless parameters:

$$\begin{cases} t^* := v_A\, t/l \ , \\ k^* := lk \ . \end{cases}$$

Note that $t^* \equiv s/l$ if the velocity v_A is constant. In that case t^* stands for arc length measured in units of length l. In the following we shall restrict ourselves to this case for convenience. This implies that the model becomes independent of v_A. Omitting the $*$ superscript the dimensionless equation of motion reads:

(3.4) $\begin{cases} \dot{\varphi} = -(\sin \varphi - \tan \alpha \varphi \cos \varphi + k) \ , \\ \varphi(0) = \varphi_0 \ . \end{cases}$

To get a feeling for the behaviour of its solution we shall analyse (3.4) for some special cases. So, we now enter modelling Steps 4 (*Reduction*) and 5 (*Asymptotic analysis*).

(i) Straight line

We first consider the case that the front wheel follows a straight line, which implies $k = 0$. At $t = 0$ the vehicle forms an angle φ_0 with this line.

For $\alpha = 0$ (3.4) reduces to

$$\dot{\varphi} = -\sin \varphi \ ,$$

with stationary points $\varphi = 0, \pi$. From the Jacobian 'matrix' $J(\varphi) := -\cos \varphi$ we directly find that $\varphi = 0$ is a stable point and $\varphi = \pi$ an unstable point. Driving backwards is an unstable activity! The solution $\varphi(t)$ is readily obtained by separation of variables:

$$\int_{\varphi_0}^{\varphi(t)} \frac{d\psi}{\sin \psi} = -\int_0^t d\tau = -t \ .$$

Thus (see [34])

$$\varphi(t) = 2 \tan^{-1}\left(e^{-t} \tan(\tfrac{1}{2} \varphi_0)\right) .$$

For $\alpha \neq 0$ we have

(3.5) $\dot\varphi = -\sin \varphi + \tan \alpha\varphi \, \cos \varphi .$

The stationary points satisfy the equation

$$\tan \varphi = \tan \alpha\varphi ,$$

i.e.

$$\sin(1 - \alpha)\, \varphi = 0 .$$

If $\alpha = 1$ each φ is stationary, but for $\alpha \neq 0$ the situation is quite complex. In that case the points $\varphi_i = i\,\pi/(1 - \alpha)\,\mathrm{mod}(2\pi)$, $i = 1, 2, \ldots$, are stationary. If α is rational, the number of φ_i is finite, but for α irrational this number is infinite.

These results show that we must *interpret* our model anew. It appears that the model – as it stands – admits quite peculiar solutions due to the fact that the front wheel may turn around without limit. We therefore add the restriction

$$|\varphi| \leq \varphi_m$$

with $\varphi_m < \pi$ a given maximum angle. Under this restriction the only stationary point is given by $\varphi = 0$. Clearly we now have for some initial conditions φ_0 with $|\varphi_0| < \varphi_m$ that the solution $\varphi(t)$ will tend to zero if $t \to \infty$, while for other values of φ_0 the solution $\varphi(t)$ will reach the upper bound φ_m and the vehicle will stop abruptly.

(ii) Circle

In this case the path of the front wheel is a circle with (constant) curvature k. If the radius of the circle is denoted by R, we have $k = l/R$.

For $\alpha = 0$ (3.4) reduces to

(3.6) $\dot\varphi = -(\sin \varphi + k) .$

No stationary point exists if $k > 1$, i.e. $l > R$. For $k < 1$ the system has two stationary points φ_1 and φ_2 where $-\pi/2 \leq \varphi_1 \leq 0$ and $\varphi_2 = 2\pi - \varphi_1$. If $|\varphi_1| < \varphi_m < |\varphi_2|$, then φ_1 is a global attractor. If $\varphi_m < |\varphi_1|$ the vehicle will get stuck at a certain moment. The cases $k = 1$ and $\varphi_1 = \varphi_2 = -\pi/2$ deserve special attention. Clearly, in this case we assume that $\varphi_m \geq \pi/2$. In the stationary situation the rear wheel remains in the centre of the circle while the front wheel turns around. The corresponding picture of the phase space is given in Fig. XII.4. Note that $\varphi = -\pi/2$ is a global attractor only if $\varphi_m = \infty$. It is not a stable point in the sense of Lyapunov (see §V.2), because an orbit starting at $\varphi_0 = -\pi/2 - \varepsilon$ $(\varepsilon > 0)$ will always leave the point $\varphi = -\pi/2$ regardless of the value of ε. The vehicle completely turns around its own front wheel and then approaches the stationary situation $\varphi = -\pi/2$.

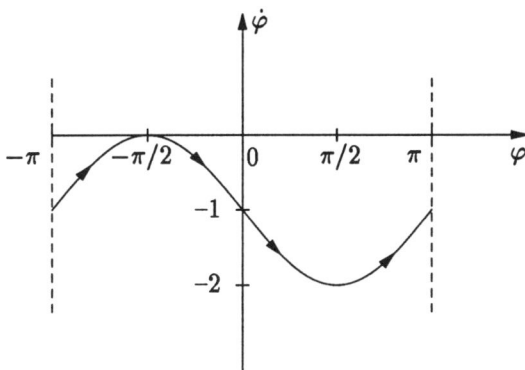

Figure XII.4   For $\alpha = 0$, $k = 1$, the vehicle follows the curve drawn in phase space.

(iii) Turning

We now consider the effect of rear-wheel steering while the vehicle is turning. For the path of the front wheel we take parts of a straight line, a circle and again a straight line respectively. This path is drawn in Fig. XII.5. At $t = 0$ we take $\varphi_0 = 0$ and the vehicle starts turning. The radius of the (circular) turn is R, measured in units of l. As mentioned earlier, the dimensionless parameter t is equal to the arc length along the front wheel path if the velocity is constant. This means that the front wheel has finished turning at $t = \pi R/2$. During turning the position $\mathbf{x}_A(t)$ of the front wheel is given explicitly by

$$\mathbf{x}_A(t) = \begin{cases} R \begin{bmatrix} \sin(t/R) \\ -\cos(t/R) \end{bmatrix}, & 0 \le t < \pi R/2 \\[2ex] \begin{bmatrix} R \\ t - (\pi R/2) \end{bmatrix}, & t \ge \pi R/2 . \end{cases}$$

For $0 \le t < \pi R/2$ the basis vectors $(\mathbf{t}, \mathbf{n})$ are represented by

$$\begin{cases} \mathbf{t}(t) = \begin{bmatrix} \cos(t/R) \\ \sin(t/R) \end{bmatrix} , \\[2mm] \mathbf{n}(t) = \begin{bmatrix} -\sin(t/R) \\ \cos(t/R) \end{bmatrix} , \end{cases}$$

and for $t \geq \pi R/2$ by

$$\begin{cases} \mathbf{t}(t) = \begin{bmatrix} 0 \\ 1 \end{bmatrix} , \\[2mm] \mathbf{n}(t) = \begin{bmatrix} -1 \\ 0 \end{bmatrix} . \end{cases}$$

During turning the angle $\varphi(t)$ is given by

(3.7a) $$\varphi(t) = \int_0^t f\big(\varphi(\tau)\big)\, d\tau ,$$

where

(3.7b) $$f(\varphi) = \begin{cases} -(\sin\varphi - \tan\alpha\varphi\ \cos\varphi + 1/R) , & 0 \leq t < \pi R/2 , \\[2mm] -(\sin\varphi - \tan\alpha\varphi\ \cos\varphi) , & t \geq \pi R/2 . \end{cases}$$

The position $\mathbf{x}_B(t)$ of the rear wheel is in terms of $\varphi(t)$ given by

$$\mathbf{x}_B(t) = \mathbf{x}_A(t) - \Big(\cos\varphi(t)\ \mathbf{t}(t) + \sin\varphi(t)\ \mathbf{n}(t)\Big) .$$

Fig. XII.5 illustrates the trajectory $\mathbf{x}_B(t)$, $-1 \leq t \leq 2.3\pi R$ for the cases $\alpha = -1.0$, -0.5, 0.0, and 0.5. For convenience we set $R = 1.0$. The numerical integration in (3.7a) is performed with the Fehlberg routine (see §III.6). From Fig. XII.5 we see that the trajectories of front and rear wheels almost coincide when the wheels are steered in opposite directions with equal amplitudes, thus if $\alpha = -1$.

We shall not work out modelling Steps 6 (*Numerical analysis*), 7 (*Verification*), 8 (*Reiteration*), and 9 (*Implementation*). The verification would imply that we had to compare measured and calculated trajectories. It might appear that the model describes the experiments very well, but only if the vehicle is very long or if α is relatively small. One might come to the conclusion that it is absolutely necessary to take into account that real trucks have not two wheels but four or more. These details would require an extension of model (3.4). The implementation might vary from case to case. It will always comprise a computer program, in which the differential equations are numerically solved. The set-up of the input and output of this program can only be organised effectively in close interaction with the customer.

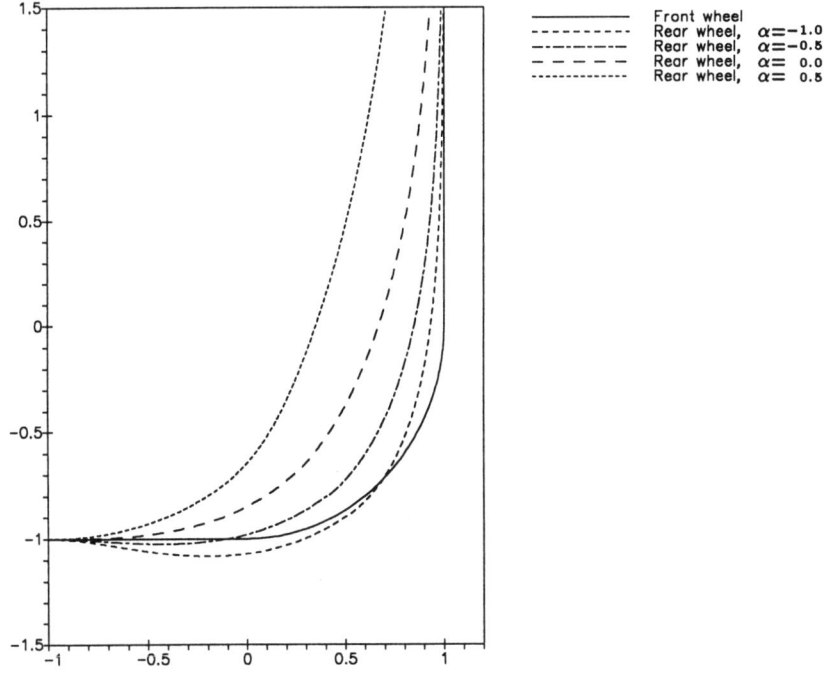

Figure XII.5 The paths of the front wheel (solid line) and the rear wheel (broken lines) while taking a turn of 90 degrees.

4. Inverse resonance

The following mechanical problem may serve as an illustration of many concepts introduced in Chapters IV and XI. The system under consideration is fairly simple, but the differential equations involved are of a very general character and occur in many applications. We consider a long needle that rotates in a vertical plane around its midpoint. Along the needle a bead may move as shown in Fig. XII.6. The construction is such that the bead can freely pass the origin. While the needle rotates, three forces are exerted on the bead: the gravitational force, the frictional force, and, due to the rotation, the centrifugal force. The last force is a so-called pseudo-force. It tends to push the bead from the needle, whereas the gravitational force may be directed both from and to the origin depending on the bead being below or above the origin respectively. The frictional force tends to reduce the bead's velocity. We shall describe the motion of the bead as a function of its initial position and

velocity and are especially interested in finding out under what conditions the
bead follows a bounded orbit on the needle and does not fly off.

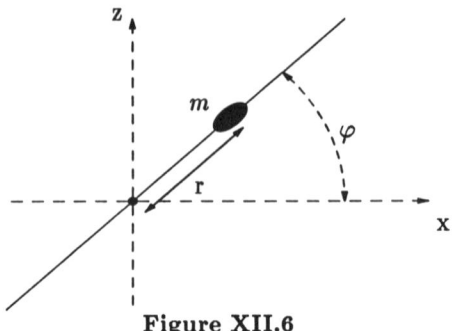

Figure XII.6

We model this problem using the theory from Chapter XI. In the bead-needle
system kinematic constraints are in force. In §XI.2 two methods are presented
to deal with mechanical systems with kinematic constraints. First we shall
apply the method of *Lagrange multipliers*. Denoting the bead's position by
Cartesian coordinates $\mathbf{r} := (x, y, z)^T$, Newton's second law reads:

$$m\ddot{\mathbf{r}} = -\alpha\dot{\mathbf{r}} + \mathbf{F}_g ,$$

where m is the mass of the bead and $\alpha > 0$ is the frictional constant.
The frictional force is modelled in the usual way, namely as being linearly
proportional to the velocity. The gravitational force $\mathbf{F}_g$ is assumed to be
constant and uniform. We choose the x- and z-axes as indicated in Fig. XII.6.
$\mathbf{F}_g$ is then given by

$$\mathbf{F}_g = \begin{bmatrix} 0 \\ 0 \\ -mg \end{bmatrix} ,$$

with g the gravitational acceleration. The first kinematic constraint is

$$f_1 := y = 0 .$$

From Fig. XII.6 we find that the second constraint is given by

$$f_2 := z - x \tan\varphi = 0 .$$

The angular velocity ω of the needle is constant. This leads to

$$\varphi(t) = \omega t + \varphi_0 .$$

Taking the gradients of the constraints we obtain

$$\mathbf{c}_1 := \frac{\partial f_1}{\partial \mathbf{r}} = (0,1,0)^T \ , \qquad \mathbf{c}_2 := \frac{\partial f_2}{\partial \mathbf{r}} = (-tan\,\varphi, 0, 1)^T \ .$$

Application of (XI.2.8) yields in the present case

$$m\ddot{\mathbf{r}} + \alpha\dot{\mathbf{r}} - \mathbf{F}_g + \lambda_1 \mathbf{c}_1 + \lambda_2 \mathbf{c}_2 = 0 \ ,$$

where λ_1 and λ_2 are the Lagrange multipliers. This differential equation together with the two kinematic constraints forms a a set of differential-algebraic equations (DAE). The DAE consists of 5 equations for the 5 unknowns x, y, z, λ_1, and λ_2. The constraint $y = 0$ leads to the trivial reduction $y \equiv 0$ and thus $\lambda_1 \equiv 0$ for all t. The remaining DAE is hard to solve analytically and requires a numerical approach. We shall not pursue this, however, since the multiplier approach is not so convenient here due to the use of Cartesian coordinates.

In the remainder of this section we shall show that use of the *Euler-Lagrange formalism* even allows for an analytical solution if we use suitably chosen generalised coordinates. In this approach we need an expression for the kinetic energy T, which is given by

$$(4.1) \qquad T = \tfrac{1}{2} m(\dot{x}^2 + \dot{y}^2 + \dot{z}^2) \ .$$

Since the bead is attached to the needle, $x, y,$ and z cannot vary independently. They satisfy the relations

$$(4.2) \qquad y = 0 \ , \qquad x = r\,\cos(\omega t + \varphi_0) \ , \qquad z = r\,\sin(\omega t + \varphi_0) \ ,$$

with r the position of the bead along the needle. Although relations (4.2) strongly resemble a transformation to polar coordinates, they should not be interpreted that way. Note that r may have both positive and negative values. The variable r is the natural candidate for the generalised coordinate of the system. Differentiation of the constraints (4.2) and substitution into (4.1) yields the kinetic energy T in terms of r:

$$(4.3) \qquad T = \tfrac{1}{2} m(\dot{r}^2 + \omega^2 r^2) \ .$$

The gravitational force $\mathbf{F}_g$ can be derived from the potential

$$(4.4) \qquad V = mg\,z = mg\,r\,\sin(\omega t + \varphi_0) \ ,$$

and is thus conservative. The latter property does not hold for the frictional force, but we shall initially ignore friction.

(i) Frictionless case

If friction is ignored, the Euler-Lagrange and Hamiltonian formalisms dealt with in §XI.2, 3 can be used. The Lagrangian L, defined as the difference between kinetic and potential energy, is given by

(4.5) $L(r, \dot{r}, t) = \frac{1}{2} m(\dot{r}^2 + \omega^2 r^2) - mg\, r\, \sin(\omega t + \varphi_0)$.

The Euler-Lagrange equation

$$\frac{d}{dt}\left(\frac{\partial L}{\partial \dot{r}}\right) - \frac{\partial L}{\partial r} = 0$$

yields the equation of motion

(4.6) $m\ddot{r} = m\omega^2 r - mg\, \sin(\omega t + \varphi_0)$.

The right-hand side contains the centrifugal force and the radial component of the gravitational force. The former is a so-called *pseudo-force*, which originates from the (time dependent) kinematic constraint. The latter is a so-called *applied force* exerted on the system from outside. This second order equation can be written as a system of two first order equations via the Hamiltonian formalism. To that end we introduce the momentum p:

(4.7) $p := \dfrac{\partial L}{\partial \dot{r}} = m\dot{r}$,

and the Hamiltonian H associated with L:

(4.8) $H(p, t) := p\dot{r} - L = \dfrac{p^2}{2m} - \frac{1}{2} m\omega^2 r^2 + mg\, r\, \sin(\omega t + \varphi_0)$.

We recognise that H is the sum of the kinetic energy $p^2/2m$, the potential energy $-\frac{1}{2} m\omega^2 r^2$ corresponding to the centrifugal force, and the potential energy corresponding to gravity. Because of the general relation

$$\frac{dH}{dt} = \frac{\partial H}{\partial t} ,$$

we conclude that the total energy of the system is not conserved. This is a consequence of the time dependence of the kinematic constraint. The Hamiltonian equations

$$\dot{r} = \frac{\partial H}{\partial p} , \quad \dot{p} = -\frac{\partial H}{\partial r}$$

read in the present case:

(4.9) $\dot{r} = \dfrac{p}{m}$, $\dot{p} = m\omega^2 r - mg\, \sin(\omega t + \varphi_0)$.

To get insight into the essential (combinations of) parameters of the model, we put it into dimensionless form. We have four parameters, viz. m, g, ω, and φ_0, but it does not make sense to vary these independently. As in §2 we find the dimensionless quantities

(4.10) $t^* := \omega t$, $r^* := r\, \dfrac{\omega^2}{g}$, $m^* := 1$.

The corresponding dimensionless momentum is

$$(4.11) \qquad p^* := m^* \frac{dr^*}{dt^*} = p \frac{\omega}{mg} ,$$

and the dimensionless equation of motion is

$$(4.12) \qquad \ddot{r}^* = r^* - \sin(\omega t^* + \varphi_0) .$$

This is a linear, inhomogeneous system which reads in standard form:

$$(4.13a) \qquad \dot{x} = A\,x + b$$

with

$$(4.13b) \qquad x := \begin{bmatrix} r^* \\ p^* \end{bmatrix} , \quad A := \begin{bmatrix} 0 & 1 \\ 1 & 0 \end{bmatrix} , \quad b := \begin{bmatrix} 0 \\ -\sin(t^* + \varphi_0) \end{bmatrix} .$$

These equations depend on one parameter only, viz. the initial angle φ_0 of the needle. They describe a wide class of linear, periodically driven, harmonic systems. The harmonic force is now centrifugal, i.e. directed away from the origin. For the mathematical pendulum, described in Examples XI.5 and XII.2, the force is centripetal, i.e. directed towards the origin. In the centripetal case the matrix A reads

$$(4.14) \qquad A := \begin{bmatrix} 0 & 1 \\ -1 & 0 \end{bmatrix} .$$

A centripetal force apparently gives rise to bounded solutions, unless *resonance* occurs, i.e. if the frequency of the driving force is close to an eigenfrequency of the system. A centrifugal force gives rise to unbounded solutions, unless the driving force prohibits this. We refer to this phenomenon as 'inverse resonance'.

From Chapter IV we know that the solution of the present linear system exists for any time and is explicitly given by

$$(4.15) \qquad x(t^*) = Y(t^*)\,x_0 + \int_0^{t^*} Y(t^*)\,Y^{-1}(s)\,b(s)\,ds ,$$

with x_0 the initial conditions (r_0^*, p_0^*) and Y the fundamental matrix

$$Y(t) := \exp(tA).$$

Using the property $A^2 = I$, we find for the matrix A in (4.13b)

$$(4.16) \qquad Y(t) = (\cosh t)I + (\sinh t)A .$$

Substitution of this expression into (4.15) directly yields that $r^*(t^*)$ is given by

$$(4.17) \qquad r^*(t^*) = c_+ \, e^{t^*} + c_- \, e^{-t^*} + \tfrac{1}{2} \sin(t^* + \varphi_0) \, ,$$

where

$$(4.18a) \quad c_+ := \tfrac{1}{2}\left(r_0^* + p_0^* - \tfrac{1}{2}\left(\sin \varphi_0 + \cos \varphi_0\right)\right) \, ,$$

$$(4.18b) \quad c_- := \tfrac{1}{2}\left(r_0^* - p_0^* - \tfrac{1}{2}\left(\sin \varphi_0 - \cos \varphi_0\right)\right) \, .$$

We recognise the sum of the general solution of the homogeneous system and a particular solution of the full equation. From expression (4.17) we directly conclude that the solution remains bounded only if $c_+ = 0$. So, inverse resonance occurs if

$$(4.19) \qquad r_0^* + p_0^* = \tfrac{1}{2}\left(\sin \varphi_0 + \cos \varphi_0\right) \, .$$

The term with c_- will die out, and after a transient period a periodic orbit remains with period $2\pi/\omega$. Note that after π/ω the bead starts repeating its motion, but on the other half of the needle. This orbit, which is expressed as $r^*(t^*) = \tfrac{1}{2}\sin(t^* + \varphi_0)$, is most easily understood in Cartesian coordinates (x, z). Using (4.2) we obtain the circle

$$(4.20) \qquad x^2 + \left(z - \frac{g}{4\omega^2}\right)^2 = \left(\frac{g}{4\omega^2}\right)^2 \, .$$

This is depicted in Fig. XII.7 for the case where $g = \omega^2$.

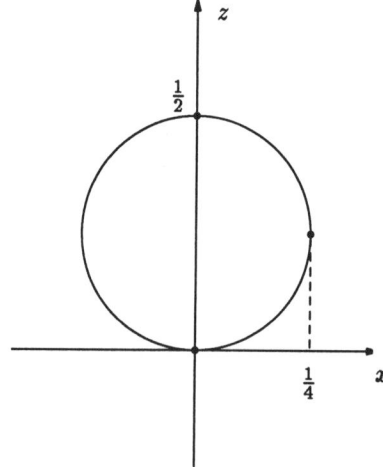

Figure XII.7   The periodic orbit of the bead in physical space if $g = \omega^2$.

(ii) The effect of friction

When friction is present the Euler-Lagrange equations (XI.2.13) are no longer applicable and the Lagrange equations (XI.2.9) have to be used. For the bead-needle system they read

(4.21) $\dfrac{d}{dt}\left(\dfrac{\partial T}{\partial \dot{r}}\right) - \dfrac{\partial T}{\partial r} = (\mathbf{F}_g + \mathbf{F}_f) \cdot \left(\dfrac{\partial \mathbf{r}}{\partial r}\right)$,

where $\mathbf{r} := (x, z)^T$, and the dot denotes the inner product. The frictional force $\mathbf{F}_f$ is taken proportional to the velocity and in the opposite direction, i.e.

(4.22) $\mathbf{F}_f := -\alpha \dot{\mathbf{r}}$, $\alpha > 0$.

Evaluation of (4.21) leads to the extended equation of motion

(4.23) $m\ddot{r} = m\omega^2 r - \alpha \dot{r} - mg \sin(\omega t + \varphi_0)$.

The introduction of friction, and thus of the extra parameter α, influences the way the system can be non-dimensionalised. The dimensionless quantities t^* and r^* are still as in (4.10), but m^* and p^* are now given by

(4.24) $m^* := m\dfrac{\omega}{\alpha}$, $p^* := p\dfrac{\omega^2}{\alpha g}$.

The dimensionless equation of motion has a similar form to (4.13), but with $\mathbf{A}$ given by

(4.25) $\mathbf{A} := \begin{bmatrix} 0 & 1 \\ 1 & -\beta \end{bmatrix}$,

where

(4.26) $\beta := \dfrac{\alpha}{m\omega}$.

In addition to φ_0, the system now depends on the dimensionless parameter β too. All bead-needle systems with the same φ_0 and β behave identically. This implies that increasing the frictional coefficient α by a certain factor has the same effect as decreasing the product $m\omega$ by the same factor.

The solution of (4.23) is the sum of a particular solution and the general solution of the homogeneous equation

(4.27) $\dot{\mathbf{x}} = \mathbf{A}\mathbf{x}$,

with $\mathbf{x}$ defined as in (4.13b) and $\mathbf{A}$ given by (4.25). Equation (4.27) has only the origin as a stationary point. Its stability is determined by the eigenvalues $\lambda_\pm$ of $\mathbf{A}$. These are given by

(4.28) $\lambda_\pm = \tfrac{1}{2}\left(-\beta \pm \sqrt{\beta^2 + 4}\right)$, with corresponding eigenvectors

$$(4.29) \qquad \mathbf{v}_\pm = \begin{bmatrix} 1 \\ \lambda_\pm \end{bmatrix} .$$

Because $\mathbf{A}$ is symmetric, the $\lambda_\pm$ are real and the $\mathbf{v}_\pm$ orthogonal. Because $\lambda_+ > 0$ and $\lambda_- < 0$, the origin is a saddle point and thus unstable, as is to be expected from the physics of the bead-needle system. A particular solution of (4.23) is easily found by substituting a linear combination of $\cos(t^* + \varphi_0)$ and $\sin(t^* + \varphi_0)$. This eventually yields the solution

$$(4.30) \qquad r^*(t^*) = c_+ \, e^{\lambda_+ t^*} + c_- \, e^{\lambda_- t^*} +$$

$$+ \frac{1}{4 + \beta^2} \{ \beta \cos(t^* + \varphi_0) + 2 \sin(t^* + \varphi_0) \} .$$

In terms of initial conditions $(r_0^*, p_0^*, \varphi_0)$ the coefficients $c_\pm$ are given by

$$(4.31) \qquad c_+ := \frac{1}{\sqrt{4 + \beta^2}} \Big(p_0^* - \lambda_- r_0^* +$$

$$+ \frac{1}{4 + \beta^2} \{ (\beta \lambda_- - 2) \cos \varphi_0 + (2\lambda_- + \beta) \sin \varphi_0 \} \Big) ,$$

$$(4.32) \qquad c_- := \frac{-1}{\sqrt{4 + \beta^2}} \Big(p_0^* - \lambda_+ r_0^* +$$

$$+ \frac{1}{4 + \beta^2} \{ (\beta \lambda_+ - 2) \cos \varphi_0 + (2\lambda_+ + \beta) \sin \varphi_0 \} \Big) .$$

The structure of solutions (4.17) and (4.30) is quite similar. For $\beta = 0$ the case (4.30) reduces to the particular case (4.17). From (4.30) we again conclude that inverse resonance occurs if $c_+ = 0$. After a transient period, in which the term with λ_- damps out, a periodic solution results, given by

$$(4.33) \qquad r(t) = \frac{g}{\omega^2} \, r^*(t^*) = a \cos(\omega t + \varphi_0) + b \sin(\omega t + \varphi_0) ,$$

with coefficients

$$(4.34) \qquad a := \frac{g}{\omega^2} \frac{\beta}{4 + \beta^2} , \qquad b := \frac{g}{\omega^2} \frac{2}{4 + \beta^2} .$$

This periodic orbit of the bead is most easily visualised in Cartesian coordinates (x, z). Using (4.1) one finds

$$(4.35) \qquad \begin{cases} x(t) = \dfrac{a}{2} (1 + \cos 2\varphi) + \dfrac{b}{2} \sin 2\varphi , \\[2mm] z(t) = \dfrac{a}{2} \sin 2\varphi + \dfrac{b}{2} (1 - \cos 2\varphi) , \end{cases}$$

with $\varphi := \omega t + \varphi_0$.

This representation shows that the periodic solutions are circles defined by

$$(4.36) \qquad \left(x - \frac{a}{2}\right)^2 + \left(z - \frac{b}{2}\right)^2 = \left(\frac{a}{2}\right)^2 + \left(\frac{b}{2}\right)^2 = \left(\frac{1}{2}\right)^2.$$

Note that they pass through the origin. It is particularly interesting to study this family of circles as a function of the friction parameter β. The midpoints $(x_0, z_0)^T$ of the circles in terms of β are given by

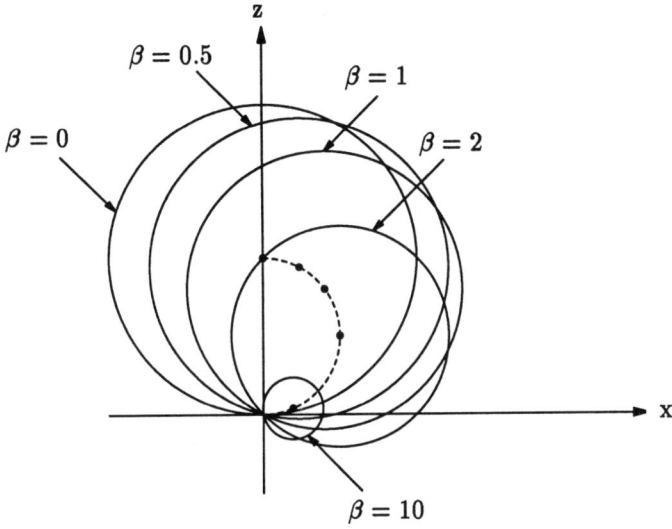

Figure XII.8   The orbits of the bead represented by (4.36) for various values of the friction parameter β.

$$(4.37) \qquad (x_0, z_0)^T := \tfrac{1}{2}(a, b)^T = \frac{g}{4\omega^2} \; \frac{1}{1 + (\beta/2)^2} \left(\frac{\beta}{2}, 1\right)^T.$$

When β varies from 0 to ∞. the midpoints themselves lie on a circle, which is given by

$$(4.38) \qquad x_0^2 + \left(z_0 - \frac{g}{8\omega^2}\right)^2 = \left(\frac{g}{8\omega^2}\right)^2.$$

In Fig. XII.8 these circles are drawn for various values of β. If $\beta = 0$, i.e. either no friction is present or the product $m\omega$ is extremely large, the periodic orbit is symmetric with respect to the z-axis and its radius is maximal. For increasing values of β the circle shrinks and its midpoint approaches the origin following a circular arc. For $\beta \to \infty$ the origin becomes a stationary point of the system.

5. Solitary water waves

In this modelling example we study a fascinating phenomenon, vividly described by J. Scott Russell in his 'Report on Waves' (1844):

> "I believe I shall best introduce this phaenomenon by describing the circumstances of my own acquaintance with it. I was observing the motion of a boat which was rapidly drawn along a narrow channel by a pair of horses, when the boat suddenly stopped – not so the mass of water in the channel which it had put in motion; it accumulated round the prow of the vessel in a state of violent agitation, then suddenly leaving it behind, rolled forward with great velocity, assuming the form of a large solitary elevation, a rounded, smooth and well-defined heap of water, which continued its course along the channel apparently without change of form or diminution of speed. I followed it on horseback, and overtook it still rolling on at a rate of some eight or nine miles an hour, preserving its original figure some thirty feet long and a foot to a foot and a half in height. Its height gradually diminished, and after a chase of one or two miles I lost it in the windings of the channel."

This solitary water wave can completely be analysed mathematically. We use it to illustrate several techniques dealt with in this book: global analysis of the phase plane (§V.6), Hamiltonian systems (§XI.3, 4), and one-step integration methods (§III.1).

(i) Modelling

In 1895 Korteweg and De Vries modelled the phenomenon described above. It is beyond the scope of this book to give a complete derivation of the equations and their analysis. We shall restrict ourselves to mentioning the assumptions used and taking the resulting equations for granted. Our goal is to answer the question whether these equations possess a solitary wave solution. To start with, Korteweg and De Vries modelled the problem as a long, straight canal taking into account one spatial dimension only. The quantity of interest is the water surface profile y. Assuming that this profile is uniform across the canal, y is a function of time t and position x along the canal. The fluid is taken to be incompressible, the internal friction is neglected, and the wave amplitude is assumed to be small compared to the depth d of the canal. Because t and x are the independent variables, the model consists of a partial differential equation for the dependent variable y:

(5.1) $\dfrac{\partial y}{\partial t} = -\sqrt{\dfrac{g}{h}}\,\dfrac{\partial}{\partial x}\left(hy + \tfrac{3}{4}y^2 + \tfrac{1}{2}\sigma\,\dfrac{\partial^2 y}{\partial x^2}\right),$

where g is the gravitational constant, h the water depth if the water is at rest, and σ given by

(5.2) $\sigma := \tfrac{1}{3}d^3 - \dfrac{\tau d}{\rho g};$

here τ is the surface tension and ρ the water density. A dimensional analysis as in §2 results in the dimensionless variables

(5.3) $y^* := y/h\,,\qquad x^* := x/h\,,\qquad t^* := t\sqrt{g/h}\,.$

The system has one dimensionless parameter:

(5.4) $\sigma^* := \sigma/d^3\,.$

In the following we shall use dimensionless quantities only, and omit the $*$ for convenience. In dimensionless form (5.1) then reads:

(5.5) $\dfrac{\partial y}{\partial t} = -\dfrac{\partial}{\partial x}\left(y + \tfrac{3}{4}y^2 + \tfrac{1}{2}\sigma\,\dfrac{\partial^2 y}{\partial x^2}\right).$

The wave amplitude is measured with respect to the water surface at rest, so by assumption $y \ll 1$. We introduce a variable $s := x - vt$, with v the (still unknown) wave velocity, and presume a solution of the form $y = y(s)$. Such a solution is called a *solitary wave*. Substitution into (5.5) reduces this equation to an ordinary differential equation:

(5.6) $\dfrac{d}{ds}\left((1-v)y + \tfrac{3}{4}y^2 + \tfrac{1}{2}\sigma\,\dfrac{d^2 y}{ds^2}\right) = 0\,.$

The expression between parentheses is thus constant. Its value follows from the boundary conditions, which state that the water surface is at rest far away from the solitary wave front. This implies that y and all its derivatives vanish for $s \to \pm\infty$. We conclude that the constant must equal zero, so that (5.6) is equivalent to

(5.7) $\sigma\,\dfrac{d^2 y}{ds^2} = by - \tfrac{3}{2}y^2$

with $b := 2(v-1)$. For $\sigma = 0$ only constant solutions are found, so we take $\sigma \neq 0$ in what follows. A simple rescaling reduces this equation to the form

(5.8) $\dfrac{d^2 y}{ds^2} = by - \tfrac{3}{2}y^2\,.$

Here, we have used the transformation $s^* := s/\sqrt{\sigma}, y^*(s^*) := y(s)$, and $b^* := b$, if $\sigma > 0$, whereas $s^* := s/\sqrt{-\sigma}, y^*(s^*) := -y(s)$, and $b^* := -b$, if $\sigma < 0$. We shall omit the $*$ superscript below.

Equation (5.8) already allows for some general conclusions. Indeed one may interpret this equation as the equation of motion of a particle of unit mass under the influence of a harmonic force, proportional to y, and a nonlinear force, proportional to y^2. The latter is directed to the origin. Since the latter dominates for large values of y, the solutions are forced to remain bounded, irrespective of the direction of the linear force determined by the sign of b. For smaller values of y, the harmonic force dominates. If $b < 0$ this force gives rise to oscillating solutions in the vicinity of $y = 0$. Because this does not agree with the condition $y = 0$ for $s \to \pm\infty$, we conclude that $b > 0$ or $v > 1$ (in dimensionless form).

We shall analyse (5.8) further along three different lines: global phase plane analysis, the Hamiltonian approach, and numerical evaluation. It is worthwhile to apply these methods in parallel, because each of them illuminates new features of the system.

(ii) Phase plane analysis

In standard form (5.8) reads:

$$(5.9) \qquad \dot{\mathbf{y}} = \mathbf{f}(\mathbf{y}) \,,$$

with

$$(5.10) \qquad \mathbf{y} := \begin{bmatrix} y_1 \\ y_2 \end{bmatrix} := \begin{bmatrix} y \\ \dfrac{dy}{ds} \end{bmatrix} \,, \qquad \mathbf{f} := \begin{bmatrix} y_2 \\ by_1 - \tfrac{3}{2} y_1^2 \end{bmatrix} \,.$$

The stationary points of the nonlinear vector field $\mathbf{f}$ are the origin $(0,0)^T$ and a point on the y_1-axis: $\mathbf{y}_0 := (2b/3, 0)^T$. Their stability properties are determined by the eigenvalues of the Jacobian matrix

$$(5.11) \qquad \mathbf{J} := \begin{bmatrix} 0 & 1 \\ b - 3y_1 & 0 \end{bmatrix} \,.$$

At the origin and $\mathbf{y}_0$ the eigenvalues $\lambda_\pm$ of J are $\pm\sqrt{b}$ and $\pm\sqrt{-b}$ respectively. For both points the eigenvectors are given by $\mathbf{v}_\pm := (1, \lambda_\pm)^T$. From §IV.7 we conclude that for $b > 0$ the origin is a saddle point and $\mathbf{y}_0$ a centre. For $b < 0$ it is the other way round, but we have already found above that in the latter case no solitary wave solution exists. This conclusion is confirmed by the following argument: a solitary wave solution is a homoclinic orbit, i.e. it converges to one and the same stationary point for $s \to \pm\infty$. In the present model the boundary conditions force the solution to approach the origin as $s \to \pm\infty$. This implies that the origin must be a saddle point and not a centre. Orbits approaching a centre point tend to oscillate around it. In that case y would become negative, which is inconsistent with the physics of the system. The existence of the sol-

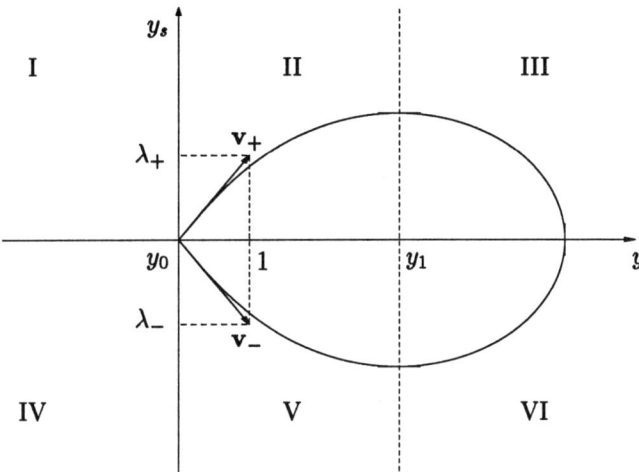

Figure XII.9 Phase plane of the solitary wave for system (5.10). The isoclines divide the plane into regions I–VI.

itary wave solutions can be shown by construction using the ideas from §V.6. First we observe that the y_1-axis is a horizontal isocline: solutions cross it perpendicularly and thus vertically. The vertical isoclines are the y_2-axis and the vertical $y_1 = 2b/3$. They can be passed only horizontally. The isoclines divide the phase plane into the regions I–VI as depicted in Fig. XII.9. In each region the signs of the components of $\mathbf{f}$ are fixed. From that observation we roughly know the directions of the solutions for increasing s values. Because the origin is a saddle point, we know that a solitary wave approaching it for $s \to \infty$ is tangent to the eigenvector $\mathbf{v}_-$ at the origin. We use this insight to start at the origin with $s = +\infty$ and follow a possible solitary wave solution $(y_1(s), y_2(s))^T$ for decreasing s values. Because $\mathbf{v}_-$ is directed into region V, the solution enters this. In region V, $y_1(s)$ increases and $y_2(s)$ decreases. As discussed above, the solution will remain bounded and will therefore cross the vertical isocline between regions V and VI. In region VI both components $y_1(s)$ and $y_2(s)$ increase. From this it follows that the solution will cross the horizontal isocline, i.e. the y_1-axis, somewhere. Because the y_1-axis is an axis of symmetry of the vector field $\mathbf{f}$, the rest of the solution is obtained by mirroring the first part with respect to the y_1-axis. For $s \to -\infty$ it will approach the eigenvector $\mathbf{v}_+$, see Fig. XII.9. This analysis yields the existence of a solitary

wave solution together with its general shape. The detailed shape will follow
from the alternative techniques used below.

(iii) Hamiltonian analysis

The forces in (5.8) can be deduced from the potential

$$(5.12) \qquad V(y_1) := \tfrac{1}{2}\left(-by_1^2 + y_1^3\right) .$$

This potential is independent of s, i.e. the system is conservative. The
Hamiltonian H is given by the sum of kinetic and potential energy,

$$(5.13) \qquad H := \tfrac{1}{2}y_2^2 + V(y_1) = \tfrac{1}{2}\left(y_2^2 - by_1^2 + y_1^3\right) ,$$

as explained in §XI.3. Because H does not depend on s explicitly, we conclude
that it is constant. Its value follows from the boundary conditions. Because
both the wave profile y_1 and its slope y_2 must vanish for $s \to \pm\infty$, the
Hamiltonian is also vanishing and we conclude that

$$(5.14) \qquad y_2^2 - by_1^2 + y_1^3 = 0 .$$

This relation determines the exact shape of the orbit in Fig. XII.9 for a given
value of the parameter b, i.e. the velocity v. From (5.14) we directly observe
that the maximum wave amplitude, which corresponds to a vanishing slope,
i.e. $y_2 = 0$, is equal to b. To obtain the solution as a function of s, we rewrite
(5.14) as

$$(5.15) \qquad y_2 = \pm\, y_1 \sqrt{b - y_1} .$$

Because $y_1 = y$ and $y_2 = dy/ds$, this is an ODE for y as a function of s. We
are free to set $s = 0$ when the wave attains its maximum value. The solitary
wave is then a solution of the IVP

$$(5.16) \qquad \begin{cases} \dot{y} = -y\sqrt{b - y} \\ y(0) = b\ (= 2(v - 1)) . \end{cases}$$

Separation of variables leads to the integral equation

$$(5.17) \qquad \int_b^y \frac{1}{z\sqrt{b - z}}\, dz = -\int_0^s dt = -s ,$$

with solution

$$(5.18) \qquad y(s) = \frac{b}{\cosh^2\!\left(\tfrac{1}{2}\sqrt{b}\, s\right)} .$$

Such a solution exists for all $v > 1$. We recall that v is the dimensionless velocity here. For the velocity itself this leads to a lower boundary of $\sqrt{g/h}$ for

the wave amplitude of $2(\sqrt{gh} - 1)$. For given, admitted velocity we conclude: the deeper the canal, the faster and the higher the wave.

(iv) Numerical analysis

Let us investigate the numerical solution of IVP (5.16). We first have to realise that (5.16) has no unique solution. In addition to solution (5.18), the constant function $y = b$ also satisfies. The uniqueness theorem in §II.1.10 does not apply in this case, because the (scalar) vector field $f(y) = -y\sqrt{b-y}$ is not Lipschitz continuous at $y = b$. Searching for a solitary wave essentially requires the solution of a BVP, not an IVP. The solution $y \equiv b$ is excluded by the conditions for $s \to \pm\infty$. Note, however, that a (multiple) shooting approach would suffer from this same problem. Here we analyse only the IVP setting in more detail. Numerically, the problem of non-uniqueness can be circumvented by choosing an initial point, $s_0 > 0$ say, slightly away from the initial point $s = 0$, where we have the maximum. The complete solution is then obtained by integrating forward and, if desired, backward from $s_0 > 0$ to 0. However, since we do not know the solution, we prefer to choose a slightly smaller initial value, y_0 say, and still take $s_0 = 0$. Since the problem is autonomous, the value of s_0 is immaterial.

We analyse the application of two simple numerical schemes. The s-axis is discretised with step length $h > 0$. Following §III.1 the Euler forward scheme reads:

$$(5.19) \qquad y_{i+1} = y_i(1 - h\sqrt{b - y_i}) =: q(y_i) \ .$$

This equation can be interpreted as a successive substitution process with the stationary points $y = b$ and $y = 0$. To visualise this substitution process we have drawn the function $q(y)$ together with the diagonal through the origin and the point $(b, b)^T$ in Fig. XII.10a. Starting at a point $y_0 < b$, a horizontal line through the point $(y_0, q(y_0))^T$ is drawn. This line crosses the diagonal at $(y_1, y_1)^T$, say. The vertical through the latter point crosses the q curve at $(y_1, q(y_1))^T$. From this point on the procedure is repeated. From the geometry it is clear that the sequence y_i, $i = 0, 1, 2, ...$, converges to the origin if $|\dot{q}(0)| < 1$, and diverges if $|\dot{q}(0)| > 1$. So we conclude: *the iteration scheme* (5.19) *approaches the correct asymptotic solution for $i \to \infty$, provided the step length $h < 2/\sqrt{b}$.*

In Fig. XII.10a,b the iteration process and the resulting solution $\{y_i\}_{i=0}^{\infty}$ are given for parameter values $b = 1$, $h = 1$. The effect of using too large a step length is illustrated in Fig. XII.11, where $h = 2.25$ is used. In that case the solution initially approaches the origin, but after a few steps becomes periodic. This is a nice example of nonlinear instability: in the linear case only geometrical or exponential growth can be observed.

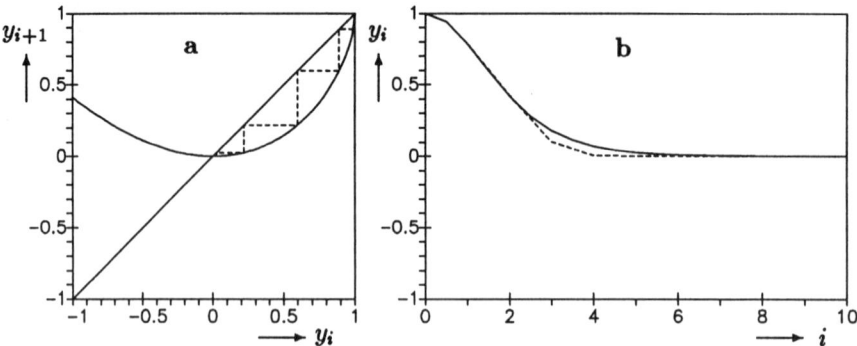

Figure XII.10   Euler forward with parameter values $b = h = 1$, $y_0 = 0.99$.

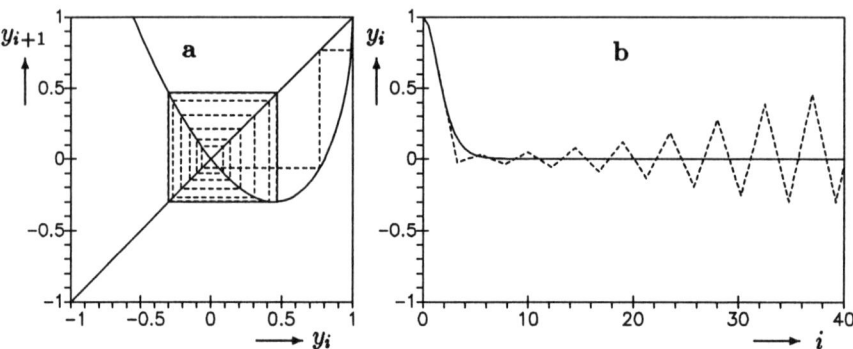

Figure XII.11   Euler forward with parameter values $b = 1$, $h = 2.25$, $y_0 = 0.99$.

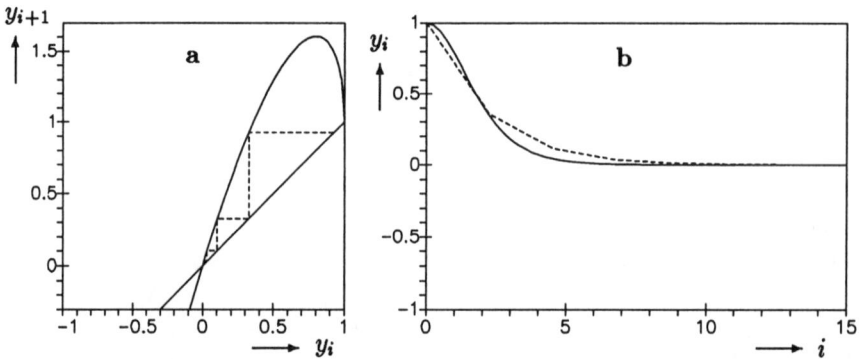

Figure XII.12   Euler backward with parameter values $b = 1$, $h = 2.25$.

The Euler backward formula for (5.16) yields the Δ-equation

$$(5.20) \qquad y_{i+1} = y_i - h\, y_{i+1}\sqrt{b - y_{i+1}} \;.$$

Collecting terms with values at the new time level, we find

$$(5.21) \qquad y_i = y_{i+1}\left(1 + h\sqrt{b - y_{i+1}}\right) =: r(y_{i+1}) \;.$$

This may be interpreted as an inverse successive substitution process. The construction of the solution $\{y_i\}_{i=0}^{\infty}$ is the inverse procedure of what has been described in the explicit case before. Starting with some value y_0, a horizontal straight line is drawn through the point $(y_0, y_0)^T$ at the diagonal in Fig. XII.12a. This line crosses the function r in, say, $(y_1, r(y_1))^T$. Next, a horizontal line is drawn through $(y_1, y_1)^T$ resulting in y_2. The procedure is then repeated. The process converges to the origin if $|\dot r(0)| > 1$, but this condition is always fulfilled. So we conclude: *the iteration scheme (5.21) approaches the correct asymptotic solution for $i \to \infty$, provided the initial value $y_0 < b$, but irrespective of the step length.*

In Fig. XII.12(a,b) the substitution process and the resulting series of y_i values are given for parameter values $b = 1$, $h = 2.25$, and $y_0 = 0.9$.

6. Infectious diseases I

We shall develop a simple, but instructive model for describing certain epidemics. Most actual epidemic problems have their own characteristics, which makes attempts to construct a generic model futile. See, e.g., [17]. Investigation of a (restricted) class of epidemic diseases centres around a specific theme. Here we shall make the following assumptions:

- The infectious disease is caused by a pathogen, say a virus, that spreads very fast in the atmosphere, e.g. by wind.

- The viruses reproduce themselves in infected individuals only.

- The incubation time is negligible, so every individual who contracts a virus becomes infective immediately afterwards.

- Every infected individual recovers after some time.

- The population remains at a fixed level in the time interval and the region under consideration.

Modelling such an infectious disease implies that we must prescribe relations for the time dependent behaviour of:

$x_1(t)$: density of viruses ,

$x_2(t)$: concentration of infected people as a percentage of
the total (constant) number of inhabitants .

The fact that x_1 does not depend on position is an implication of the assumption that the spread of viruses is fast, so that the virus concentration is uniform in space. In the subsequent section a model is studied in which the diffusion of pathogens, and thus of the disease, is a dominant factor.

A model consistent with the assumptions listed above is given by the ODE

(6.1)
$$\begin{cases} \dot{x}_1 = a_{11}x_1 + a_{12}x_2 , \\ \dot{x}_2 = f(x_1) + a_{22}x_2 . \end{cases}$$

The terms on the right-hand side are explained as follows:

- $a_{11}x_1$, with $a_{11} < 0$: the viruses die out if they do not find appropriate places – susceptible individuals – to settle and reproduce themselves.

- $a_{12}x_2$, with $a_{12} > 0$: infected individuals act as a source of viruses.

- $f(x_1)$, the infection rate: the function f represents the probability that individuals become infected as a function of virus concentration. This term is an approximation of the product $f(x_1)(1 - x_2)$, i.e. infection rate times the concentration of sound individuals, but the factor $1 - x_2$ is neglected under the assumption $x_2 \ll 1$.

- $a_{22}x_2$, with $a_{22} < 0$: infected individuals recover and the number of recovering people is taken to be linearly proportional to the concentration of infected people.

We now investigate model (6.1) for different infection rates.

(i) Linear infection rate

A trivial choice is to take f linear:

(6.2) $f(x_1) = a_{21}x_1 ,\qquad a_{21} > 0 .$

This seems a reasonable way to model the infection rate, but we shall show that the consequences do not reflect realistic epidemics. The linear model has the standard form

(6.3) $\dot{\mathbf{x}} = \mathbf{A}\mathbf{x} ,$

with $\mathbf{x} := (x_1, x_2)^T$ and $\mathbf{A}$ a constant 2×2 matrix. The only stationary point is the origin. Its stability properties follow from the eigenvalues $\lambda_\pm$ of $\mathbf{A}$ (see §IV.3):

(6.4) $\lambda_\pm := \frac{1}{2}\left(e \pm \sqrt{e^2 - 4d}\right)$,

with $d := \det(\mathbf{A}) = a_{11}a_{22} - a_{12}a_{21}$, and $e := \mathrm{Tr}(\mathbf{A}) = a_{11} + a_{22}$. The discriminant $e^2 - 4d$ can be written as

(6.5) $e^2 - 4d = (a_{11} - a_{22})^2 + 4a_{21}a_{12}$.

Because the latter expression is positive, the eigenvalues are real. We distinguish between three cases, depending on the sign of d. For convenience, we introduce

(6.6) $\beta := \dfrac{a_{11}a_{22}}{a_{12}a_{21}} = 1 + \dfrac{d}{a_{12}a_{21}}$.

The cases sign $d >, =, < 0$ then correspond to $\beta >, =, < 1$ respectively. Following §IV.3 and §V.1 we have

– $\beta > 1$ and thus $\lambda_+, \lambda_- < 0$,   *asymptotic stability*
– $\beta = 0$ and thus $\lambda_+ = 0$, $\lambda_- < 0$,   *(Lyapunov) stability*
– $\beta < 1$ and thus $\lambda_+ > 0$, $\lambda_- < 0$,   *instability, saddle point.*

So, for $\beta > 1$ no epidemic will occur. In the linear case asymptotic stability is equivalent to global attractivity: however large the numbers of viruses and/or infected individuals initially may be, these numbers will gradually decrease as time evolves. In practice, $\beta = 0$ will never be satisfied exactly, so this case is not considered any further. For $\beta < 1$ the origin is unstable. Any small perturbation will cause the system to leave the origin in the direction of the eigenvector $\mathbf{v}_+$ corresponding to the positive eigenvalue λ_+. The eigenvectors are given by

(6.7) $\mathbf{v}_\pm := \begin{bmatrix} 1 \\ (-a_{11} + \lambda_\pm)/a_{12} \end{bmatrix}$.

Note that $\mathbf{v}_+$ is directed into the first quadrant. The perturbed system will follow $\mathbf{v}_+$ and approach $+\infty$ for $t \to \infty$. The number of infected individuals thus becomes unlimited.

Modelling the infection rate in a linear way results in a model predicting the occurrence of either no epidemic or unlimited growth of the epidemic. We conclude that model (6.1) with (6.2) apparently is not adequate to describe reality.

(ii) Limited infection risk

The unboundedness of the solutions for linear infection rates stems from the unboundedness of the infection rate f in (6.2). To repair this shortcoming we introduce an expression for f which models a saturation for larger values of x_1:

(6.8) $f(x_1) = \dfrac{a_{21}x_1}{1 + \alpha x_1}$,

with $a_{21} > 0$ and $\alpha > 0$. For $\alpha x_1 \ll 1$ this function behaves linearly, while for larger αx_1 it approaches the constant a_{21}/α. The origin is still a stationary point of (6.1) with (6.8) and the same stability analysis as given above applies. We find that it is a saddle point if $\beta < 1$. Under this condition a second stationary point $(\hat{x}_1, \hat{x}_2)^T$ arises, given by

(6.9) $\hat{x}_1 = \dfrac{1}{\alpha} \dfrac{1 - \beta}{\beta}$, $\hat{x}_2 = \left(\dfrac{-a_{11}}{a_{12}}\right) \hat{x}_1$.

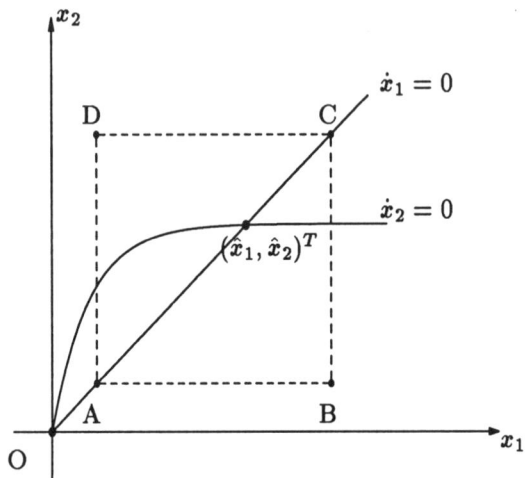

Figure XII.13   Phase plane of (6.1) with (6.8).

In Fig. XII.13 the horizontal and vertical isoclines, determined by the equations $\dot{x}_1 = 0$ and $\dot{x}_2 = 0$, respectively, are sketched for $\beta < 1$. The stationary points are the points of intersections of the isoclines. The stability properties of $(\hat{x}_1, \hat{x}_2)^T$ follow from the eigenvalues of the Jacobian matrix

(6.10) $\mathbf{J} = \mathbf{J}(x_1) := \begin{bmatrix} a_{11} & a_{12} \\[2mm] \dfrac{df}{dx_1} & a_{22} \end{bmatrix}$,

evaluated at the stationary point $\hat{x}_1$. For f, given by (6.8), we have

(6.11) $\dfrac{df}{dx_1} = \dfrac{a_{21}}{(1 + \alpha x_1)^2}$.

The eigenvalues $\lambda_{\pm}$ at $\hat{x}_1$ are

(6.12) $\lambda_{\pm} := \frac{1}{2} \left(e \pm \sqrt{e^2 - 4\beta d} \right)$.

Because $\beta < 1$ the discriminant is positive, just as in (6.3), and the $\lambda_{\pm}$ are real and negative, so that $(\hat{x}_1, \hat{x}_2)^T$ is asymptotically stable. This result stems from the linearisation around the stationary point and is valid only locally. To establish the basin of attraction we may apply the method introduced in §V.6, using rectangles $ABCD$ as drawn in Fig. XII.13. The vertices A and C may shift along the $\dot{x}_1 = 0$ isocline. The isoclines divide the first quadrant into regions with fixed signs of $\dot{x}_1$ and $\dot{x}_2$. These signs globally determine the direction of the vector field in each region. We consider rectangles with A between $(0,0)^T$ and $(\hat{x}_1, \hat{x}_2)^T$, and C on the other side of $(\hat{x}_1, \hat{x}_2)^T$ along $\dot{x}_1 = 0$. Then the vector field is directed inwards from everywhere on the boundary of the rectangle, except for the points A and C. From §V.6 we conclude that the attractor $(\hat{x}_1, \hat{x}_2)^T$ has the entire first quadrant as basin of attraction, because A may approach the origin arbitrarily close and C may shift to infinity. Even the positive x_1- and x_2-axes belong to the basin of attraction of $(\hat{x}_1, \hat{x}_2)^T$, except for the origin, of course.

In the case $\beta > 1$ a similar construction with rectangles proves that then the origin is a global attractor.

Model (6.1) with infection rate (6.8) predicts that for $\beta < 1$ an epidemic will break out, however small the initial number of pathogens may be. The system will tend to a situation with a fixed number of infected individuals: the number of recovering individuals precisely balances the number of individuals getting infected. To remedy the epidemic, one should try to get $\beta > 1$. The parameter a_{11} is specific for the virus under consideration and cannot be changed. The parameter a_{12} can probably be diminished by isolating the infected individuals. The parameter a_{22} can be increased by application of medicines, if available. For influenza, for example, no appropriate medicine is known. The parameter a_{21} decreases if the condition of the individuals is improved so much that they are able to resist infections more successfully.

The parameter α, introduced in (6.8), does not have any influence on the occurrence of an epidemic, but determines, together with the other parameters, the value of $\hat{x}_2$, and thus the gravity of the epidemic.

(iii) Threshold infection rate

For most infectious diseases the infection rate shows a threshold behaviour: the pathogen does not form a threat, unless the pathogen concentration exceeds a threshold value. The reason is that the immune system of healthy individuals is able to fight low virus concentrations successfully. The threshold phenomenon can be incorporated into model (6.1). To that end the infection rate f has to be adjusted such that it (nearly) vanishes for small values of x_1. The essential feature is appropriately described by an expression of the form

$$(6.13) \qquad f(x_1) = \frac{a_{21}x_1^2}{(1 + \alpha x_1^2)} \ .$$

For larger values of x_1 this choice behaves similarly to (6.8), whereas for small x_1-values its slope nearly vanishes. The number of stationary points of (6.1) with (6.13) depends on the parameter values. With the definition

$$(6.14) \qquad \delta_\pm := 1 \pm \sqrt{1 - 4\alpha\beta^2} \ ,$$

the possible cases are:

$- 4\alpha\beta^2 > 1 : (0,0)^t$,

$- 4\alpha\beta^2 = 1 : (0,0)^T$ and $(\hat{x}_1, \hat{x}_2)^T := \left(\dfrac{1}{2\alpha\beta}, \dfrac{a_{21}}{2\alpha} \right)^T$,

$- 4\alpha\beta^2 < 1 : (0,0)^T, \ (\hat{x}_1, \hat{x}_2)^T := \left(\dfrac{\delta_-}{2\alpha\beta}, \dfrac{a_{21}\delta_-}{2\alpha} \right)^T$, and

$$(x_1^*, x_2^*)^T := \left(\frac{\delta_+}{2\alpha\beta}, \frac{a_{21}\delta_+}{2\alpha} \right)^T .$$

Because the cases $4\alpha\beta^2 \geq 1$ strongly resemble the results obtained for infection rate (6.8), we omit the analysis, and only recall that for $4\alpha\beta^2 > 1$ the origin is a global attractor. The situation for $4\alpha\beta^2 < 1$ is sketched in Fig. XII.14 and needs further investigation. Linearisation around the stationary points reveals that $(0,0)^T$ and $(x_1^*, x_2^*)^T$ are asymptotically stable, whereas $(\hat{x}_1, \hat{x}_2)^T$ is a saddle point.

To estimate the basins of attraction of the two former points we employ contracting rectangles. Consider the rectangles $ABCD$ and $OEFG$ as denoted in Fig. XII.14. Careful analysis of the regions in which the isoclines divide the first quadrant clarifies why the vector field is directed inwards along their boundaries. This holds as long as F shifts along the $\dot{x}_1 = 0$ isocline between the points $(0,0)^T$ and $(\hat{x}_1, \hat{x}_2)^T$, where A is between $(\hat{x}_1, \hat{x}_2)^T$ and $(x_1^*, x_2^*)^T$, and C beyond $(x_1^*, x_2^*)^T$. We conclude that the basin of attraction of $(0,0)^T$ is at least $\{(x_1, x_2)^T \mid 0 \leq x_1 \leq \hat{x}_1, 0 \leq x_2 \leq \hat{x}_2\}$, with the exception of $(\hat{x}_1, \hat{x}_2)^T$ itself. For $(x_1^*, x_2^*)^T$ the basin is minimal $\{(x_1, x_2)^T \mid x_1 \geq \hat{x}_1, x_2 \geq \hat{x}_2\}$, also with the exception of $(\hat{x}_1, \hat{x}_2)^T$. About the remaining parts of the first quadrant we may draw conclusions too. Because $(\hat{x}_1, \hat{x}_2)^T$ is a saddle point, it lies on a stable manifold or so-called *separatrix*, as explained in §V.4. In the present, planar case this is a curve Γ. The two parts Γ_1 and Γ_2 of Γ on both sides of $(\hat{x}_1, \hat{x}_2)^T$ are orbits which approach the stationary point for $t \to \infty$. Because Γ is situated at one side of both isoclines, the x_2-value will monotonically decrease if the x_1-value increases along this curve. For the same reason Γ_2

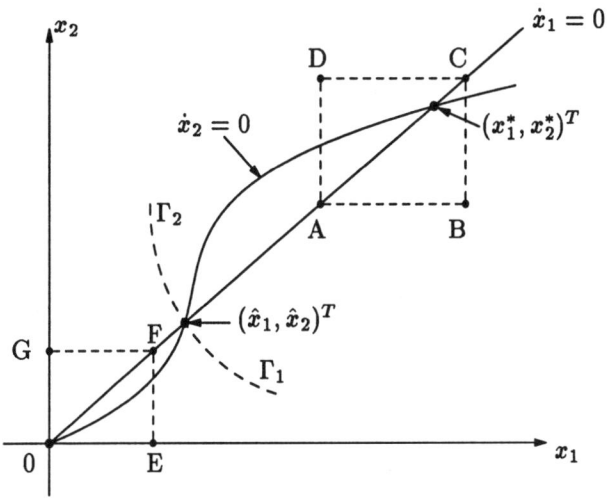

Figure XII.14 Phase plane of (6.1) with (6.13).

has a monotonically increasing x_2-value if its x_1-value decreases. In fact, the separatrix Γ is just the boundary between the basins of attraction of $(0,0)^T$ and $(x_1^*, x_2^*)^T$. Γ itself is the (one-dimensional) basin of attraction of the saddle point.

It is an interesting question whether Γ_1 cuts the x_1-axis or approaches it asymptotically. The same question can be asked for Γ_2 and the x_2-axis. A somewhat delicate analysis proves that both the x_1- and the x_2-axis are indeed cut. The proof does not give useful general insights, and we shall restrict ourselves to the numerical treatment therefore.

Because the problem is autonomous, we can easily obtain an ODE for the orbits, and in particular for the separatrix Γ. They satisfy

(6.15) $$\frac{dx_2}{dx_1} = \frac{a_{21}x_1^2 + (1 + \alpha x_1^2)a_{22}x_2}{(1 + \alpha x_1^2)(a_{11}x_1 + a_{12}x_2)} \ .$$

So we can find x_2 as a function of x_1 by integrating this equation, as long as the denominator is sufficiently bounded away from zero.

The only point we know on Γ is $(\hat{x}_1, \hat{x}_2)^T$. This point is inappropriate to be used as the initial value, because it is a stationary point, so that the right-hand side of (6.15) is not defined there. To circumvent this in a practical way, we can use the insight that at $(\hat{x}_1, \hat{x}_2)^T$ Γ is tangent to the eigenvector $\mathbf{v}_-$ in (6.7).

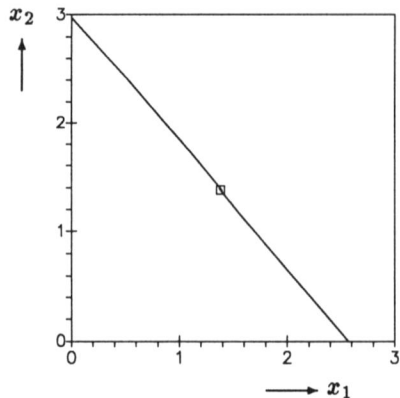

Figure XII.15   Saddle point $(1.38, 1.38)$, $\varepsilon = 0.01$, $\alpha = 0.2$.

Let us denote this (normalised) eigenvector by $(\nu_1, \nu_2)^T$. Approximations for the initial values are then

$$(6.16) \qquad x_2 \doteq \hat{x}_2 \pm \varepsilon \nu_2 \qquad \text{for } x_1 \doteq \hat{x} \pm \varepsilon \nu_1 \,, \qquad \varepsilon \ll 1 \,,$$

with the $+$ and $-$ signs referring to Γ_1 and Γ_2 respectively. For Γ_1 (6.15) has to be integrated for increasing argument until x_2 vanishes. For Γ_2 (6.15) has to be integrated for decreasing argument from $\hat{x}_1$ to 0. In Fig. XII.15 an example of a calculated Γ curve is given based on Fehlberg (ABSTOL $= 10^{-2}$, RELTOL $= 10^{-3}$), see §III.6.

We conclude that the use of (6.13) for the infection rate leads to a considerably more complex behaviour of the model than the use of (6.8). In the case of (6.13) no epidemic will break out if $4\alpha\beta^2 > 1$. But even if this condition is not fulfilled, no starting point below the Γ curves in Fig. XII.14 and Fig. XII.15 will give rise to an epidemic. Also, in these cases the population of viruses will die out.

7. Infectious diseases II

The epidemic model in §6 is suitable for infectious diseases with a uniform spatial distribution. Here we study the effect when this distribution is non-homogeneous. One may think of diseases penetrating a population via contacts between individuals. Examples are influenza, rabies, pest, and Aids. For convenience, the following assumptions are being made:

– The incubation time is negligibly short.

- The pathogens can only survive and reproduce themselves while residing in infected individuals.

- Every infected individual recovers or dies after some time. Recovered individuals become immune.

The model has to describe the time and space dependence of:

$y_1(t, \mathbf{x})$: concentration of individuals not yet infected and thus susceptible individuals,

$y_2(t, \mathbf{x})$: concentration of infected individuals.

In fact a third class y_3 plays a rôle in the model, namely the class of recovered and thus immune individuals. Because of the relation $y_1 + y_2 + y_3 = 1$ it is not necessary to introduce separate formulae for y_3.

The concentrations are taken with respect to the population density $\bar{y}$, averaged over the region and time period under consideration. In view of the considerations above, we describe the interactions between the two classes y_1 and y_2 by the following equations:

(7.1)
$$\begin{cases} \dfrac{\partial y_1}{\partial t} = -\alpha y_1 y_2 \ , \\[2ex] \dfrac{\partial y_2}{\partial t} = \alpha y_1 y_2 - \beta y_2 + \gamma \Delta y_2 \ . \end{cases}$$

The rationale of the different terms is as follows:

- $-\alpha y_1 y_2$ with rate of infection $\alpha > 0$. The product $y_1 y_2$ is a measure of the probability that a healthy individual meets an infected one. This term also appears in the second equation with opposite sign, because a decrease of y_1 leads to an increase of y_2.

- $-\beta y_2$ with rate of recovery $\beta > 0$. Recovered individuals become immune and leave the classes y_1 and y_2 once and for all.

- $\gamma \Delta y_2$ with Laplace operator $\Delta := \partial^2/\partial x^2 + \partial^2/\partial y^2$ and rate of diffusion $\gamma \geq 0$. This term describes that at least a subset of the infected individuals is still mobile.

The concentrations y_1 and y_2 are already dimensionless, but it is worthwhile to make the complete model non-dimensional. We shall apply the methods of §2. The dimensions of the variables and parameters are: $[t] = [T]$, $[\mathbf{x}] = [L]$, $[\alpha] = [L^2/T]$, $[\beta] = [1/T]$, $[\gamma] = [L^2/T]$, and $[\bar{y}] = [1/L^2]$. Appropriate dimensionless quantities are

(7.2) $\quad t^* := \alpha \bar{y} t \ , \quad \mathbf{x}^* := \sqrt{\dfrac{\alpha \bar{y}}{\gamma}} \mathbf{x} \ , \quad \beta^* := \dfrac{\beta}{\alpha \bar{y}} \ .$

The corresponding dimensionless version of model (7.1) is

(7.3a) $\dfrac{\partial y_1^*}{\partial t^*} = -y_1^* y_2^*$

(7.3b) $\dfrac{\partial y_2^*}{\partial t^*} = y_1^* y_2^* - \beta^* y_2^* + \Delta y_2^*$.

The model thus depends on only one parameter, namely β^*. In the following we shall use dimensionless quantities but omit the superscript $*$.

Partial differential equations, such as (7.3), are outside the scope of this book. However, the spatial propagation of the disease is appropriately studied if we use the fact that in many cases epidemics travel with a constant speed from one side of a continent to the opposite side with a more or less plane wave front; examples of such epidemics are rabies and influenza. Taking the direction in which such a *solitary wave* moves as the x-axis, the model becomes one-dimensional in space. The partial differential equation (7.3b) reduces further to an ordinary one if we introduce the variable $s = x + vt$. The velocity v is unknown in advance. The Laplace operator reduces to the second derivative with respect to s and model (7.3) transforms into the following set of first order equations:

(7.4)
$$
\begin{cases}
\dot{y}_1 = -y_1 y_2 / v \\[4pt]
\dot{y}_2 = y_3 \\[4pt]
\dot{y}_3 = v y_3 - y_1 y_2 + \beta y_2 ,
\end{cases}
$$

with the derivatives taken with respect to s. Here we have defined $y_3 := \dot{y}_2$ to have the system in standard (first order) form.

The analysis of a system with three components, such as in (7.4), is much more complicated than that of a system with two components. In the latter case the vector field can easily be visualised. It is therefore worthwhile to seek a reduction of the number of equations via constants of motion. Such a quantity is conserved along orbits in the phase space. Let us assume that $H(y_1, y_2, y_3)$ is a constant of motion. Then it must hold that

(7.5) $\dfrac{dH}{ds} = \dfrac{\partial H}{\partial y_1} \dot{y}_1 + \dfrac{\partial H}{\partial y_2} \dot{y}_2 + \dfrac{\partial H}{\partial y_3} \dot{y}_3 = 0$,

with the $\dot{y}_i$, $i = 1, 2, 3$, given by (7.4). This equation is satisfied by the expression

(7.6) $H = y_1 - \beta \ln y_1 + y_2 - \dfrac{y_3}{v}$.

Orbits of the system in phase space are thus contour lines, on which H has a constant value. The value of H in case of an epidemic must follow from the boundary conditions imposed for $s \to \pm\infty$. The cases $s = \pm\infty$ correspond to the situations long before and after the epidemic occurred. Long before the epidemic, everybody is healthy but susceptible, so $y_1(-\infty) = 1$. After the epidemic all individuals are (again) healthy, but some of them are still susceptible, because they did not get infected, while the others are immune. The number of susceptibles after the epidemic is unknown and is denoted by a. The wave profile approaches zero for $s \to \pm\infty$, so y_2 and its derivative y_3 vanish in these limits. In summary, the boundary conditions are

$$(7.7) \qquad y_1(-\infty) = 1, \quad y_1(+\infty) = a, \quad y_2(\pm\infty) = 0, \quad y_3(\pm\infty) = 0.$$

Hence (7.4) and (7.7) constitute a BVP, with four conditions for the four unknowns y_1, y_2, y_3, and v; note that it can be brought into standard form by adding the ODE $\dot{v} = 0$ to (7.4). Substituting the conditions at $s = -\infty$ into (7.6) we find that a solitary wave corresponds to the value $H = 1$. The conditions at $s = +\infty$ lead to the relation

$$(7.8) \qquad a - \beta \ln a = 1.$$

This is an implicit equation for a as a function of the (dimensionless) parameter β. Because $a < 1$ we have

$$(7.9) \qquad a < \beta < 1.$$

By substituting $H = 1$, we may eliminate one of the components, say y_3, after which (7.4) reduces to

$$(7.10) \qquad \begin{cases} \dot{y}_1 = -y_1 y_2/v \\ \dot{y}_2 = -v(1 - y_1 - y_2 + \beta \ln y_1). \end{cases}$$

The stationary points are $(a, 0)^T$ and $(1, 0)^T$. At $(a, 0)^T$ the eigenvalues $\lambda_\pm$ and eigenvectors $\mathbf{v}_\pm$ of the Jacobian matrix are given by

$$(7.11) \qquad \lambda_\pm := \tfrac{1}{2}\left(v \pm \sqrt{v^2 - 4(a - \beta)}\right)$$

$$(7.12) \qquad \mathbf{v}_\pm := \left(1, -\frac{a}{v\lambda_\pm}\right)^T.$$

The corresponding quantities at $(1, 0)^T$ are obtained by setting $a = 1$ in these expressions. Because y_1 and y_2 are concentrations, we have $y_1, y_2 \geq 0$. This implies that the eigenvalues $\lambda_\pm$ cannot be complex; for if they were complex, solutions approaching one of the stationary points would start to circle around it, thus leaving the first quadrant. This argument leads to the following condition on the velocity:

$$(7.13) \qquad v \geq 2\sqrt{1 - \beta}.$$

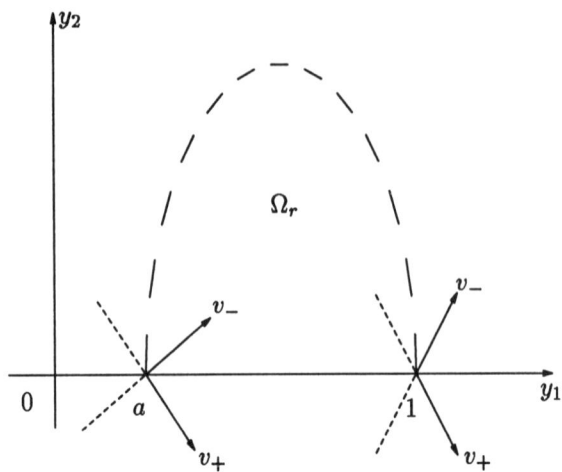

Figure XII.16   Phase plane of the system (7.10).

With this condition $(a,0)^T$ is a saddle point and $(1,0)^T$ an unstable node. A possible solitary wave solution will start in $(1,0)^T$ for $s = -\infty$ and finish in $(a,0)^T$ for $s = +\infty$. To show the existence of such a wave we focus on the region Ω, with $0 < r \leq 1$ defined by

$$(7.14) \qquad \Omega_r = \Big\{ (y_1, y_2) \mid a \leq y_1 \leq 1, \ 0 \leq y_2 \leq r(1 - y_1 + \beta \ln y_1) \Big\},$$

which is sketched in Fig. XII.16. The vector field points inwards in Ω_r along the y_1-axis between $y_1 = a$ and $y_1 = 1$. This is also the case along the upper boundary of Ω_r if r is chosen appropriately. Within Ω_r, $\dot{y}_1$ has a negative sign: all solutions within Ω_r will move in the direction of decreasing y_1. Let us follow an orbit starting at $(1,0)^T$ for $s = -\infty$ and leaving this point such that it enters Ω_r. The orbit cannot leave Ω_r and will monotonically move to the left. The only possibility is that it eventually converges to the utmost left point of Ω_r, i.e. $(a,0)^T$.

According to §V.4 the solitary wave solution approaches the saddle point $(a,0)^T$ along the stable manifold, or separatrix, which is tangent to $\mathbf{v}_-$ at $(a,0)^T$. In the present two-dimensional case the stable manifold coincides with the solitary wave we are looking for. Its numerical calculation is quite straightforward, because we can use the fact that the sign of $\dot{y}_1$ is constant along the orbit, so that y_1 is a monotonic function of t. This implies that we can use y_1 as independent variable in the solution instead of t. Eliminating t from (7.10) we obtain

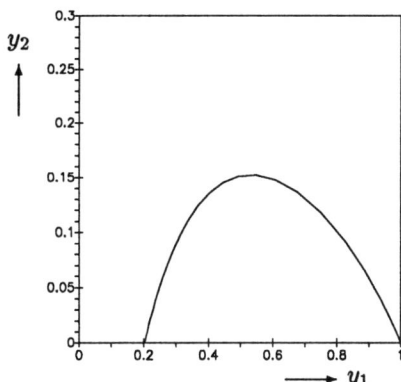

Figure XII.17 $\varepsilon = 10^{-6}$, $\beta = 0.5$, $v = 1.5$, $a = 0.2032$.

$$(7.15) \qquad \frac{dy_2}{dy_1} = \frac{v^2(1 - y_1 + \beta \ln y_1 - y_2)}{y_1 y_2} .$$

Considering v as a parameter, this equation is easily integrated, e.g. by Fehlberg. We cannot start at $(a, 0)^T$, because this point is stationary. As explained in §6, this problem can be circumvented by choosing the initial conditions $(a, 0)^T + \varepsilon v_-$ with $0 < \varepsilon \ll 1$. The resulting orbit in the phase plane is given in Fig. XII.17 for $\beta = 0.5$, $v = 1.5$, and $a = 0.2032$. To obtain y_1 as a function of t, one substitutes this solution $y_2(y_1)$ into the first equation of (7.10) and integrates. The time dependence of $y_2(t)$ follows immediately from $y_2(y_1(t))$. The resulting $y_1(t)$ and $y_2(t)$ curves are given in Fig. XII.18.

We conclude that solitary wave solutions exist for all velocities satisfying (7.13), provided that $\beta < 1$. To prevent the occurrence of an epidemic, one should make sure that

$$(7.16) \qquad \beta^* = \frac{\beta}{\alpha \bar{y}} \geq 1 .$$

Of course, if we really wish to know the actual concentrations (i.e. the solution of (7.4), (7.7)), we have to employ a nonlinear BVP solver and provide it with a good estimate; such an estimate may be found from a qualitative analysis as above.

In practice it is very hard to influence the recovery rate β for this type of disease. In some cases the infection rate α can be diminished by vaccination. If this is not possible one has to resort to keeping the average concentration of individuals below a critical value. A striking example of this strategy is given

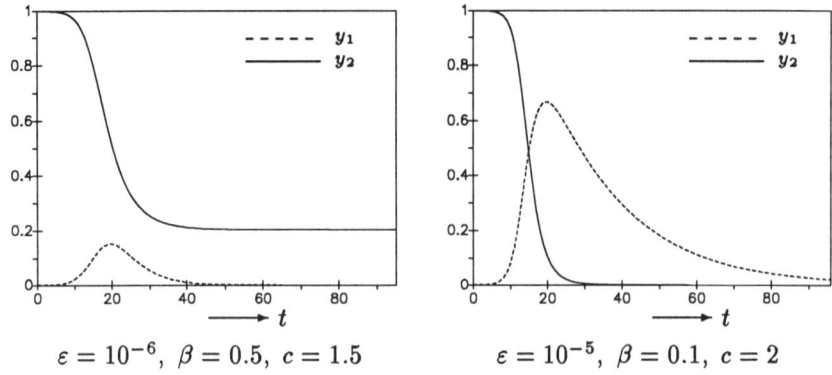

$$\varepsilon = 10^{-6}, \quad \beta = 0.5, \quad c = 1.5 \qquad\qquad \varepsilon = 10^{-5}, \quad \beta = 0.1, \quad c = 2$$

Figure XII.18

in [9]. To fight the spread of rabies among foxes, corridors have been created
in which the number of foxes is small because of intense hunting. Corridors
of about 15 km turned out to be effective barriers against rabies propagating
through the fox population.

8. Nerve conduction

Nerves transmit signals through electric pulses, which propagate along the
nerves at high speed. This phenomenon has been extensively studied. The
physiological processes involved are quite complex, but mainly based on the
transport of K^+ and Na^+ ions across the nerve membrane. In 1952 a model
was published by [47], which highly stimulated the research in this field. The
model described a lot of the characteristics of processes of this kind quite well.
A special version of the model was proposed by Fitzhugh and Nagumo in 1961.
See [55]. The latter is the subject of this section. We shall derive the model
equations, analyse some special cases, and outline the results of a full analysis.

The essential part of the nerve is the membrane. We model this part as an
infinitely long, hollow cylinder. The voltage difference V between the outside
and the inside of the cylinder is the central quantity in the model. The currents
through the membrane are very small, but the resistance is quite high, so that
V can still be measured accurately. The potential V depends on time t and
position x along the membrane. Because of cylinder symmetry the system
is in fact one-dimensional. The electrical properties of the membrane can be
described in terms of elementary components: resistor, capacitor, and inductor.

The relations between the current I through and the voltage drop V over these components are

(8.1)
$$\begin{cases} V = IR & \text{(linear resistor)} \\[2mm] I = C\dfrac{dV}{dt} & \text{(capacitor)} \\[2mm] V = L\dfrac{dI}{dt} & \text{(inductor)} \end{cases}$$

with R the resistivity, C the capacity, and L the inductivity. The spatial coordinate x is continuous. For modelling purposes, however, we discretise it, cutting the membrane into small parts of width h. We assume that between neighbouring parts only axial currents, and within a part only radial currents occur.

In the Fitzhugh-Nagumo model the membrane is described by the electric circuit drawn in Fig. XII.19. The lower horizontal wire represents the outside of the membrane. Outside the membrane the potential is taken to be the same everywhere. The upper horizontal wire represents the interior of the nerve, which has electrical resistivity R_i in the axial direction. A voltage drop V gives rise to three types of currents through the membrane. The fast transport of ions is denoted as current I_4; it is in general a nonlinear function of V, say $f(V)$. In what follows we shall investigate two expressions for $f(V)$. The slow transport of ions is represented by the current I_5. This branch consists of an inductor with inductivity L, and a linear resistor with resistivity R_a. The capacitor branch with current I_3 represents the ability of the membrane to accumulate a number of ions temporarily.

We shall first derive the equations for the discrete model in Fig. XII.19. After that the limit $h \downarrow 0$ will be taken. It is important to note that the currents I_3, I_4, I_5, and the parameters C, L and R_a are defined per unit area. The identical parts of the membrane have surface area $2\pi a h$. The currents I_1 and I_2 and the parameter R_b are defined per unit length. Applying the basic laws (8.1) we obtain the following equations for the currents $I_1, ..., I_4$:

(8.2)
$$\begin{cases} I_1 = -\dfrac{V_i - V_{i-1}}{h\,R_b}\,, \\[3mm] I_2 = -\dfrac{V_{i+1} - V_i}{h\,R_b}\,, \\[3mm] I_3 = 2\pi a h\,C\,\dfrac{\partial V_i}{\partial t}\,, \\[3mm] I_4 = 2\pi a h\,f(V_i)\,, \end{cases}$$

where V_i is short for $V(t, x_i)$. Note that all voltages and currents depend on time and position.

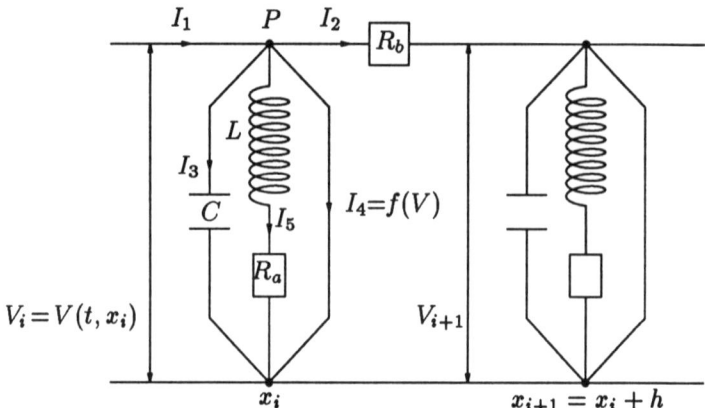

Figure XII.19 Electric circuit representing the currents and voltages involved in nerve conduction.

Application of Kirchhoff's first law, which states that the sum of all currents in a point of a circuit vanishes, to the point P in Fig. XII.19 yields the differential-difference equation

(8.3) $\qquad C\dfrac{\partial V_i}{\partial t} = \dfrac{1}{2\pi a\, R_b}\ \dfrac{V_{i+1} - 2V_i + V_{i-1}}{h^2} - I_5 - f(V_i)\ .$

Taking the limit $h \downarrow 0$ we obtain the (partial) differential equation:

(8.4) $\qquad C\dfrac{\partial V_i}{\partial t} = \dfrac{1}{2\pi a\, R_b}\ \dfrac{\partial^2 V_i}{\partial x^2} - I_5 - f(V)\ .$

The I_5 branch yields a separate equation. Because I_5 is the only current which plays a rôle, we shall omit the index and write $I := I_5$ in what follows. The sum of the voltage drops over the inductor and the resistor must be equal to V:

(8.5) $\qquad V = L\dfrac{\partial I}{\partial t} + I\, R_a\ .$

Equations (8.4) and (8.5) establish our model. The number of model parameters is quite large and it certainly pays to put the model into dimensionless form. To that end we need characteristic values for the voltage drop and the current. It is appropriate to relate these values to properties of the function f which governs the current I_4. Let us consider the system in the rest situation. Taking both the time and the spatial derivatives in (8.4) and (8.5) equal to zero, we find that the (uniform) equilibrium potential V_E satisfies the equation

(8.6) $V_E + f(V_E)R_a = 0$.

For the functions f to be studied below, this equation has a unique non-vanishing solution.

A characteristic value R_E for the resistor is provided by

(8.7) $R_E = \left[\left(\dfrac{df}{dV}\right)_{V=V_E}\right]^{-1}$.

In terms of V_E and R_E the dimensionless quantities are:

(8.8)
$$
\begin{cases}
t^* := \left(\dfrac{1}{C\,R_E}\right) t \ , \\[2mm]
x^* := \left(\dfrac{R_E}{2\pi a\, R_b}\right)^{1/2} x \ , \\[2mm]
V^* := \left(\dfrac{1}{V_E}\right) V \ , \\[2mm]
I^* := \left(\dfrac{R_E}{V_E}\right) I \ , \\[2mm]
f^*(V^*) := -\left(\dfrac{R_E}{V_E}\right) f(V) \ .
\end{cases}
$$

Omitting the superscript $*$, the dimensionless model reads:

(8.9)
$$
\begin{cases}
\dfrac{\partial V}{\partial t} = \dfrac{\partial^2 v}{\partial x^2} - I + f(V) \ , \\[2mm]
\dfrac{\partial I}{\partial t} = \varepsilon(V - \gamma I) \ .
\end{cases}
$$

The system is thus governed by two dimensionless parameters:

(8.10)
$$
\begin{cases}
\varepsilon := \dfrac{C\,R_E^2}{L} \ , \\[2mm]
\gamma := \dfrac{R_a}{R_E} \ .
\end{cases}
$$

Substitution of typical, measured values for C, L, R_a, and R_E yields that $\varepsilon \ll 1$ and $\gamma \approx 1$. Hence (8.9) is a singularly perturbed system. This motivates us to investigate the model first under the assumption $\varepsilon = 0$, i.e. the reduced problem.

(i) The reduced problem

When $\varepsilon = 0$, the current I is constant and the model consists of one equation only:

(8.11) $\dfrac{\partial V}{\partial t} = \dfrac{\partial^2 V}{\partial x^2} + f(V)$.

Nerve conduction models have to describe at least one characteristic aspect found from measurements, namely that electrical pulses may travel along the nerve at constant speed. To investigate whether (8.11) possesses solitary wave solutions, we assume V to be of the form $V(t, x) = V(s)$ with $s = x + vt$. The wave velocity v is not known in advance. Because the system has no preferred direction we may take $v > 0$. Under this assumption and with the notation $\mathbf{y} := (y_1, y_2)^T := (V, dV/ds)^T$ we write (8.11) as

$$(8.12) \quad \dot{\mathbf{y}} = \begin{bmatrix} y_2 \\ v\,y_2 - f(y_1) \end{bmatrix} .$$

The Jacobian matrix of this system is

$$(8.13) \quad \mathbf{J}(y_1) := \begin{bmatrix} 0 & 1 \\ -f'(y_1) & v \end{bmatrix} ,$$

with eigenvalues

$$(8.14) \quad \lambda_\pm := \frac{v}{2} \pm \sqrt{\left(\frac{v}{2}\right)^2 - f'(y_1)} ,$$

and eigenvectors

$$(8.15) \quad \mathbf{v}_\pm := \begin{bmatrix} 1 \\ \lambda_\pm \end{bmatrix} .$$

The behaviour of the solutions of (8.12) depends on the form of the current function f. We consider two nonlinear expressions. Linear models do not possess solitary wave solutions.

(ii) Fisher's f function

Fisher, see [55], has suggested taking

$$(8.16) \quad f(y_1) = y_1 (1 - y_1) .$$

The corresponding stationary points of (8.12) are $A := (0, 0)^T$, $B := (1, 0)^T$. In both points y_2, and thus the slope dV/ds, vanishes, but the values of V are different. Evaluation of $\lambda_\pm$ at A shows that the origin is unstable for all velocity values v: for $v \geq 2$ it is an unstable node, for $v < 2$ an unstable spiral point. Evaluation of $\lambda_\pm$ at B shows that this is a saddle point. These observations have important consequences. The measured travelling wave pulses along nerves correspond to homoclinic solitary waves. Such a wave becomes completely flat and vanishing for $s \to \pm\infty$. This implies that the corresponding orbit in the plane approaches the origin for $s \to \pm\infty$.

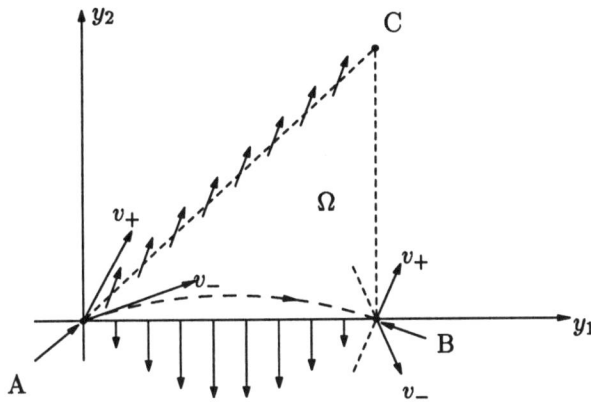

Figure XII.20   Phase plane of the reduced system (8.12) with (8.16).

Because the origin is unstable in the Fisher model, no such solitary wave exists. However, we shall show that the model has a heteroclinic solitary wave solution starting at the origin A for $s = -\infty$ and finishing in B for $s = +\infty$. For $v < 2$ the origin is a spiral point. Solutions leaving such a point tend to spiral around it. In the present case such a solution would then leave the first quadrant. Because this would correspond to concentrations becoming negative, we impose the condition

$$v > 2 \ .$$

Let us focus on the right-angled triangle Ω in Fig. XII.20 with vertices A, B, and $C := (1, b)^T$. The value of b can be chosen such that the vector field of (8.12) with (8.16) points outward along the hypotenuse AC. The slope of the hypotenuse is equal to b and the slope of the vector field along this line is $v - (1 - y_1)/b$. In the interval $0 \le y_1 \le 1$ the latter value is bigger than the former for $0 \le y_1 \le 1$ if $b \ge \lambda_-$. So we choose b such that the slope of AC is steeper than that of $\mathbf{v}_-$, as shown in Fig. XII.20. Along AB the vector field points downward and thus outward from Ω, too.

Let us follow a possible solitary wave in the reverse direction. This implies that the direction field also reverses. We start at $s = +\infty$ in B and leave this point in the direction of v_- so that the orbit enters Ω. Because the sign of $\dot{y}_1$ is constant (and negative for reversed s) in the upper half-plane, as seen from (8.12), the orbit we follow has to move to the left. This orbit cannot pass the line segments AB and AC because of the direction of the vector field

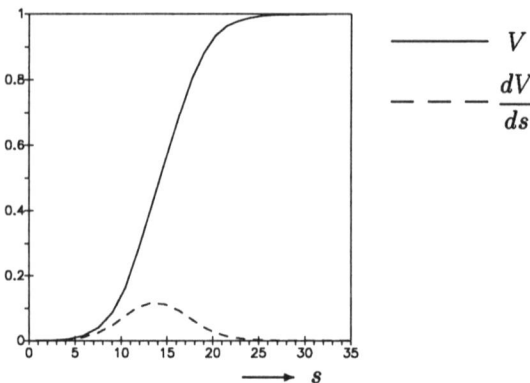

Figure XII.21 Solitary wave solution of (8.12) with (8.16) for velocity $v = 2.1$.

along these sides. We conclude that this orbit converges to A for $s \to -\infty$. As discussed in §7, the solitary wave, whose existence is shown in this way, is precisely the separatrix which leaves the saddle point B tangent to $\mathbf{v}_-$.

We find that (8.12) with (8.16) has solitary wave solutions for all velocities $v > 2$. These waves are heteroclinic: the voltages in front of and behind the wavefront are different. In Fig. XII.21 both V and dV/ds are given as functions of s for $v = 2.1$.

(iii) Nagumo's f function

Nagumo, see [55], suggested the following refinement:

$$(8.17) \qquad f(y_1) := y_1(1 - y_1)(c - y_1) , \qquad 0 < c < 1/2 .$$

Model (8.12) with (8.17) has three stationary points: $A := (0,0)^T$, $B := (1,0)^T$ and $E := (a,0)^T$. They are shown in Fig. XII.22.

The points A and B are saddle points and E is an unstable node or spiral point. Because the sign of $\dot{y}_1$ is fixed and positive in the first quadrant, all orbits move to the right as s increases. If a solitary wave exists it can only be a heteroclinic one, starting at A and finishing at B. To show its existence we shall use a technique reminiscent of single shooting. The solitary wave we are looking for will be tangent to $\mathbf{v}_+$ at A. According to (8.15), the slope of $\mathbf{v}_+$ is determined by the value of λ_+ at A. This value depends on the still unknown velocity v:

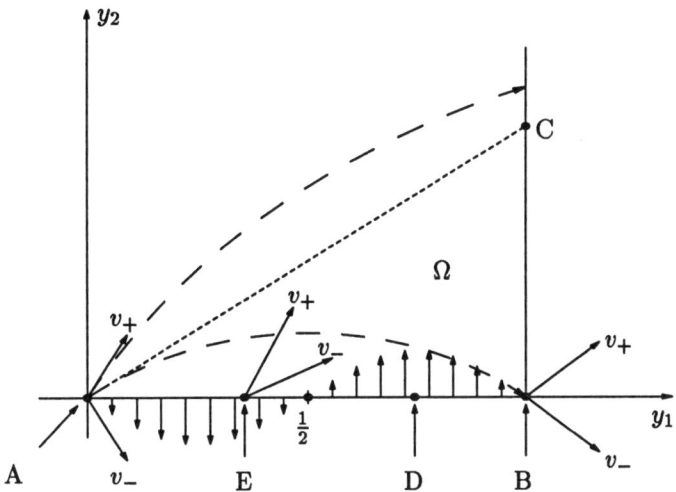

Figure XII.22 Phase plane of the reduced system (8.12) with (8.17).

$$(8.18) \qquad \lambda_+(A) := \frac{v}{2} + \sqrt{\left(\frac{v}{2}\right)^2 + c} \, .$$

Hence we see that λ_+, and thus the slope of $\mathbf{v}_+$ at A, is a monotonically increasing function of v. Let us investigate the two cases that v is very large and very small. We first select an arbitrary value for b. For large values of v the slope of v_+ at A, and thus of the solitary wave we are looking for, will be steeper than the slope b of AC. See Fig. XII.22. In fact the value of v can be chosen such that the orbit of the solitary wave lies above the line segment AC in the interval $0 \le y_1 \le 1$. The line segment AC is defined by $y_2 = by_1$, $0 \le y_1 \le 1$. The slope r of the vector field along AC is given by

$$(8.19) \qquad r(y_1) := v - \frac{(y_1 - c)(1 - y_1)}{b} \, .$$

From this we see that v can be chosen such that $r(y_1) > b$ for $0 \le y_1 \le 1$. So, an orbit leaving A above the line segment AC will remain above AC and hit the vertical line through B and C above C.

Next, we consider the case $v = 0$. Then, ODE (8.12) reduces to

$$(8.20) \qquad \ddot{y}_1 = -f(y_1) \, .$$

This ODE is of the Hamiltonian type, see §XI.3. The corresponding *potential* is given by

$$(8.21) \qquad P(y_1) := \int\limits_0^{y_1} f(y)\, dy \; ,$$

where the lower bound in the integral is arbitrarily chosen, because only the derivative of P is of importance. The Hamiltonian, in this case given by the sum of the kinetic and potential energy, is a constant of motion. Thus

$$(8.22) \qquad H := \tfrac{1}{2} y_2^2 + P(y_1) \; .$$

The value of the energy follows from the initial conditions of the orbit under consideration. For a solitary wave starting at $A = (0,0)^T$ we have $H \equiv 0$ and thus

$$(8.23) \qquad y_2(y_1) = \sqrt{-2P(y_1)} \; .$$

The solitary wave starts at $A = (0,0)^T$ and moves into the first quadrant in the direction of increasing y_1 as follows from (8.12). From the Nagumo expression for f, see (8.17), we directly see that $P(y_1)$ is positive for small y_1 initially. However, for some y_1 in the interval $0 \le y_1 \le 1$, $\bar{y}_1$ say, we have

$$(8.24) \qquad P(\bar{y}_1) = \int\limits_0^{\bar{y}_1} f(y)\, dy = 0 \; .$$

So, $y_2(\bar{y}_1) = 0$, from which we conclude that the solitary wave corresponding to $v = 0$ hits the y_1-axis at $(\bar{y}_1, 0)^T$. Now we may apply a continuity argument. Consider the curve consisting of the line segments DB and BC in Fig. XII.22. For $v = 0$, the orbit leaving A in the direction of $\mathbf{v}_+$ hits one end point D. For a certain v value, $v_0 > 0$ say, this orbit hits the other end point C. When v varies from 0 to v_0 the intersection point of the orbit with the line segments DB and BC will move monotonically from D to C, because the orbits corresponding to different values of v cannot cross. We conclude that for some value of v the saddle point $B = (1,0)^T$ is hit, and this is precisely the heteroclinic, solitary wave we are looking for. In B this orbit must be tangent to $\mathbf{v}_-$.

Further analysis of the nerve conduction model (8.9) with $\varepsilon \ne 0$ involves many other technical details which are not treated here; see, e.g., [54]. We conclude with some remarks.

(i) In the analysis above we set $\varepsilon = 0$. This implies that the capacitor branch in Fig. XII.19, representing the ability to store electric charge temporarily, is ignored. Both the Fisher (8.16) and the Nagumo (8.17) equations for $f(V)$ possess solitary wave solutions. In both models only heteroclinic waves are allowed, for which the voltage drop V has different values before and after the wavefront. This does not agree with experiments. The electrical pulses, travelling along nerves, are homoclinic. Far from the pulse in the tails of the travelling wave, V approaches zero and its derivative tends to vanish. The approximation $\varepsilon = 0$ is apparently too crude.

(ii) The Fisher model (8.16) allows for solitary wave solutions at infinitely many velocities, whereas the Nagumo model (8.17) has only one such solution with a uniquely determined velocity. In experiments it has been observed that a specific nerve geometry corresponds to a unique wave velocity.

(iii) The solitary wave solutions of the Fisher model are easy to calculate numerically. Starting at $(1, 0)$ and following the wave in the negative s direction nearly each numerical integration scheme will yield a solution which converges to $(0, 0)$ because this point is an attractor.

(iv) The solitary wave solution of the Nagumo model, however, is hard to calculate. The velocity of the wave is not known in advance, but has to be estimated, e.g. using a shooting method. See, e.g., [54]. The direction in which the solutions leave $(0, 0)$ depends on the still unknown value of v. The shooting process turns out to be highly unstable. Reversion of the direction of s does not improve this, because the solution to be calculated travels from saddle point to saddle point. One can also reformulate the problem as a boundary value problem by adding the equation $\dot{v} = 0$.

(v) The complete Fitzhugh-Nagumo model consists of (8.9) with $\varepsilon \neq 0$ and the Nagumo expression (8.17). This model is three-dimensional with the origin as saddle point. It can be shown that the complete model has a homoclinic solitary wave solution starting and finishing at the origin. It describes the phenomena observed in nerve conduction quite well.

9. Torsion in a crank shaft

In compressors the crank shaft is a relatively vulnerable part. Via the crank the rotation of the shaft is transformed into an oscillating motion of the piston (or vice versa as in a combustion engine). In this a periodic force is exerted on the shaft, which may give rise to torsional oscillations in the shaft. If the amplitudes of these oscillations are too large, the axis is in danger of getting ruptured. We study this phenomenon using a mathematical model, which, though simple, still contains the essential features of realistic compressors. The model describes a shaft with one crank and one piston. The shaft is driven by a powerful engine and we assume that this happens with constant angular velocity ω, irrespective of the load of the crank. In practice this assumption is quite reasonable, because most compressors are equipped with a heavy flywheel. Friction will be ignored; in practice the losses due to friction are very small. The system is sketched in Fig. XII.23.

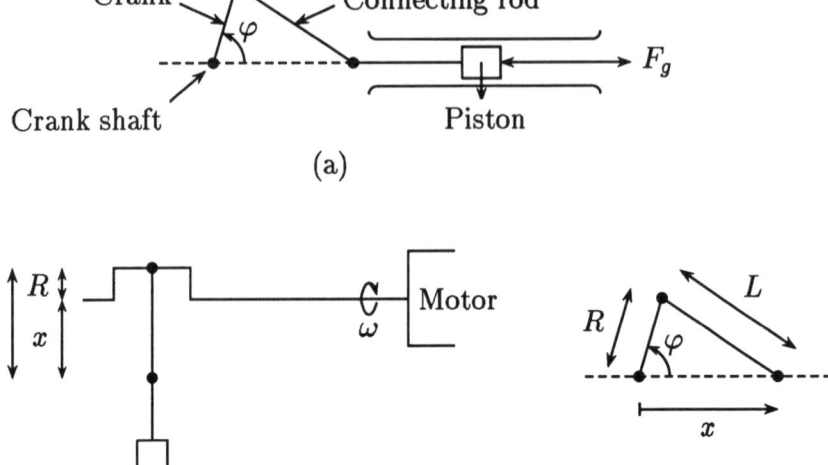

Figure XII.23 Schematic diagrams of a compressor: (a, c) side views, (b) top view.

Clearly this is a mechanical system with kinematic constraints, so we have a DAE associated with it. The system is conveniently described within the Lagrangian formalism dealt with in §XI.2. We denote the rotation angle of the crank by $\varphi(t)$. The torsion $\varphi(t)$ may deviate from the rotation angle ωt of the engine. For the derivation of the equation of motion we need an expression for the kinetic energy. It is the sum of the kinetic energies T_c of the crank, T_{cs} of the crank shaft, T_p of the piston, and T_{cr} of the connecting rod:

(9.1) $T = T_c + T_{cs} + T_p + T_{cr}$.

The crank and the crank shaft are rigid bodies rotating around a fixed axis with angular velocity $\dot\varphi$. For these components we have

(9.2) $T_c + T_{cs} = \frac{1}{2}\left(I_c + I_{cs}\right)\dot\varphi^2$

with I_c and I_{cs} the constant moments of inertia of crank and crank shaft respectively. Though the piston and connecting rod do not rotate, T_p and T_{cr} can be cast into a similar form provided that φ-dependent moments of inertia $I_p(\varphi)$ and $I_{cr}(\varphi)$ are introduced. We show this for the piston. Denoting the position of the piston by x (see Fig. XII.23), its kinetic energy is given by

(9.3) $T_p = \frac{1}{2} m_p \, \dot{x}^2$,

with m_p the piston mass. From the geometry it follows that x and φ are related by

(9.4) $x(\varphi) = L(\lambda \cos \varphi + \sqrt{1 - \lambda^2 \sin^2 \varphi})$

where λ is defined as $\lambda := R/L$. By differentiation of (9.4) we find an expression for $\dot{x}$ in terms of $\dot{\varphi}$. Substitution into (9.3) then yields

(9.5)
$$\begin{cases} T_p = \frac{1}{2} I_p(\varphi) \, \dot{\varphi}^2 \, , \\ I_p(\varphi) = m_p \, R^2 \, f^2(\varphi) \, , \end{cases}$$

where the dimensionless function f is defined by

(9.6) $f(\varphi) = (1 + g(\varphi, \lambda)) \sin \varphi$,

with

(9.7) $g(\varphi, \lambda) = \dfrac{\lambda \cos \varphi}{\sqrt{1 - \lambda^2 \sin^2 \varphi}}$.

To obtain an expression for I_{cr}, we note that the motion of the connecting rod is the sum of the linear motion of its centre of mass and the rotation with respect to its centre of mass. Omitting details we mention that I_{cr} can be expressed in terms of the mass m_{cr} of the connecting rod, its moment of inertia I_{cr} with respect to its centre of mass, and a third dimensionless function h given by

(9.8) $h^2(\varphi) := (1 - p)^2 \cos^2 \varphi + \left(1 + p\, g(\varphi, \lambda)\right)^2 \sin^2 \varphi$.

The parameter p, with $0 < p < 1$, denotes the position of the centre of mass. We have $L_1 = pL$ and $L_2 = (1-p)L$ with L_1 and L_2 as indicated in Fig. XII.24. The expression for $I_{cr}(\varphi)$ reads:

(9.9) $I_{cr}(\varphi) = m_{cr} \, R^2 \, f^2(\varphi) + I_{cr} \, h^2(\varphi)$.

In practice, one often has a geometry with $\lambda \ll 1$. Then we may approximate up to first order in λ:

(9.10)
$$\begin{cases} f(\varphi) & = \; \sin \varphi + \frac{1}{2} \lambda \sin(2\varphi) \\ g(\varphi) & = \; \lambda \cos \varphi \\ h^2(\varphi) & = \; 1 + p(p - 2) \cos^2 \varphi + \lambda p \sin \varphi \sin 2\varphi \, . \end{cases}$$

So, in that case we have $g^2 \ll h^2$.

We assume the torsion $\varphi - \omega t$ to vary linearly along the crank shaft between crank and engine and take for the torsional moment at the crank position a harmonic repulsive force:

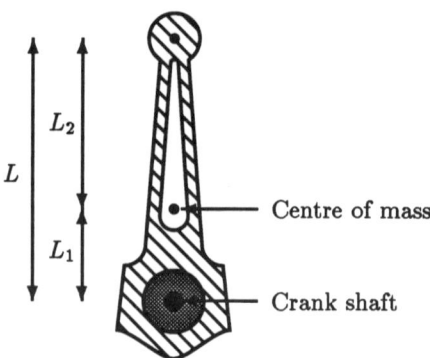

Figure XII.24   The connecting rod and its centre of mass.

(9.11) $M_{\text{torsion}}(\varphi) = -k(\varphi - \omega t)$, $k > 0$.

To the gas force F_{gas} exerted on the piston there is a corresponding gas moment given by

(9.12) $M_{\text{gas}}(\varphi) = F_{\text{gas}} \dfrac{dx}{d\varphi} = F_{\text{gas}} \, R \, f(\varphi)$.

Substitution of the expressions for T, M_{torsion}, and M_{gas} into the Lagrange equation (see §XI.2)

(9.13) $\dfrac{d}{dt}\left(\dfrac{\partial T}{\partial \dot\varphi}\right) - \dfrac{\partial T}{\partial \varphi} = M_{\text{torsion}} + M_{\text{gas}}$,

yields the equation of motion

(9.14) $I\,\ddot\varphi + \tfrac{1}{2}\,\hat I\,\dot\varphi^2 + k(\varphi - \omega t) = M_{\text{gas}}(\varphi)$,

with $\hat I := dI/d\varphi$. This equation can be put into dimensionless form by division through $I(\varphi)$ (which is always positive), and introduction of the dimensionless variable

(9.15) $\tau := \omega t$.

We define $\varphi^*(\tau) \equiv \varphi(t)$ and omit the $*$ index in the following. The resulting equation reads

(9.16) $\ddot\varphi + \dfrac{1}{2}\left(\dfrac{\hat I}{I}\right)\dot\varphi^2 + \dfrac{k}{\omega^2 I}\,(\varphi - \tau) = \dfrac{1}{\omega^2 I}\,M_{\text{gas}}(\varphi)$,

where differentiation is meant with respect to τ. We are interested in the stability properties of the torsion angle

(9.17) $y(\tau) := \varphi(\tau) - \tau$,

rather than in the angle φ itself. We note that after the transition from φ to y the moments become functions of both y and τ:

$$I = I(y + \tau) \ , \qquad M_{gas} = M_{gas}(y + \tau).$$

In terms of y the equation of motion reads

(9.18) $\ddot{y} + \dfrac{\hat{I}}{I}\dot{y} + \dfrac{1}{2}\dfrac{\hat{I}}{I}\dot{y}^2 + \dfrac{k}{\omega^2 I}y = \dfrac{1}{\omega^2 I}M_{gas} - \dfrac{1}{2}\dfrac{\hat{I}}{I}$,

with $\hat{I}(y + \tau) := \partial I(y + \tau)/\partial y$.

In practice the torsion angle y has to remain very small. This suggests that we linearise (9.18) around $y = 0$. The equation is first written into the standard form. Next, the Jacobian matrix of the vector field in $\mathbf{y} = 0$ is evaluated. Substitution of this matrix into the linearised equation then yields

(9.19) $\ddot{y} + a(\tau)\dot{y} + b(\tau)y = F(\tau)$,

with

(9.20)
$$\begin{cases} a(\tau) = \dfrac{\hat{I}(\tau)}{I(\tau)} \ , \\[2mm] b(\tau) = \dfrac{k}{\omega^2 I(\tau)} + \tfrac{1}{2}\dot{a}(\tau) - \dfrac{1}{\omega^2 I(\tau)}\dot{M}_{gas}(\tau) \ , \\[2mm] F(\tau) = \dfrac{1}{\omega^2 I(\tau)}M_{gas}(\tau) - \tfrac{1}{2}a(\tau) \ , \end{cases}$$

(with $\dot{a} := da/d\tau$), etc. The coefficients $a(\tau)$ and $b(\tau)$ and the driving force $F(\tau)$ are 2π-periodic in τ.

The term $b(\tau)y$ in (9.19) represents a harmonic force with time dependent spring constant. In practice the values of k and ω are chosen such that, for a given geometry, the sign of b is negative; the harmonic force is then repulsive and the solutions bounded, unless resonance occurs. The sign of $a(\tau)$ is not fixed and the term $a(\tau)\dot{y}$ in (9.19) cannot be interpreted as a friction force.

For convenience we rewrite (9.19) into a form without the $\dot{y}$ term. To that end we apply the Liouville transformation (see §IV.6)

(9.21) $x(\tau) := y(\tau)\exp\left(\tfrac{1}{2}\displaystyle\int_0^\tau a(s)\,ds\right)$.

This leads to

(9.22) $\ddot{x} + c(\tau)x = G(\tau)$,

with

(9.23)
$$\begin{cases} c(\tau) := b(\tau) - \tfrac{1}{2}\dot{a}(\tau) - \tfrac{1}{4}a^2(\tau) , \\[2mm] G(\tau) := F(\tau)\exp\left(\tfrac{1}{2}\displaystyle\int_0^\tau a(s)\,ds\right) . \end{cases}$$

We note that the stability properties of x and y for $\tau \to \infty$ are the same. This follows from the fact that

$$(9.24)\qquad \int_0^{2\pi} a(\tau)\,d\tau = 0 ,$$

because $a(\tau) = d\log I/d\tau$, and I is 2π-periodic. Thus x and y coincide for $\tau = 2\pi n$, $n \in \mathbb{N}$. Between these points their difference has an upper bound independent of τ.

In standard form (9.22) reads:

$$(9.25)\qquad \dot{\mathbf{x}} = \mathbf{A}(\tau)\,\mathbf{x} + \mathbf{b}(\tau)$$

with $\mathbf{x} := (x,\dot{x})^T$, $\mathbf{b} := (0,G)^T$, and

$$(9.26)\qquad \mathbf{A}(\tau) := \begin{bmatrix} 0 & 1 \\ -c(\tau) & 0 \end{bmatrix} .$$

The general solution of equations like (9.25) is given in §IV.3. Because $\mathbf{A}$ is periodic, we may apply the Floquet theorem, see §IV.5. An important rôle is played by the fundamental matrix $\mathbf{Y}(\tau)$. It is the solution of the initial value problem

$$(9.27)\qquad \begin{cases} \dot{\mathbf{Y}} = \mathbf{A}(\tau)\,\mathbf{Y} , \\[2mm] \mathbf{Y}(0) = \mathbf{I} . \end{cases}$$

The columns of $\mathbf{Y}$ are thus calculated by integrating the homogeneous differential equation $\dot{\mathbf{x}} = \mathbf{A}(\tau)\,\mathbf{x}$ starting from standard basis vectors $(1,0)^T$ and $(0,1)^T$. The stability of solutions of the homogeneous part of (9.25) is determined by the eigenvalues of $\mathbf{Y}(2\pi)$. These eigenvalues $\lambda_\pm$ are called *characteristic multipliers*. Let us introduce the notation

$$(9.28)\qquad \begin{cases} d = \det\left(\mathbf{Y}(2\pi)\right) , \\[2mm] s = \tfrac{1}{2}\operatorname{Tr}\left(\mathbf{Y}(2\pi)\right) . \end{cases}$$

The eigenvalues $\lambda_\pm$ of $\mathbf{Y}(2\pi)$ are solutions of

$$(9.29)\qquad \lambda^2 - 2s\lambda + d = 0 .$$

Since $\text{Tr}(\mathbf{A}) = 0$ it follows from Theorem IV.2.14 that $d = 1$. The $\lambda_\pm$ are thus given by

$$(9.30) \qquad \lambda_\pm = s \pm \sqrt{s^2 - 1}\,.$$

We distinguish three cases:

(i) $|s| > 1$: λ_+, λ_- real; one of them has an amplitude larger than one, and the other has an amplitude smaller than one.

(ii) $|s| = 1$: $\lambda_+ = \lambda_- = +1$ or $\lambda_+ = \lambda_- = -1$.

(iii) $|s| < 1$: λ_+, λ_- is a complex conjugate pair of unit magnitude.

If $|s| > 1$, one of the λ has magnitude larger than one, which gives rise to unbounded solutions. For $|s| \leq 1$, both λ have unit magnitude so that the solutions will be bounded and periodic. The period is $2\pi n$, with n the smallest integer for which both

$$(9.31) \qquad \lambda_+^n = 1 \quad \text{and} \quad \lambda_-^n = 1\,.$$

Some special cases are $n = 1$ if $s = 1$, $n = 2$ if $s = -1$, $n = 4$ if $s = 0$, and $n = \infty$ if s and thus $\lambda_\pm$ are irrational numbers. In the latter case the solutions are *quasi-periodic*.

The full inhomogeneous equation (9.25) may have unbounded solutions if

- the homogeneous version shows this behaviour, thus if $|s| > 1$.

- the Fourier spectrum of the driving force $G(\tau)$ contains a component with the same frequency as a periodic solution of the homogeneous part, and resonance occurs. The driving force G in (9.23) is 2π-periodic in τ. The frequencies in its spectrum are thus the positive integers $1, 2, 3, \dots$. The frequencies of the periodic solutions of the homogeneous part of (9.25) are $1/(2\pi n)$, with n integer and ≥ 1. From this we directly conclude that resonance occurs only if $n = 1$, thus if $s = +1$. In all other cases the frequencies of the periodic solutions lie below the lowest frequency in the Fourier spectrum of G.

In summary, we conclude that, in terms of the stability definitions given in §V.1, the origin is unstable if $|s| > 1$, and Lyapunov stable if $|s| < 1$. We note that the origin is not a stationary point. So, a solution starting at the origin will not stay there. Its distance to the origin will remain small if the system is in a stable mode.

As an example we evaluate the formalism, dealt with above, for a compressor with geometrical parameters $I_c = 10$ kg m^2, $R = 0.15$ m, $L = 1.0$ m, $m_p = 900$ kg, and $k = 5.10^6$ Nm. We assume the influences of crank shaft and connecting rod to be negligible and set $I_{cs} = I_{cr} = 0$. To demonstrate

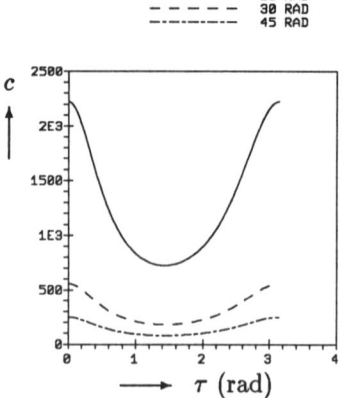

Figure XII.25   The function $c(\tau)$ for three values of the angular velocity ω.

resonance behaviour it suffices to study the compressor without load and we set $F_{gas} = 0$. The matrix **A** in (9.26) has one variable element $c(\tau)$. Under the conditions mentioned above $c(\tau)$ is π-periodic. In Fig. XII.25 $c(\tau)$ is given for the angular velocity values $\omega = 15, 30$, and 45 rad/s. From this figure it is clear that $c(\tau)$ and thus **A** depend very smoothly on τ.

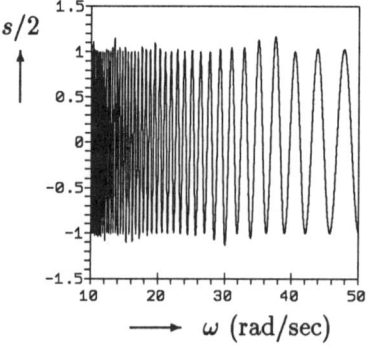

Figure XII.26   The trace s as a function of the angular velocity ω.

To calculate s, the trace of the fundamental matrix evaluated after one period, we integrate (9.25) for $0 \leq \tau \leq \pi$ starting at the vectors $(1,0)^T$ and $(0,1)^T$ respectively. The resulting vectors at $\tau = \pi$ form the columns of the fundamental matrix. The resulting s-value as a function of ω is given in

Fig. XII.26 for the range $10 \leq \omega \leq 50$ rad/s.

We observe that $s(\omega)$ behaves more or less like a sine with varying frequency. Nearly everywhere we have $|s| < 1$, so stability is assured for those ω-values. However, s takes on values equal to or slightly larger than 1 for ω-values which accumulate as $\omega \to 0$. In these points resonance is to be expected as discussed above. We conclude that, when the compression is started, the angular velocity ω of the engine must be increased as fast as possible until a value is reached that is not close to a value for which $s(\omega) \geq 1$. For the compressor under consideration we see from Fig. XII.26 that the ω-value of the engine in the stationary situation should be larger than about 40 rad/s. Above that value there are points with $s(\omega) = 1$, but they do not lie very dense on the ω-axis.

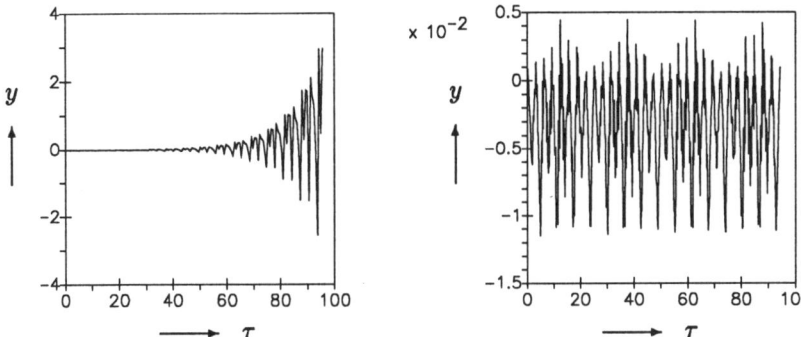

Figure XII.27   Resonant (left) and non-resonant (right) behaviour of the torsion in a crank shaft.

To demonstrate the different types of behaviour, we have calculated the torsion for $\omega = 37.6$ and 38.3 The corresponding s-values are $s(37.6) = 1.16$ and $s(38.3) = 0.0$. The results are given in Fig. XII.27a,b. The difference between resonant and non-resonant behaviour is very pronounced in this system, because all friction is neglected. In practice even a small amount of friction will temper the enhancement of the torsion amplitude considerably.

In this section we showed how the stability properties can be estimated if the compressor is very simple and contains only one crank. Extension to advanced systems including more cranks and frictional effects is not hard, because the general ideas will remain the same. By calculating s, the trace of the fundamental matrix after one period, as a function of the system parameters, one can (numerically) establish which regions in parameter space correspond to (un)stable solutions. These insights are of great importance for designers of compressors, enabling them to avoid quite a lot of trial-and-error experiments.

10. The dripping faucet

In this section we analyse a daily-life system which is very convenient for illustrating many aspects of chaotic behaviour. Chaotic systems show sensitive dependence on the initial conditions. This sensitivity can generally be switched on and off by adjusting a parameter in the model. This kind of bifurcation is most easily understood for scalar Δ-equations, as shown in Chapter VI. A fascinating alternative example, suggested by [70], is the dripping behaviour of a faucet. Experiments show that the dripping pattern varies with the flow rate in a remarkable way. In the experiments the intervals between the falling drops could be measured by laser equipment with a photoelectric sensor.

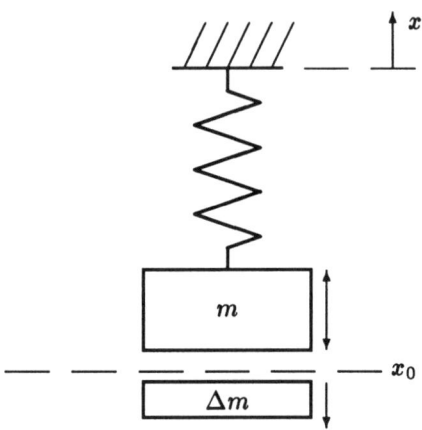

Figure XII.28   Model of the dripping faucet, consisting of a mass-spring system with time dependent mass. The lower mass Δm represents a newly formed drop.

We model the dripping faucet in quite a simple way using only three differential equations. As explained in Chapter VI, this is the minimum number for which chaotic behaviour can occur in continuous time systems. Although the model neglects nearly all physical details of the faucet, it reproduces many of the observed aspects. The dynamics of a drop when still clinging to the faucet is described by a mass-spring system with a gradually increasing mass. At some moment a part of the mass is suddenly separated from the rest. Such a part represents a newly formed drop. In Fig. XII.28 the mass-spring system is depicted. The amount of water attached to the faucet is presented by the time dependent mass $m(t)$. If the faucet is open, $m(t)$ will increase linearly with time:

(10.1) $m(t + t_i) = \beta t + m(t_i),$

where β measures the flow rate and t_i is the last dripping moment. The mass $m(t)$ is a discontinuous function of t, because $m(t)$ is instantaneously decreased at the dripping moments. Relation (10.1) therefore holds between successive dripping moments only. In the following we shall use the corresponding differential equation

(10.2) $\quad \dot{m} = \beta$.

The spring represents the adhesive forces and the surface tension of the water. It has spring constant $k > 0$. The gravitational force exerted on the mass is $-m(t)\,g$, with g the gravitational acceleration. The third force which has to be taken into account is friction. We assume it to be proportional to the velocity $v(t)$ of the mass. The frictional constant will be denoted by $\gamma > 0$. The position of the centre of gravity of the mass is denoted by $x(t)$ and measured as indicated in Fig. XII.28. Newton's second law is expressed here as

(10.3) $\quad \dfrac{d}{dt}(mv) = -mg - kx - \gamma v$.

In addition to equations (10.2) and (10.3) we have the relation

(10.4) $\quad \dot{x} = v$.

At this point we have to make a choice for the state vector characterising the system. Dripping is an instantaneous decrease of mass, during which only mass and not position or velocity is influenced. This suggests taking the vector $(x, v, m)^T$ as state vector, so that only one component changes at the moment of dripping. Summarising the model contained in equations (10.2), (10.3) and (10.4), we obtain

(10.5) $\quad \begin{bmatrix} \dot{x} \\ \dot{v} \\ \dot{m} \end{bmatrix} = \begin{bmatrix} v \\ -g - k\dfrac{x}{m} - (\beta + \gamma)\dfrac{v}{m} \\ \beta \end{bmatrix}$.

The ODE (10.5) describes the system between two successive moments of dripping. This three-dimensional system is autonomous and nonlinear. It can be reduced using the relation (10.1), resulting in a two-dimensional, linear, and non-autonomous system. This transformation has little practical use, and we shall consider only (10.5) in the following.

The solution of (10.5) will in general be oscillating. Because $m(t)$ steadily increases, the average position will become lower and lower. We fix a position $-x_0$ with $x_0 > 0$, as indicated in Fig. XII.28, and assume that dripping occurs when $x(t)$ passes $-x_0$ in the downward direction. At that moment a part Δm of m is instantaneously cut off, while x and v are left unchanged. The effect will be that the gravitational force is instantaneously diminished, so that the remaining mass will reverse its velocity shortly after the moment of dripping, and jump upwards. The drop, separating from the faucet at time t_i, has mass $\Delta m(t_i)$. For $\Delta m(t_i)$ we choose a function of the velocity $v(t_i)$, such that $\Delta m(t_i)$ is small if $v(t_i)$ is small, and large if $v(t_i)$ is large. A convenient choice is

$$(10.6) \qquad \Delta m(t_i) = m(t_i) \exp\left(\frac{\alpha}{v(t_i)}\right), \qquad i = 1, 2, \dots .$$

Note that $v(t_i)$ is not positive. The parameter $\alpha > 0$ measures the rate of mass reduction. In the exceptional 'grazing' case of $v(t_i)$ vanishing we set $\Delta m = 0$, so that no drop is formed.

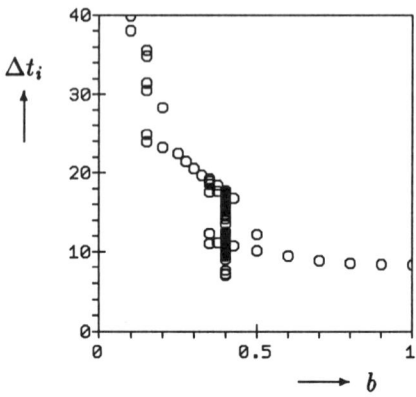

Figure XII.29   Values attained by Δt_i as a function of parameter b. The transient values for $i = 1, ..., 100$ are discarded.

The complete model consists of (10.5) with (10.6). It contains the parameters g, k, x_0, α, β, γ. Dimensional analysis, as discussed in §2, shows that only three non-dimensional parameters are essential. One way to non-dimensionalise the model is by introducing the dimensionless variables

$$(10.7) \qquad t^* := \frac{k}{\beta+\gamma} t , \quad x^* := \frac{x}{x_0} , \quad v^* := \frac{\beta+\gamma}{k\,x_0} v , \quad m^* := \frac{k}{(\beta+\gamma)^2} m .$$

The corresponding dimensionless parameters are

$$(10.8) \qquad a := \frac{(\beta+\gamma)\,\alpha}{k\,x_0} , \quad b := \frac{\beta}{\beta+\gamma} , \quad c := \frac{(\beta+\gamma)^2\,g}{k^2 x_0} , \quad x_0^* := 1 .$$

The corresponding dimensionless version of (10.5) is given by the ODE

$$(10.9) \qquad \begin{bmatrix} \dot{x}^* \\[2mm] \dot{v}^* \\[2mm] \dot{m}^* \end{bmatrix} = \begin{bmatrix} v^* \\[2mm] -c - \dfrac{x^* + v^*}{m^*} \\[2mm] b \end{bmatrix} .$$

The dimensionless version of (10.6) is

$$(10.10) \quad \Delta m^*(t_i^*) = m^*(t_i^*) \exp\left(-\frac{a}{v^*(t_i^*)}\right) .$$

We conclude that the model depends on three independent parameters only. In the following the * will be omitted, unless confusion might arise. When model (10.9) with (10.10) is evaluated numerically using, e.g., the routine Fehlberg, a large variety of solutions are obtained for different sets of (a, b, c). It turns out that many interesting aspects can be studied by varying b, while a and c are kept constant. For the latter we use the values $a = 0.01$ and $c = 0.15$; the solutions are then bounded. Much information is provided by the time intervals

$$\Delta t_i := t_{i+1} - t_i , \qquad i = 1, 2, \dots .$$

After a transient period the series $\{\Delta t_i\}, i = 1, 2, \dots$, attains the values given in Fig. XII.29 as a function of parameter b. The limit values are independent of the initial conditions, but show a strong dependence on the value of b. For most b-values the system becomes periodic. The period observed depends on b. If b is varied in the interval $[0, 1]$ many bifurcations are observed, at which the period switches from one value to another. Note that no period doubling is seen (as is characteristic for the logistic equation depicted in Fig. VI.3).

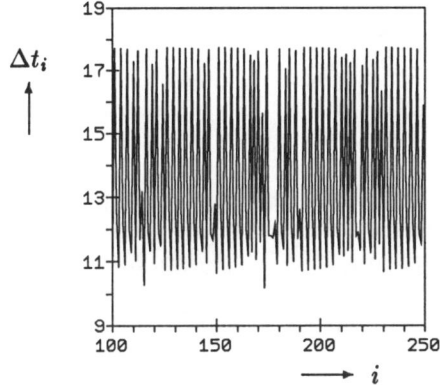

Figure XII.30   Behaviour of the Δt_i-series for $b = 0.4$.

For b close to the value 0.4 a different behaviour is found. Then the solution does not become periodic, and the values attained form a continuum. In Fig. XII.30 the Δt_i-series is given as a function of i for $b = 0.4$. This series seems to be generated at random. Fig. XII.31 shows that this is not the case. In this *reconstruction plot* Δt_{i+1} is drawn versus Δt_i. If the series in Fig. XII.30 had been stochastic, we would have found a cloud of points in

Fig. XII.31. Instead a curve seems to be found. A closer look at this figure reveals a fine structure. In Fig. XII.32 a detail is shown. It shows that the figure actually consists of at least two lines. Further refinement reveals that these lines themselves also consist of lines. In fact the figure appears to consist of a cascade of lines. This fine structure is typical for chaotic attractors. Because

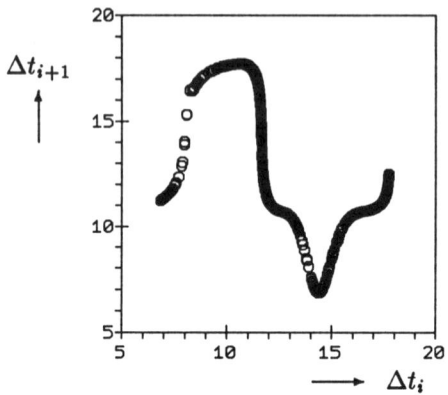

Figure XII.31   Reconstruction plot of the Δt_i-series.

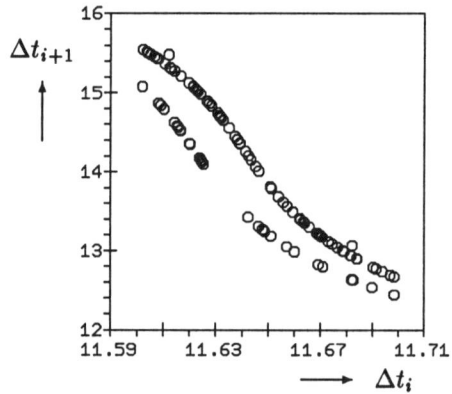

Figure XII.32   Magnification of a part of **Figure XII.31**.

the attractor is very thin, the Δt_i-series can be considered to be generated in first approximation by the one-step Δ-equation

(10.11) $\Delta t_{i+1} = f(\Delta t_i),$

where the function $f(\Delta t)$ follows the figure in Fig. XII.31 in an average way. In this figure we see that $f(\Delta t)$ has one point in common with the straight line

$\Delta t_{i+1} = \Delta t_i$. This implies that equation (10.11) has one stationary solution Δt^*. This solution corresponds to a period-1 orbit for which the falling drops are equidistant. This orbit is unstable, because $|f'(\Delta t^*)| > 1$. Period-2 orbits are found from plotting Δt_{i+2} against Δt_i. This yields the composed function $f \circ f$:

$$(10.12) \quad \Delta t_{i+2} = f\Big(f(\Delta t_i)\Big) =: f \circ f(\Delta t_i) .$$

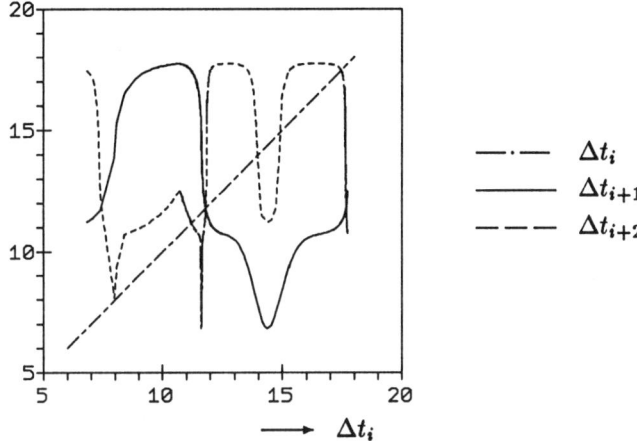

$$\begin{array}{ll} -\cdot- & \Delta t_i \\ \underline{\qquad} & \Delta t_{i+1} \\ --- & \Delta t_{i+2} \end{array}$$

$$\longrightarrow \Delta t_i$$

Figure XII.33

The functions f and $f \circ f$ are given in Fig. XII.33. From this figure we see that equation (10.12) has seven stationary points. One of them equals the stationary point of (10.11). The others correspond to three period-2 orbits of the dripping faucet. The latter orbits are unstable, because $|f'| > 1$ at the stationary points. We can repeat this procedure and study composed functions $f \circ ... \circ f$ of any order. The conclusion is that, in the chaotic regime, the dripping faucet has periodic orbits of all possible periods, and all these orbits are unstable. This observation is typical for chaotic attractors. They usually contain infinitely many unstable periodic orbits of all possible periods.

Let us now turn to the state space spanned by (x, v, m). Orbits in the state space turn to converge to an attractor, independent of the initial conditions. Its structure resembles a band cut through by the plane $x = 1$. In Fig. XII.34 the cross-sections of the attractor with the plane $x = 1$ are given.

The orbits arrive in the upper part of Fig. XII.34 at moments of dripping t_i. From there they instantaneously jump to the lower part. After that they travel along the attractor to the upper part again in the period Δt_i. Following the orbits during a cycle over the attractor, we may construct a *Poincaré map* of the attractor given by

$$(10.13) \quad \Big(1, v(t_i), m(t_i)\Big) \rightarrow \Big(1, v(t_{i+1}), m(t_{i+1})\Big) ,$$

with all the values in (10.13) taken before (or after) dripping.

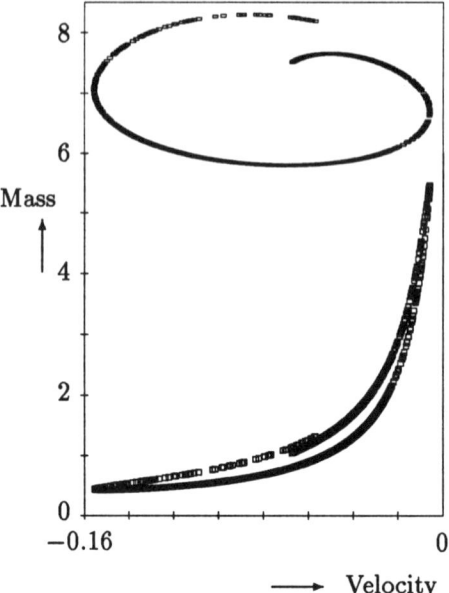

Figure XII.34 Poincaré sections of the attractor of the dripping faucet in phase
space with the plane $x = 1$.

Although the two curves in Fig. XII.34 suggest that the attractor is two-
dimensional, this is not true. If one zooms in at these curves, a complex
structure of layers is found. In Fig. XII.35 a magnification of a part of
Fig. XII.34 is shown.

It is to be expected that the dimension of the chaotic attractor will be fractal,
but quite close to 2. Here the *capacity* or *box-counting* dimension D_0 is meant.
This dimension does not take into account that some regions on the attractor
are visited more frequently than others. In practice usually the correlation
dimension D_2 and not D_0 is estimated, because efficient standard software is
available for this [74]. The correlation dimension indeed takes into account the
visiting frequencies.

In Fig. XII.36 an estimate of the correlation integral $P_2(l)$, defined in
(VI.5.5), is given as a function of l on a log-log scale. For this estimate we
have used 10000 points on the attractor. The slope of this plot yields the
estimate

$$D_2 = 1.65 .$$

The fact that the estimates for D_0 and D_2 differ considerably indicates that

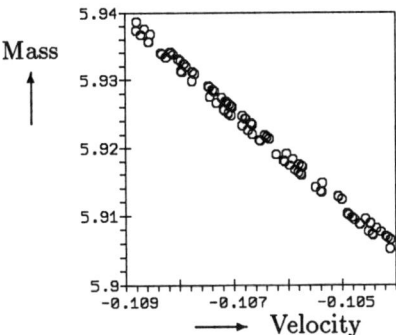

Figure XII.35   Magnification of a part of **Figure XII.34**.

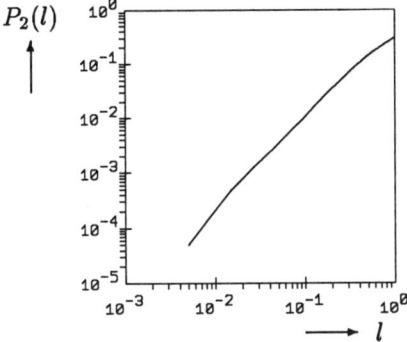

Figure XII.36   Two-point correlation function $P_2(l)$ as a function of the mesh size l.

the visiting frequency is far from being uniform over the attractor. This is indeed observed from the calculated solutions.

The attractor is chaotic if, by definition, the largest Lyapunov exponent λ_1 is positive. Then the orbits on the attractor tend to diverge on average. As explained in §VI.3, the estimation of the Lyapunov exponents requires the evaluation of the linearised version of the equations of motion along a typical orbit. The time evolution of the present system consists of a continuous time part, given by (10.9), and the discrete map (10.10). This implies that we would have to combine Definitions VI.3.13 and VI.3.21. This implementation is not straightforward. We prefer to estimate λ_1 through a simpler approach, based on the information contained in the Δt_i-series. Each of the bands in Fig. XII.34 is a Poincaré section of the attractor. Starting at a point at the upper band we can follow the corresponding orbit until it reaches this band again. This yields the Poincaré map

$$P_1: \quad (v_i, m_i) \to (v_{i+1}, m_{i+1}) , \quad i = 1, 2, 3, \dots.$$

It maps the upper band in Fig. XII.34 onto itself, and contains all the information about the dynamics of the system. Poincaré maps are discrete, and therefore easier to handle than continuous time evolutions in most cases. Another useful map is provided by the time evolution of the Δt_i. The map

$$P_2: \quad \Delta t_i \to \Delta t_{i+1} , \quad i = 1, 2, 3, \dots,$$

is closely related to the map P_1. This follows from the fact that a point (v_i, m_i) at one of the bands in Fig. XII.34 contains all information to calculate the period Δt_i in a smooth way. This implies that if P_1 shows chaotic behaviour, this also holds for P_2, and vice versa. So, if P_2 has a positive λ_1, this also holds for P_1. It is quite simple to obtain an estimate for the Lyapunov exponent of P_2. We have already seen that P_2 is approximately given by the map in (10.11) with f depicted in Fig. XII.31. Following the theory in §VI.3 we can estimate λ_1 from

$$\lambda_1 = \lim_{n \to \infty} \frac{1}{n} \sum_{i=0}^{n-1} \ln \|f'(x_i)\| .$$

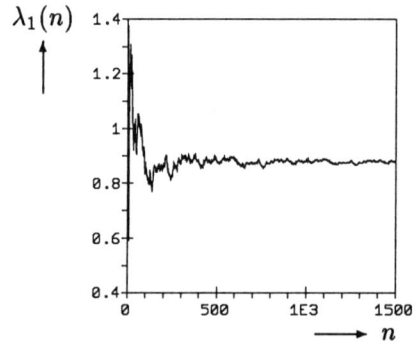

Figure XII.37 Convergence of the estimate for the Lyapunov exponent λ_1.

The behaviour of the estimate $\lambda_1(n)$ as a function of n is given in Fig. XII.37. The estimated value is clearly positive and given by

$$\lambda_1 = 0.9 .$$

From this we conclude that the time evolution of the attractor for $b = 0.4$ is chaotic. The value of λ_1 indicates that inaccuracies are magnified by a factor of $exp(\lambda_1 t)$ for each cycle, roughly speaking. This implies that prediction of the evolution is reliable only within a time horizon. From (VI.7.2) we know

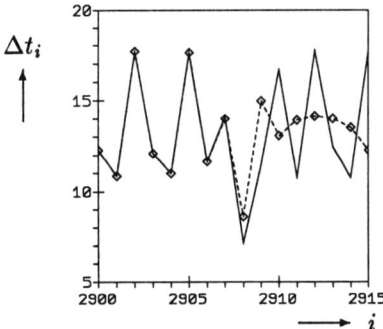

Figure XII.38   True (solid line) and predicted (dotted line) values for the
Δt_i-series.

that a rough estimate of the time horizon is given by

$$T_{hor} = -\frac{1}{\lambda_1}\ln \epsilon_0 \; ,$$

where ϵ_0 is the inaccuracy of the data. To show this we apply the prediction
method presented in §VI.5 to the Δt_i-series. The series has length 3000. The
first 2900 data points are used to predict the values for $i = 2901, 2902, ...$
The data used are given with an accuracy of 5 decimals, so $\epsilon_0 = 5 \; 10^{-6}$. The
corresponding true values are also available. They are numerically obtained
with an accuracy which is much larger than ϵ_0, so that the reliability of the
predictions can be estimated by comparing them to these data. In Fig. XII.38
the true and predicted values are given. It is seen that the predictions follow the
true values closely for about the first 10 prediction steps. After about 15 steps
they turn out to be completely unreliable. This observation hardly depends on
the particular value the prediction started, and appears to be typical for the
data. Substituting the values for ϵ_0 and λ_1 into the expression for T_{hor} yields
the estimate

$$T_{hor} = 13.5 \; ,$$

which is in good agreement with the observations.

Appendix A

Interpolation and quadrature

Let a function $f(t)$ be specified at $n+1$ points t_i, $i = 0, ..., n$. A straightforward way to interpolate $f(t)$ between the known values makes use of polynomials. The *Lagrange basis polynomials* $L_i(t)$ are defined by

$$L_i(t) := \frac{(t - t_0) ... (t - t_{i-1})(t - t_{i+1}) ... (t - t_n)}{(t_i - t_0) ... (t_i - t_{i-1})(t_i - t_{i+1}) ... (t_i - t_n)} .$$

One can easily see that they satisfy $L_i(t_i) = 1$, $i = 0, ..., n$, and $L_i(t_j) = 0$ if $i \neq j$. From this it is directly clear that the polynomial

$$p(t) := \sum_{i=0}^{n} L_i(t) f(t_i)$$

indeed attains the desired values at the grid points t_i. The degree of $p(t)$ is n, which is is in general the minimum value for polynomial interpolation at $n + 1$ points. It can be proven that the Lagrange interpolation error can be expressed as

$$p(t) - f(t) = (t - t_0) ... (t - t_n) \frac{f^{(n+1)}(\xi)}{(n + 1)!} ,$$

for some ξ in the smallest interval containing t_0, t_n and t.

If not only function values but also derivatives are prescribed, interpolation can be performed in a similar way, using the *Hermite basis polynomials* $H_{1,i}$ and $H_{2,i}$. The first one is defined as

$$H_{1,i}(t) := \left(1 - 2L_i'(t_i)(t - t_i)\right) L_i^2(t) .$$

The second one is defined as

$$H_{2,i}(t) := (t - t_i) L_i^2(t) .$$

The basis polynomials $H_{1,i}(t)$ and $H_{2,i}(t)$ satisfy the properties $H_{1,i}(t_j) = \delta_{i,j}$, $H_{1,i}'(t_j) = 0$, $H_{2,i}(t_j) = 0$, and $H_{2,i}'(t_j) = \delta_{ij}$ ($\delta_{ij} = 0$ if $i \neq j$ and $\delta_{ii} = 1$). From this we can directly find the interpolation polynomial

$$p(t) = \sum_{i=0}^{n} \{H_{1,i}(t) \, f(t_i) + H_{2,i}(t) \, f'(t_i)\} \; ,$$

which is of degree $2n + 1$. The Hermite interpolation error can be expressed as

$$p(t) - f(t) = (t - t_0)^2 \ldots (t - t_n)^2 \, \frac{f^{(2n+2)}(\xi)}{(2n + 2)!} \; .$$

The integral

$$\int_{a}^{b} f(t) \, dt$$

can be approximated using these interpolation polynomials. Let $n + 1$ grid points t_i be chosen within the interval $[a, b]$ with $t_0 = a$ and $t_n = b$ and let $f(t)$ be known at the grid points. Lagrange interpolation then leads to the approximation

(A.1) $$\int_{a}^{b} f(t) \, dt \; \doteq \; \sum_{i=0}^{n} w_i \, f(t_i) \; ,$$

with

$$w_i := \int_{a}^{b} L_i(t) \, dt \; .$$

Quadrature formula (A.1) yields exact results if $f(t)$ is a polynomial of order $\leq n$. Hermite interpolation yields the quadrature formula

(A.2) $$\int_{a}^{b} f(t) \, dt \; \doteq \; \sum_{i=0}^{n} \{w_{1,i} \, f(t_i) + w_{2,i} \, f'(t_i)\}$$

with

$$w_{1,i} := \int_{a}^{b} H_{1,i}(t) \, dt$$

and

$$w_{2,i} := \int_{a}^{b} H_{2,i}(t) \, dt \; .$$

The grid points t_i can be chosen such that $w_{2,i} = 0$ for all i. In that case (A.2) reduces to the form (A.1) with the $H_{1,i}$, and thus the $w_{1,i}$, evaluated with respect to this special grid; actually one can show that $w_{1,i} = \int_a^b L_i(t)$. The resulting formula is called *Gaussian quadrature*. The method is clearly exact for polynomials of order $\leq 2n + 1$.

If the grid points t_i are equispaced, a simple relation exists between interpolating polynomials and backward difference operators. For completeness we repeat the basic definitions of the latter. The backward difference operators ∇^k are for $k = 0, 1, 2$ defined as

$$\nabla^0 f_n = f_n , \quad \nabla^1 f_n = f_n - f_{n-1} , \quad \nabla^2 f_n = f_n - 2f_{n-1} + f_{n-2} .$$

In general we have the recurrence relation

$$\nabla^k f_n = \nabla^{k-1} f_n - \nabla^{k-1} f_{n-1} .$$

From this we find

$$(A.3) \qquad f_{n-k} = \sum_{j=0}^{k} (-1)^j \binom{k}{j} \nabla^j f_n .$$

The interpolating polynomial at equispaced grid points $t_0, \cdots, t_n$ with grid size h then reads in terms of the ∇^k:

$$(A.4) \qquad p(t) = \sum_{j=0}^{n} (-1)^j \binom{-s}{j} \nabla^j f_n , \qquad s := \frac{t - t_n}{h} .$$

Now consider the ODE

$$\dot{x} = f(t, x) .$$

Then

$$x(t_{i+1}) - x(t_i) = \int_{t_i}^{t_{i+1}} f(\tau, x(\tau))\, d\tau .$$

If we apply (A.4) on the interval (t_{i-k+1}, t_i), i.e. approximate $f(t, x(t))$ by such a polynomial p there, we obtain

$$(A.5) \qquad x(t_{i+1}) - x(t_i) \doteq \sum_{j=0}^{k-1} \hat{\gamma}_j \nabla^j f_i ,$$

where

$$\hat{\gamma}_j := (-1)^j \int_0^1 \binom{-\tau}{j} d\tau .$$

This is the k-step *Adams-Bashforth* formula.
The $\hat{\gamma}_j$ are simply calculated by recursion:

$$\hat{\gamma}_0 = 1$$

$$\hat{\gamma}_j = 1 - \frac{1}{j+1}\,\hat{\gamma}_0 - \frac{1}{j}\,\hat{\gamma}_1 - \ldots - \frac{1}{2}\,\hat{\gamma}_{j-1}\,, \qquad j \geq 1\,.$$

If we apply (A.4) on the interval (t_{i-k+1}, t_{i+1}), we obtain

$$(A.6) \qquad x(t_{i+1}) - x(t_i) \doteq \sum_{j=0}^{k} \bar{\gamma}_j \, \nabla^j f_{i+1}\,,$$

where

$$\bar{\gamma}_j = (-1)^j \int_{-1}^{0} \binom{-s}{j} \, ds\,.$$

This is the k-step *Adams-Moulton* formula.
From this we find

$$\bar{\gamma}_0 = 1$$

$$\bar{\gamma}_j = -\frac{1}{j+1}\,\bar{\gamma}_0 - \frac{1}{j}\,\bar{\gamma}_1 - \ldots - \frac{1}{2}\,\bar{\gamma}_{j-1}\,, \qquad j \geq 1\,.$$

The coefficients $\hat{\gamma}_j$ and the $\bar{\gamma}_j$ are independent of k. They are related by

$$\sum_{j=0}^{k} \bar{\gamma}_j = \hat{\gamma}_k\,.$$

Appendix B

Norms

First, we consider norms of vectors $\mathbf{x}$ in a linear vector space V. A norm on V, denoted by $\|\cdot\|$, has to satisfy the following four conditions ($\mathbf{x}, \mathbf{y} \in V$, $\lambda \in \mathbb{R}$ (or $\mathbb{C}$)):

(i) $\|\mathbf{x}\| \in \mathbb{R}$ and $\|\mathbf{x}\| \geq 0$

(ii) $\|\mathbf{x}\| = 0$ if and only if $\mathbf{x} = 0$

(iii) $\|\lambda \mathbf{x}\| = |\lambda| \, \|\mathbf{x}\|$

(iv) $\|\mathbf{x} + \mathbf{y}\| \leq \|\mathbf{x}\| + \|\mathbf{y}\|$ (*triangular inequality*).

If $V = \mathbb{R}^n$, the norm is often a so-called *Hölder norm*, defined by

$$\|\mathbf{x}\|_p := \left(\sum_{i=1}^{n} |x_i|^p \right)^{\frac{1}{p}}, \qquad 1 \leq p \leq \infty \, ,$$

with x_i the components of $\mathbf{x}$. For $p = 1, 2$, and ∞ we have

$$\|\mathbf{x}\|_1 = \sum_{i=1}^{n} |x_i| \, ,$$

$$\|\mathbf{x}\|_2 = \left(\sum_{i=1}^{n} |x_i|^2 \right)^{\frac{1}{2}} \qquad (\textit{Euclidean norm}) \, ,$$

$$\|\mathbf{x}\|_\infty = \max_i |x_i| \, ,$$

respectively.

Two norms $\|\cdot\|_\alpha$ and $\|\cdot\|_\beta$ are called *equivalent* if two constants c_1, c_2 exist, such that for all $\mathbf{x} \in V$

$$c_1 \, \|\mathbf{x}\|_\alpha \leq \|\mathbf{x}\|_\beta \leq c_2 \, \|\mathbf{x}\|_\alpha \, .$$

From this we have that a series convergent with respect to one norm is also convergent with respect to an equivalent norm. If $V = I\!R^n$ this implies that convergence considerations are norm independent because any two norms on $I\!R^n$ are equivalent. See, e.g., [?].

Next matrix norms are considered. Let V be the linear space consisting of matrices. A norm on V satisfies the conditions similar to (i), $\cdots$, (iv). A vector norm induces a matrix norm in a natural way as follows:

$$(B.1) \qquad \|\mathbf{A}\| := \max_{\mathbf{x} \neq 0} \frac{\|\mathbf{A}\mathbf{x}\|}{\|\mathbf{x}\|} \ .$$

Often such a norm is called an *associated* matrix norm. As one can easily verify we have

$$\|\mathbf{A}\| = \max_{\mathbf{x}=1} \|\mathbf{A}\mathbf{x}\| \ .$$

On top of (i), $\cdots$, (iv) such an associated norm apparently also has a *multiplicativity* property

$$(v) \qquad \|\mathbf{A}\mathbf{B}\| \leq \|\mathbf{A}\| \, \|\mathbf{B}\| \ .$$

The most well known matrix norms associated to Hölder norms are

$$\|\mathbf{A}\|_1 = \max_j \sum_{i=1}^{n} |a_{ij}| \ ,$$

$$\|\mathbf{A}\|_\infty = \max_i \sum_{j=1}^{n} |a_{ij}| \ ,$$

$$\|\mathbf{A}\|_2 = \left(\rho(\mathbf{A}^T \mathbf{A}) \right)^{\frac{1}{2}} \ ,$$

where $\rho(\mathbf{B})$ is the absolutely largest eigenvalue of $\mathbf{B}$, see Appendix C.

If V is the linear space of scalar functions $x(t)$ defined on an interval $[a, b]$, we can introduce analogues of the Hölder norm for the continuous case:

$$(B.2) \qquad \|x\|_p := \left(\int_a^b |x(t)|^p \, dt \right)^{\frac{1}{p}} \ .$$

Clearly, if V is a space of vector functions one has to replace the modulus by a suitable vector norm in (B.2). For $p = \infty$ we have

$$\|x\|_\infty = \sup_{t \in [a,b]} |x(t)| \ .$$

Each norm on a linear space V generates a *metric* d on V by the definition

(B.3) $d(\mathbf{x}, \mathbf{y}) := \|\mathbf{x} - \mathbf{y}\|$, $\mathbf{x}, \mathbf{y} \in V$.

This metric has the property of *translation invariance*:

$$d(\mathbf{x} + \mathbf{z}, \mathbf{y} + \mathbf{z}) = d(\mathbf{x}, \mathbf{y}) .$$

Appendix C

Jordan matrices

Any $n \times n$ matrix $\mathbf{A}$ can be brought onto *Jordan form* via a similarity transformation. This implies that a nonsingular matrix $\mathbf{S}$ exists such that the matrix $\mathbf{J} := \mathbf{S}^{-1} \mathbf{A} \mathbf{S}$ has the form

$$\mathbf{J} = \begin{bmatrix} \mathbf{J}_1 & & & \emptyset \\ & \mathbf{J}_2 & & \\ & & \ddots & \\ \emptyset & & & \mathbf{J}_p \end{bmatrix} .$$

The matrices $\mathbf{A}$ and $\mathbf{J}$ have the same eigenvalues $\lambda_1, ..., \lambda_p$ $(p \leq n)$. Each *Jordan block* $\mathbf{J}_j$ is characterised by its size r_j and its eigenvalue λ_j $(\in \mathbb{R}$ or $\mathbb{C})$. The $r_j \times r_j$ Jordan blocks have the form

$$\mathbf{J}_j = \begin{bmatrix} \lambda_j & 1 & & \emptyset \\ & \lambda_j & 1 & \\ & & \ddots & 1 \\ \emptyset & & & \lambda_j \end{bmatrix} , \qquad j = 1, ..., p .$$

Different Jordan blocks may have equal eigenvalues and/or size. The number of Jordan blocks with one and the same eigenvalue is called the *geometric multiplicity* of that eigenvalue. The geometric multiplicity is equal to the number of linearly independent eigenvectors corresponding to that eigenvalue. This can be understood from the fact that a Jordan block of size r_j has exactly one eigenvector given by $(1, 0, ..., 0)^T$ with length r_j.

The *algebraic multiplicity* of an eigenvalue is equal to the number of times that the eigenvalue occurs on the main diagonal of $\mathbf{J}$. This multiplicity is thus given by the sum of the dimensions r_j of the corresponding Jordan blocks. An eigenvalue of algebraic multiplicity m is an m-fold root of the characteristic polynomial

$$\det(\mathbf{A} - \lambda\mathbf{I}) = 0,$$

which has the same roots as the polynomial $\det(\mathbf{J} - \lambda\mathbf{I}) = 0$.

As an example we consider a matrix with Jordan form

$$\mathbf{J} = \begin{pmatrix} 1 & 1 & 0 \\ 0 & 1 & 0 \\ 0 & 0 & 1 \end{pmatrix}.$$

This matrix has a three-fold eigenvalue $\lambda = 1$, i.e. its algebraic multiplicity is three; however, its geometric multiplicity is 2.

Note that the columns of $\mathbf{S}$ are just eigenvectors of $\mathbf{A}$ if $\mathbf{J}$ is a diagonal matrix. If any of the sizes r_j is larger than one, the matrix $\mathbf{A}$ is called *defect*. In this case the vectors in $\mathbf{S}$ which are not eigenvectors are sometimes called *principal vectors*.

In solving ODE we naturally encounter the matrix function $e^{\mathbf{A}t}$. We can also transform this to its Jordan form:

$$\mathbf{S}^{-1} e^{\mathbf{A}t} \mathbf{S} = e^{\mathbf{J}t} = \exp\left(\text{diag}(\mathbf{J}_1 t, ..., \mathbf{J}_p t)\right),$$

as can easily be checked from the definition of $e^{\mathbf{A}t}$ as a series expansion. The matrix $e^{\mathbf{J}t}$ has as main diagonal blocks the matrices $e^{\mathbf{J}_j t}$, i.e.

$$e^{\mathbf{J}_j t} = \mathbf{I} + t\mathbf{J}_j + \frac{t^2}{2!}(\mathbf{J}_j)^2 + \frac{t^3}{3!}(\mathbf{J}_j)^3 + ... \,.$$

For $r_j > 1$ the Jordan block $\mathbf{J}_j$ can be written as

$$\mathbf{J}_j = \lambda_j \mathbf{I} + \mathbf{D},$$

where $\mathbf{I}$ is the $r_j \times r_j$ unit matrix and $\mathbf{D}$ is the $r_j \times r_j$ matrix with ones on the diagonal directly above the main diagonal. Because $\mathbf{I}$ and $\mathbf{D}$ commute, i.e. $\mathbf{ID} = \mathbf{DI}$, we may write

$$e^{\mathbf{J}_j t} = e^{\lambda \mathbf{I} t} \cdot e^{\mathbf{D}t} = e^{\lambda t}\left(1 + t\mathbf{D} + \frac{t^2}{2!}\mathbf{D}^2 + \frac{t^3}{3!}\mathbf{D}^3 + ...\right).$$

The series expansion of $e^{\mathbf{D}t}$ breaks off, because $\mathbf{D}^i = \mathbf{0}$, if $i > r_j$. From this we find

$$e^{\mathbf{J}_j t} = e^{\lambda_j t} \begin{bmatrix} 1 & t & \frac{t^2}{2!} & \cdots & \frac{t^{r_j-1}}{(r_j-1)!} \\ & 1 & \ddots & & \\ & & \ddots & \ddots & \\ & & & \ddots & t \\ \varnothing & & & & 1 \end{bmatrix}.$$

We conclude that $\|e^{\mathbf{A}t}\|$ is uniformly bounded for $t \to \infty$, if $\text{Re}(\lambda_j) \leq 0$, $j = 1, ..., p$, where $\text{Re}(\lambda_j) = 0$ is only allowed if the corresponding Jordan blocks have size 1.

For (linear) difference equations we naturally have the matrix function $\mathbf{A}^l$, for integers $l \geq 0$. This l-th power of $\mathbf{A}$ can also be transformed to Jordan form via

$$\mathbf{S}^{-1} \mathbf{A}^l \mathbf{S} = \mathbf{J}^l .$$

In a way similar to the treatment of $e^{\mathbf{J}t}$ we find for a Jordan block $\mathbf{J}$ with eigenvalue λ and size $r > 1$

$$\mathbf{J}^l = \lambda^{l-r+1} \begin{bmatrix} \lambda^{r-1} & l\,\lambda^{r-2} & \cdots & \cdots & \binom{l}{r-1}\lambda^0 \\ & \lambda^{r-1} & \ddots & & \vdots \\ & & \ddots & \ddots & \vdots \\ & & & \ddots & l\,\lambda^{r-2} \\ \emptyset & & & & \lambda^{r-1} \end{bmatrix} .$$

From this form we conclude that $\|\mathbf{A}^l\|$ is uniformly bounded for $t \to \infty$ if $|\lambda_j| \leq 1$, $j = 1, ..., p$, where $|\lambda_j| = 1$ is only allowed if the corresponding Jordan blocks have size 1.

Appendix D

Linear difference equations with constant coefficients

Consider the linear k-step difference equation

$$(D.1) \qquad x_i = a_1 x_{i-1} + a_2 x_{i-2} + \dots + a_k x_{i-k-1} \ , \qquad i \geq k - 1 \ ,$$

with constant coefficients $a_1, \dots, a_{k-1}$. The solution is uniquely determined if initial conditions $x_0, x_1, \dots, x_k - 1$ are given. The general form of the solutions can be found from the insight that $x_i = \lambda^i$ is a solution for some $\lambda \ (\in \mathbb{R} \text{ or } \mathbb{C})$ if λ satisfies the equation

$$\lambda^k = a_1 \lambda^{k-1} + \dots + a_k \ .$$

If this polynomial has k different roots $\lambda_1, \dots, \lambda_k$, the general solution of (D.1) is given by

$$x_i = \sum_{j=1}^{k} c_j \, (\lambda_j)^i \ , \quad i \geq 0 \ ,$$

with the coefficients c_j determined from the initial values. If some of the λ_j are r_j-fold, the solution of (D.1) is

$$(D.2) \qquad x_i = \sum_{j=1}^{p} \sum_{l=0}^{r_j-1} c_{jl} \, i^l (\lambda_j)^i \ , \quad i \geq 0 \ ,$$

with $p \leq k$, and $\displaystyle\sum_{j=1}^{p} r_j = k.$

Appendix E

Order symbols

Here, we specify the meaning of the order symbols $O(\cdot)$ and $o(\cdot)$. If some function $\mathbf{f}(x)$ is expanded into a series around a fixed value x_0, one is usually interested in a few lower order terms of the series only. The higher order terms are summarised in a rest term $\mathbf{R}(x)$, of which the limiting behaviour is known if $x \to x_0$. The order symbol

$$o(x^n)$$

is a short-hand notation for

$$\lim_{x \to x_0} \frac{\|\mathbf{R}(x)\|}{x^n} = 0 \ .$$

The order symbol

$$O(x^n)$$

stands for

$$\lim_{x \to x_0} \frac{\|\mathbf{R}(x)\|}{x^n} = C < \infty$$

for some constant C. The order symbols must always be interpreted in terms of a limit, and the context must make clear which limit is meant. Quite often x_0 is 0, 1, ∞.

For example, for a continuous function $f(x)$ we have

$$f(x) = f(0) + o(1) \ .$$

Appendix F

Contractive mappings

Consider the mapping

(F.1) $f(x) = x$

where f maps of some subset of a metric space into itself. A point x which satisfies (F.1) is called a *stationary point* of the mapping f. A possible method for solving equations of the form (F.1) is to use *iteration*, i.e. to solve the discrete IVP

(F.2) $\begin{cases} x_{n+1} = f(x_n) , & n = 0, 1, 2, \ldots , \\ x_0 \text{ given} . \end{cases}$

The motivation for this method is the following: if the mapping f is continuous and if the sequence $\{x_n\}$ converges, i.e. $x_n \to x$, then

$$x = \lim_n x_{n+1} = f\left(\lim_n x_n\right) = f(x) .$$

For convergence we need f to be contractive.

Definition F.3. *Suppose S is a subset of a metric space X with distance function d. A mapping $f : S \to X$ is called a contraction, with contraction constant L, if*

$$d(f(x), f(y)) \leq L\, d(x, y)$$

for all $x, y \in S$, where $0 < L < 1$.

Note that if f is a linear mapping defined on a normed linear space, then f is a contraction if and only if $\|f\| < 1$ (see Appendix B). It is also worth pointing out that a contraction may have at most one stationary point. Indeed, if x and y are both stationary points of the contraction f, then

$$d(x, y) = d(f(x), f(y)) \leq L\, d(x, y) < d(x, y) ,$$

which is a contradiction. The basic theorem on contractive mappings is called the *contractive mapping principle* (also called the contraction theorem of Banach). This theorem asserts that contractions on complete metric spaces have stationary points. For a proof see [?]

Theorem F.4. *Suppose S is a closed subset of a complete metric space (X, d) and $\mathbf{f} : S \to S$ is a contraction with contraction constant L. Then $\mathbf{f}$ has a unique stationary point $\mathbf{x}^* \in S$. If $\mathbf{x}_0$ is any point of S and $\{\mathbf{x}_n\}$ is defined by (F.2), then $\mathbf{x}_n \to \mathbf{x}^*$ for $n \to \infty$ and we have*

$$d(\mathbf{x}_n, \mathbf{x}^*) \leq L^n (1 - L)^{-1} d(\mathbf{x}_0, \mathbf{x}_1) \ .$$

It should be emphasised that the error bound given in the theorem, is an 'a priori' bound. That is, once the initial approximation $\mathbf{x}_0$ has been chosen, a contraction constant L is found and the approximation $\mathbf{x}_1$ is computed, the errors in successive approximations can be bounded *before* the approximations themselves are computed. This can be used for estimating the number of iterations sufficient to obtain a solution within a prescribed tolerance.

Appendix G

Matrix decompositions

Let $\mathbf{A} \in I\!\!R^{n \times n}$ be a matrix. Then there exists an orthogonal matrix $\mathbf{Q}$ and an upper triangular matrix $\mathbf{U}$, i.e. $\mathbf{U} = (u_{ij})$ with $u_{ij} = 0$, $i > j$, such that

$$(G.1) \qquad \mathbf{A} = \mathbf{Q} \, \mathbf{U} \, .$$

One can find such a decomposition from the *Gram-Schmidt process* by viewing the columns of $\mathbf{Q}$ as vectors. The columns of $\mathbf{Q}$ are successively formed by orthonormalisation, whereas the upper triangular $\mathbf{U}$ expresses that the space spanned by the first l columns of $\mathbf{A}$ $(l \leq n)$ is the same as the space spanned by the first l orthonormal columns of $\mathbf{Q}$. This decomposition also holds for rectangular matrices. In practice one rather computes $\mathbf{Q}^T$ $(= \mathbf{Q}^{-1})$ in

$$(G.2) \qquad \mathbf{Q}^T \mathbf{A} = \mathbf{U} \, ,$$

by Householder's method or Givens' method, cf. [?]. Here the columns of $\mathbf{A}$ below the diagonal are successively swept by an orthogonal matrix (much like in Gaussian elimination).

We can fairly easily show now that a symmetric matrix has an orthonormal system of eigenvectors. Indeed, let

$$\mathbf{A} = \mathbf{T} \, \mathbf{J} \, \mathbf{T}^{-1} \, ,$$

be the Jordan form, then write $\mathbf{T} = \mathbf{Q} \, \mathbf{U}$ according to Gram-Schmidt. Hence we find

$$\mathbf{A} = \mathbf{Q} \, \mathbf{V} \, \mathbf{Q}^T \, ,$$

where $\mathbf{V}$ is upper triangular. $\mathbf{A}$ is symmetric and so is $\mathbf{V}$; hence $\mathbf{V}$ must be a diagonal matrix. The eigenvalues of $\mathbf{A}$ are given by the diagonal elements of $\mathbf{V}$, and the eigenvectors are the corresponding columns of $\mathbf{Q}$.

Using this decomposition one can show the existence of a *singular value decomposition* (SVD). For this consider the symmetric matrix $\mathbf{A}^T \mathbf{A}$, which has positive eigenvalues, and its Jordan form,

$$A^T A = T\,S\,T^{-1}\ ,$$

where S is a positive diagonal matrix and T is orthogonal. Write $T = Q_2$ and $S = \Sigma^2$, where Σ is again a positive diagonal matrix. Hence

$$Q_2^T\,A^T A\,Q_2 = \Sigma^2\ .$$

In other words, $A\,Q_2$ is a matrix with orthogonal columns, having Euclidean norm equal to the diagonal elements of Σ. We thus find

$$A\,Q_2 = Q_1\,\Sigma\ ,$$

where Q_1 is orthonormal, i.e. having orthogonal columns and rows of unit length. Hence we may write

(G.3) $$A = Q_1\,\Sigma\,Q_2^T\ ,$$

which is called the SVD. This decomposition also exists when A is singular.

References

[1] Abramowitz M. and Stegun I.A. (1965) *Handbook of Mathematical Functions*, Dover, New York.

[2] Agarwal R.P. (1992) *Difference Equations and Inequalities: Theory, Methods and Applications*, Dekker, Basel.

[3] Aiken R.C. (ed.) (1985) *Stiff Computation*, Oxford University Press, Oxford.

[4] Ascher U.M., Mattheij R.M.M. and Russel R.D. (1995) *Numerical Solution of Boundary Value Problems for Ordinary Differential Equations*, SIAM, Philadelphia, PA.

[5] Arnold V.I. (1973) *Ordinary Differential Equations*, MIT Press, Cambridge, MA.

[6] Arnold V.I. (1978) *Mathematical Methods of Classical Mechanics*, Springer-Verlag, New York.

[7] Arrowsmith D.K. and Place C.M. (1982) *Ordinary Differential Equations*, Chapman and Hall, London.

[8] Ascher U. and Petzold L.R. (1991) *Projected Implicit Runge-Kutta Methods for Differential Algeraic Equations*, SIAM J. Numer. Anal. **128** 1097–1120.

[9] Bailey N.T.J. (1975) *The Mathematical Theory of Infectious Diseases*, Griffin, London.

[10] Barenblatt G.I. (1979) *Similarity, Self-Similarity, and Intermediate Asymptotics*, Consultants Bureau, New York.

[11] Baumgarte J. (1972) *Stabilization of Constraints and Integrals of Motion in Dynamical Systems*, Comp. Math. Appl. Mech. Engng **1**, 1–16.

[12] Bender E.A. (1978) *An Introduction to Mathematical Modeling*, Wiley-Interscience, Chichester.

[13] Birkhoff G. and Rota G.C. (1978) *Ordinary Differential Equations*, Wiley, Chichester.

[14] Bowen R. (1970) *Markov Partitions for Axiom A Diffeomorphisms*, Amer. J. Math. **92**, 725–747.

[15] Brenan K., Campbell S. and Petzold L. (1989) *Numerical Solution of Initial-Value Problems in Differential-Algebraic Equations*, North-Holland.

[16] Butcher J.C. (1987) *The Numerical Analysis of Ordinary Differential Equations: Runge-Kutta and General Linear Methods*, Wiley, Chichester.

[17] Capasso V. (1993) *Mathematical Structures of Epidemic Systems*, Springer-Verlag, Berlin.

[18] Clements R.R. (1989) *Mathematical Modelling*, Cambridge University Press, Cambridge.

[19] Coddington E.A. and Levinson N. (1955) *Theory of Ordinary Differential Equations*, McGraw-Hill, New York.

[20] Curtis C.F. and Hirschfelder J.O. (1952) *Integration of Stiff Equations*, Proc. Nat. Acad. Sci. **38**, 235–243.

[21] Dahlquist G. (1956) *Stability and Error Bounds in the Numerical Integration of Ordinary Differential Equations*, Trans. Royal Inst. Technol., Stockholm, No. **130**.

[22] Dekker K. and Verwer J.G. (1984) *Stability of Runge-Kutta Methods for Stiff Nonlinear Differential Equations*, North-Holland, Amsterdam.

[23] Derrida B., Gervois A. and Pomeau Y. (1979) J. Phys. **12A**, 269.

[24] Devaney R.L. (1987) *An Introduction to Chaotic Dynamical Systems*, Addison-Wesley, New York.

[25] van Dyke M. (1964) *Perturbation Methods in Fluid Dynamics*, Academic Press, New York.

[26] Fehlberg E. (1964) *New Higher-Order Runge-Kutta Formulas with Step-Size Control for Systems of First and Second Order Differential Equations*, ZAMM **44**, Sonderheft.

[27] Feigenbaum M.J. (1983) *Universal Behavior in Non-linear Systems*, Physica **D7**, 16–39.

[28] Gear C.W. (1971) *Numerical Initial Value Problems in Ordinary Differential Equations*, Prentice-Hall, Englewood Cliffs, NJ.

[29] Ghantmacher F.R. (1974) *Matrix Theory II*, Chelsea, New York.

[30] Gleick J. (1987) *Chaos, Making a New Science*, Penguin, New York.

[31] Glendinning P. (1994) *Stability, Instability and Chaos: An Introduction to the Theory of Nonlinear Differential Equations*, Cambridge University Press, Cambridge.

[32] Goldstein H. (1959) *Classical Mechanics*, Addison-Wesley, Reading, MA.

[33] Golub G.H. and van Loan C.F. (1989) *Matrix Computations*, Johns Hopkins University Press, Baltimore, MD.

[34] Gradshteyn I.S. and Ryzhik I.M. (1994) *Table of Integrals, Series, and Products*, Academic Press, London.

[35] Grassberger P. and Procaccia I. (1983) *Measuring the Strangeness of Strange Attractors*, Physica **9D**, 189–208.

[36] Griepentrog E. and März R. (1986) *Differential-Algebraic Equations and their Numerical Treatment*, Treubner Texte zur Math., Leipzig.

[37] Grigorieff R.D. (1977) *Numerik gewöhnlicher Differentialgleichungen 2*, Teubner, Stuttgart.

[38] Grimshaw R. (1990) *Nonlinear Ordinary Differential Equations*, Blackwell Scientific, Oxford.

[39] Groetsch C.W. (1980) *Elements of Applicable Functional Analysis*, Dekker, New York.

[40] Guckenheimer J. and Holmes P. (1983) *Nonlinear Oscillators, Dynamical Systems, and Bifurcations of Vector Fields*, Springer Verlag, New York.

[41] Hairer E., Nørsett S.P. and Wanner G. (1993), *Solving Ordinary Differential Equations I*, Springer-Verlag, Berlin.

[42] Hairer E. and Wanner G. (1991), *Solving Ordinary Differential Equations II*, Springer-Verlag, Berlin.

[43] Hall G. and Watt J.M. (ed.) (1976) *Modern Numerical Methods for Ordinary Differential Equations*, Clarendon Press, Oxford.

[44] Hammel S.M., Yorke J.A. and Grebogi C. (1988) *Numerical Orbits of Chaotic Processes represent True Orbits*, Bull. Amer. Math. Soc. 19, 465.

[45] Hartman P. (1964) *Ordinary Differential Equations*, Wiley, New York.

[46] Hentschel H.G.E. and Procaccia I. (1984) *Relative Diffusion in Turbulent Media: the Fractal Dimension of Clouds*, Phys. Rev. A. 29, 1461–1470.

[47] Hodgkin A.L. and Huxley A.F. (1952) *A Quantitative Description of Membrane Current and its Application to Conduction and Excitation in Nerve*, J. Physiol. 117, 500–544.

[48] Ipsen D.C. (1960) *Units, Dimensions and Dimensionless Numbers*, McGraw-Hill, New York.

[49] Jordan D.W. and Smith P. (1977) *Nonlinear Ordinary Differential Equations*, Oxford University Press, Oxford.

[50] Kapitaniak T. (1990) *Chaos in Systems with Noise*, World Scientific, Singapore.

[51] Kibble T.W.B. (1966) *Classical Mechanics*, McGraw-Hill, London.

[52] Lakshmikantham V. and Trigiante D. (1988) *Theory of Difference Equations: Numerical Methods and Applications*, Academic Press, Boston, MA.

[53] Milne W.E. (1970) *Numerical Solution of Ordinary Differential Equations*, Dover, New York.

[54] Miura R.M. (1982) *Accurate Computation of the Stable Solitary Wave for the Fitzhugh-Nagumo Equations*, J. Math. Biol. 13, 247–269.

[55] Murray J.D. (1993) *Mathematical Biology*, Springer-Verlag, Berlin.

[56] Newhouse S.E. (1980) *Lectures on Dynamical Systems*, in: Dynamical Systems, CIME Lectures, Bressanone, Italy, pp. 1–114, Progress in Mathematics No. 8, Birkhäuser, Boston, MA.

[57] Nordsieck A. (1962) *On Numerical Integration of Ordinary Differential Equations*, Math. Comp. 16, 22–49.

[58] Nyström E.J. (1925) *Ueber die Numerische Integration von Differentialgleichungen*, Acta Soc. Sci., Feun 50, 1–54.

[59] O'Malley R.E. (1991) *Singular Perturbation Methods for Ordinary Differential Equations*, Springer-Verlag, New York.

[60] Ortega J.M. and Rheinboldt W.C. (1970) *Iterative Solution of Nonlinear Equations in Several Variables*, Academic Press, New York.

[61] Ott E. (1993) *Chaos in Dynamical Systems*, Cambridge University Press, Cambridge.

[62] Packard N.H., Crutchfield J.P., Farmer J.D. and Shaw R.S. (1980) *Geometry from the Time Series*, Phys. Rev. Lett. **45**, 712.

[63] Pryce J.D. (1993) *Numerical Solution of Sturm-Liouville Problems*, Oxford University Press, Oxford.

[64] Rasband S.N. (1990) *Chaotic Dynamics of Nonlinear Systems*, Wiley, New York.

[65] Runge C. (1895) *Ueber die Numerische Auflösung von Differentialgleichungen*, Math. Ann. **46**, 167–178.

[66] Sano M. and Sawady Y. (1985) *Measurement of the Lyapunov Spectrum from Chaotic Time Series*, Phys. Rev. Lett. **55**, 1082.

[67] Schlögl F. (1980) *Stochastic Measures in Nonequilibrium Thermodynamics*, Phys. Rep. **62**, 267–376.

[68] Shampine L.F. and Gordon M.K. (1975) *Computer Solution of Ordinary Differential Equations, the Initial Value Problem*, Freeman, San Francisco, CA.

[69] Shampine L.F. and Watts H.A. (1979) *The Art of Writing a Runge-Kutta Code II*, Appl. Math. Comp. **5**, 93–121.

[70] Shaw R. (1984) *The Dripping Faucet as a Model Chaotic System*, Aerial Press, Santa Cruz, CA.

[71] Sparrow C. (1982) *The Lorentz Equations*, Springer-Verlag, New York.

[72] Stetter H.J. (1973) *Analysis of Discretization Methods for Ordinary Differential Equations*, Springer-Verlag, Berlin.

[73] Takens F. (1981) *Detecting Strange Attractors in Turbulence*, in: Dynamical Systems and Turbulence, Warwick, Lecture Notes in Math. 898, Eds. D.A. Rand and L.S. Young, Springer-Verlag, Berlin.

[74] Theiler J. (1987) *Efficient Algorithm for Estimating the Correlation Dimension from a Set of Discrete Points*, Phys. Rev. A. **36**, 4456–4462.

[75] Vasileva AB, Butuzov V.F. and Kalachev L.V. (1995) *The Boundary Function Method for Singular Perturbation Problems*, SIAM.

[76] Walter W. (1971) *An Elementary Proof of Peano's Existence Theorem*, Am. Math. Monthly **78**, 170–173.

[77] Wilson H.K. (1971) *Ordinary Differential Equations*, Addison-Wesley, New York.

Index